Advances in Nanogels

Advances in Nanogels

Editors

**Filippo Rossi
Chien-Chi Lin
Emanuele Mauri**

MDPI • Basel • Beijing • Wuhan • Barcelona • Belgrade • Manchester • Tokyo • Cluj • Tianjin

Editors

Filippo Rossi
Politecnico di Milano
Milan
Italy

Chien-Chi Lin
Indiana University-Purdue
University Indianapolis
Indianapolis, IN
USA

Emanuele Mauri
Università Campus
Bio-Medico
Rome
Italy

Editorial Office
MDPI
St. Alban-Anlage 66
4052 Basel, Switzerland

This is a reprint of articles from the Special Issue published online in the open access journal *Gels* (ISSN 2310-2861) (available at: https://www.mdpi.com/journal/gels/special_issues/nanogels).

For citation purposes, cite each article independently as indicated on the article page online and as indicated below:

LastName, A.A.; LastName, B.B.; LastName, C.C. Article Title. *Journal Name* **Year**, *Volume Number*, Page Range.

ISBN 978-3-0365-6420-3 (Hbk)
ISBN 978-3-0365-6421-0 (PDF)

Cover image courtesy of Filippo Rossi

© 2023 by the authors. Articles in this book are Open Access and distributed under the Creative Commons Attribution (CC BY) license, which allows users to download, copy and build upon published articles, as long as the author and publisher are properly credited, which ensures maximum dissemination and a wider impact of our publications.

The book as a whole is distributed by MDPI under the terms and conditions of the Creative Commons license CC BY-NC-ND.

Contents

About the Editors vii

Preface to "Advances in Nanogels" ix

Chien-Chi Lin, Emanuele Mauri and Filippo Rossi
Editorial on the Special Issue "Advances in Nanogels"
Reprinted from: *Gels* **2022**, *8*, 835, doi:10.3390/gels8120835 1

Emanuele Mauri, Sara Maria Giannitelli, Marcella Trombetta and Alberto Rainer
Synthesis of Nanogels: Current Trends and Future Outlook
Reprinted from: *Gels* **2021**, *7*, 36, doi:10.3390/gels7020036 3

Brielle Stawicki, Tyler Schacher and Hyunah Cho
Nanogels as a Versatile Drug Delivery System for Brain Cancer
Reprinted from: *Gels* **2021**, *7*, 63, doi:10.3390/gels7020063 27

Simona Campora, Reham Mohsen, Daniel Passaro, Howida Samir, Hesham Ashraf, Saif El-Din Al-Mofty, et al.
Functionalized Poly(N-isopropylacrylamide)-Based Microgels in Tumor Targeting and Drug Delivery
Reprinted from: *Gels* **2021**, *7*, 203, doi:10.3390/gels7040203 43

Abdul Qadir, Samreen Jahan, Mohd Aqil, Musarrat Husain Warsi, Nabil A. Alhakamy, Mohamed A. Alfaleh, et al.
Phytochemical-Based Nano-Pharmacotherapeutics for Management of Burn Wound Healing
Reprinted from: *Gels* **2021**, *7*, 209, doi:10.3390/gels7040209 61

Tisana Kaewruethai, Chavee Laomeephol, Yue Pan and Jittima Amie Luckanagul
Multifunctional Polymeric Nanogels for Biomedical Applications
Reprinted from: *Gels* **2021**, *7*, 228, doi:10.3390/gels7040228 81

Shadab Md, Nabil A. Alhakamy, Thikryat Neamatallah, Samah Alshehri, Md Ali Mujtaba, Yassine Riadi, et al.
Development, Characterization, and Evaluation of α-Mangostin-Loaded Polymeric Nanoparticle Gel for Topical Therapy in Skin Cancer
Reprinted from: *Gels* **2021**, *7*, 230, doi:10.3390/gels7040230 99

Gaurav Kant Saraogi, Siddharth Tholiya, Yachana Mishra, Vijay Mishra, Aqel Albutti, Pallavi Nayak and Murtaza M. Tambuwala
Formulation Development and Evaluation of Pravastatin-Loaded Nanogel for Hyperlipidemia Management
Reprinted from: *Gels* **2022**, *8*, 81, doi:10.3390/gels8020081 125

Françoise Chuburu, Volodymyr Malystkyi, Juliette Moreau, Maïté Callewaert, Céline Henoumont, Cyril Cadiou, et al.
Synthesis and Characterization of Conjugated Hyaluronic Acids. Application to Stability Studies of Chitosan-Hyaluronic Acid Nanogels Based on Fluorescence Resonance Energy Transfer
Reprinted from: *Gels* **2022**, *8*, 182, doi:10.3390/gels8030182 141

About the Editors

Filippo Rossi

Prof. Dr. Filippo Rossi received his MSc and then PhD in chemical engineering from Politecnico di Milano in 2007 and 2011, respectively. He then spent research periods at Uppsala University (2012) and Mario Negri Institute for Pharmacological Research (2013–2015) as a PostDoc and at Keio University (2018) as a Visiting Associate Professor. In 2015, he joined the Department of Chemistry, Materials and Chemical Engineering "Giulio Natta" of Politecnico di Milano as an Assistant Professor and has been working there as an Associate Professor of Applied Physical Chemistry since 2019. His group's research program is focused on developing novel technologies to solve problems associated with health and the environment using functionalized polymers in bulk (gels) and colloidal forms.

Chien-Chi Lin

Prof. Dr. Chien-Chi Lin is the Thomas J. Linnemeier Guidant Foundation Endowed Chair & Professor of Biomedical Engineering at Indiana University—Purdue University Indianapolis (IUPUI). He received his PhD in Bioengineering from Clemson University in 2007 and completed his postdoc training at the University of Colorado Boulder in 2010. The primary research focus of his research group is on designing multifunctional polymeric biomaterials, particularly hydrogels, with spatiotemporally tunable physicochemical properties to mimic the extracellular matrix (ECM) in the tumor microenvironment and stem cell niche. His team is also interested in developing innovative hydrogel chemistries for 3D bioprinting and for fabricating unique hydrogels suitable for releasing therapeutically relevant agents. He has been recognized by the NSF CAREER Award and the Abraham Max Distinguished Professor Award from the Purdue School of Engineering & Technology; as an Honorary Scientist and Advisor on Agricultural Science & Technology for the Rural development administration of Republic of Korea; as an invitee for the US Frontiers of Engineering Symposium by the US National Academy of Engineering; the Young Investigator Award from Georgia Institute of Technology; and the American Institute of Chemists Postdoctoral Award from the University of Colorado.

Emanuele Mauri

Dr. Emanuele Mauri is an Assistant Professor in Applied Physical Chemistry at Politecnico di Milano. He graduated in Chemical Engineering and received a Ph.D. with honors in Industrial Chemistry and Chemical Engineering at Politecnico di Milano. He was a visiting Assistant Professor at ETH Zurich, and he spent four years as an Assistant Professor at Università Campus Bio-Medico di Roma (Italy) working on the design of nanogels for drug delivery. His main research fields are now related to the synthesis and characterization of polymer-based soft materials for tissue engineering and controlled drug delivery systems, in particular the formulation of hydrogels for 3D printing and tissue engineering applications, continuous in-flow synthesis of nanoparticles through microfluidics and surface decoration of nanoparticles for target therapy in tumoral scenarios. He has been awarded with international awards such as the European Doctorate Award by European Society of Biomaterials (ESB) and Nanoinnovation's Got Talent at NanoInnovation 2019 in Rome.

Preface to "Advances in Nanogels"

In recent decades, the rise of nanotechnology has led to the design of innovative nano-biomaterials for alternative approaches in pharmacological therapies and diagnostic investigations. Among them, nanogels represent one of the most promising nanocarriers for tunable drug release and selective cell interactions.

Nanogels are composed of physically/chemically cross-linked polymers organized in a three-dimensional nanostructure, and their fundamental feature is the ability to reproduce, at a nanoscale, the typical swelling behavior of bulk hydrogels. This enhances the dispersion stability and bioavailabioty of the loaded drugs or proteins, as well as theit interactions with physiological compartments.

Moreover, other unique properties are closely correlated to the chance of fine-tuning their porosity, hydrophilic nature, stability, size, and charge by varying the chemical composition or grafting of additional functionalities. In particular, functionalization with specific chemical linkages or molecules promotes the controlled release of the payload according to specific external stimuli, extending the curative benefits over time. At the same time, the configuration of core–shell nanogels and the development of coating layers are two potential alternatives to modulate nanosystem–cell interactions and cellular uptake, with a view to the selective cell internalization approach, which represents one major challenge for the clinical administration of the nanosystems.

Starting from the contribution of Prof. Akiyoshi (1993) and Prof. Vinogradov (1999), whom designed, respectively, the first physical and chemical nanogels, innovative proposals have been studied, combining in vitro and in vivo applications in very heterogeneous disorders. This Special Issue will provide an overview of the different potential approaches to synthesize smart nanogels for controlled drug delivery, with a special emphasis on the functionalization strategies of these nanocarriers and their application in in vitro and/or in vivo models.

Filippo Rossi, Chien-Chi Lin, and Emanuele Mauri
Editors

Editorial

Editorial on the Special Issue "Advances in Nanogels"

Chien-Chi Lin [1,*], Emanuele Mauri [2,3,*] and Filippo Rossi [3,*]

1. Department of Biomedical Engineering, Indiana University-Purdue University Indianapolis, Indianapolis, IN 46202, USA
2. Department of Engineering, Università Campus Bio-Medico di Roma, Via Álvaro del Portillo 21, 00128 Rome, Italy
3. Department of Chemistry, Materials and Chemical Engineering "G. Natta", Politecnico di Milano, Via Mancinelli 7, 20131 Milan, Italy
* Correspondence: lincc@iupui.edu (C.-C.L.); e.mauri@unicampus.it (E.M.); filippo.rossi@polimi.it (F.R.)

In recent decades, the rise of nanotechnology has led to the design of innovative nano-biomaterials which are used to improve pharmacological therapies and assist with disease diagnosis. Among them, nanogels represent one of the most promising nanocarriers for tunable drug release and selective cell targeting. In general, nanogels are composed of physically/chemically cross-linked polymer chains organized in a three-dimensional nanostructure. Like their bulk gel counterpart, nanogels are capable of imbibing a large quantity of water and undergoing reversible swelling/deswelling. This enhances the dispersion stability, the interactions with physiological compartments and the bioavailability of the loaded drugs or proteins. Other unique properties include highly tunable porosity, hydrophilicity, stability, size, and charge by means of modulating chemical compositions of the polymers or grafting additional functionalities on the nanogel surface. In particular, functionalization with specific chemical linkages or molecules promotes controlled release of the payload according to specific external stimuli (e.g., redox agents, enzymatic activities, pH and temperature variations), extending the curative benefits over time. At the same time, the configuration of core–shell nanogels or the development of coating layers are two potential alternatives to modulate the nanosystem–cell interactions. Altering these properties affects cellular uptake and provides a means for selective cell internalization, which represents a major challenge for the clinical administration of nanosystems.

Starting with the contributions of Prof. Akiyoshi (1993) and Prof. Vinogradov (1999) that designed, respectively, the first physical and chemical nanogels, many innovative proposals have been studied in recent years, combining in vitro and in vivo applications in a range of medical fields [1,2].

Nanogels can be prepared by natural polymers, such as hyaluronic acid (HA) and chitosan. Malytskyi and coworkers [3] obtained stable nanogels from HA and chitosan using tripolyphosphate as a cross-linker, taking advantage of the ionic gelation mechanism. The nanogels were highly stable even in presence of enzymes, which is advantageous in biomedical applications. Chitosan was also used by Saraogi and coworkers [4] to produce nanogels using ionization methods that can then be loaded with pravastatin for hyperlipidemia treatment. The sustained release, together with the compatibility with excipients and high hemolytic activity, support the prospective use of orally administered pravastatin-loaded nanogel as an effective and safe nano-delivery system in hyperlipidemia treatment.

Nanogels are commonly prepared by synthetic polymers. Md and coworkers [5] used an emulsion–diffusion–evaporation technique with a three-level, three-factor Box–Behnken design to produce poly (D, L-lactic-co-glycolic acid)-based nanogels. The developed formulations showed excellent flux across the skin layer in the skin permeation study, and skin probe fluorescent dye in confocal microscopy revealed significant penetration of NPs into the skin, suggesting their possible use in skin-cancer treatment. Functionalization strategies can then be added to preformed nanogels for specific and selective cancer-cell targeting. Campora and coworkers [6] decorated poly(N-isopropylacrylamide) colloids with acrylic acid and coupled them with folic acid, targeting the folate receptors overexpressed by cancer cells and the chemotherapeutic drug doxorubicin. This approach supports their possible use in selective cell-cancer treatment, with consequent amelioration of chemotherapeutic strategies.

The high versatility on nanogels is also described in the four review papers in this Special Issue: Mauri et al. [7] proposed an interesting focus on the recent techniques used in nanogel design, highlighting key upgrades in terms of methodologies, microfluidics, and 3D printing. Macromolecules and biomolecules can indeed be combined to create ad hoc nanonetworks according to the final curative goals, maintaining the criteria of biocompatibility and biodegradability. From a materials perspective, as described by Kaewruethai and coworkers [8], nanogels were designed as core–shell structures that demonstrated efficient responsiveness to different stimuli such as temperature, pH, reductive environment, or radiation. Stawicki et al. [9] focused on the use of nanogels as vehicles to cross the blood–brain barrier, providing local and systemic drug-delivery systems in the treatment of brain cancer. Indeed, the current ongoing development allows patient-centered treatment that can be considered a promising tool for the management of brain cancer. Moreover, other targets can be addressed, such as wound healing, in which the ability of nanogels to carry and release drugs is extremely promising, as discussed by Qadir and coworkers [10].

In summary, the present Special Issue provides an overview of the different potential approaches to synthetizing smart nanogels for controlled drug delivery, with particular emphasis on the functionalization strategies of these nanocarriers and their application in in vitro and/or in vivo models. We hope that this Special Issue will offer good insight into recent studies on nanogels and their applications. We thank all the research teams that contributed to "Advances in Nanogels".

Funding: This research received no external funding.

Conflicts of Interest: The authors declare no conflict of interest.

References

1. Vinogradov, S.; Batrakova, E.; Kabanov, A. Poly(ethylene glycol)-polyethyleneimine NanoGel (TM) particles: Novel drug delivery systems for antisense oligonucleotides. *Colloids Surf. B Biointerfaces* **1999**, *16*, 291–304. [CrossRef]
2. Pinelli, F.; Perale, G.; Rossi, F. Coating and Functionalization Strategies for Nanogels and Nanoparticles for Selective Drug Delivery. *Gels* **2020**, *6*, 6. [CrossRef] [PubMed]
3. Malytskyi, V.; Moreau, J.; Callewaert, M.; Henoumont, C.; Cadiou, C.; Feuillie, C.; Laurent, S.; Molinari, M.; Chuburu, F. Synthesis and Characterization of Conjugated Hyaluronic Acids. Application to Stability Studies of Chitosan-Hyaluronic Acid Nanogels Based on Fluorescence Resonance Energy Transfer. *Gels* **2022**, *8*, 182. [CrossRef] [PubMed]
4. Saraogi, G.K.; Tholiya, S.; Mishra, Y.; Mishra, V.; Albutti, A.; Nayak, P.; Tambuwala, M.M. Formulation Development and Evaluation of Pravastatin-Loaded Nanogel for Hyperlipidemia Management. *Gels* **2022**, *8*, 81. [CrossRef] [PubMed]
5. Md, S.; Alhakamy, N.A.; Neamatallah, T.; Alshehri, S.; Mujtaba, M.A.; Riadi, Y.; Radhakrishnan, A.K.; Khalilullah, H.; Gupta, M.; Akhter, M.H. Development, Characterization, and Evaluation of α-Mangostin-Loaded Polymeric Nanoparticle Gel for Topical Therapy in Skin Cancer. *Gels* **2021**, *7*, 230. [CrossRef] [PubMed]
6. Campora, S.; Mohsen, R.; Passaro, D.; Samir, H.; Ashraf, H.; Al-Mofty, S.E.-D.; Diab, A.A.; El-Sherbiny, I.M.; Snowden, M.J.; Ghersi, G. Functionalized Poly(N-isopropylacrylamide)-Based Microgels in Tumor Targeting and Drug Delivery. *Gels* **2021**, *7*, 203. [CrossRef] [PubMed]
7. Mauri, E.; Giannitelli, S.M.; Trombetta, M.; Rainer, A. Synthesis of Nanogels: Current Trends and Future Outlook. *Gels* **2021**, *7*, 36. [CrossRef] [PubMed]
8. Kaewruethai, T.; Laomeephol, C.; Pan, Y.; Luckanagul, J.A. Multifunctional Polymeric Nanogels for Biomedical Applications. *Gels* **2021**, *7*, 228. [CrossRef] [PubMed]
9. Stawicki, B.; Schacher, T.; Cho, H. Nanogels as a Versatile Drug Delivery System for Brain Cancer. *Gels* **2021**, *7*, 63. [CrossRef] [PubMed]
10. Qadir, A.; Jahan, S.; Aqil, M.; Warsi, M.H.; Alhakamy, N.A.; Alfaleh, M.A.; Khan, N.; Ali, A. Phytochemical-Based Nano-Pharmacotherapeutics for Management of Burn Wound Healing. *Gels* **2021**, *7*, 209. [CrossRef]

Synthesis of Nanogels: Current Trends and Future Outlook

Emanuele Mauri [1], Sara Maria Giannitelli [1], Marcella Trombetta [1] and Alberto Rainer [1,2,*]

[1] Department of Engineering, Università Campus Bio-Medico di Roma, via Álvaro del Portillo 21, 00128 Rome, Italy; e.mauri@unicampus.it (E.M.); s.giannitelli@unicampus.it (S.M.G.); m.trombetta@unicampus.it (M.T.)

[2] Institute of Nanotechnology (NANOTEC), National Research Council, via Monteroni, 73100 Lecce, Italy

* Correspondence: a.rainer@unicampus.it; Tel.: +39-06225419214; Fax: +39-06225419419

Abstract: Nanogels represent an innovative platform for tunable drug release and targeted therapy in several biomedical applications, ranging from cancer to neurological disorders. The design of these nanocarriers is a pivotal topic investigated by the researchers over the years, with the aim to optimize the procedures and provide advanced nanomaterials. Chemical reactions, physical interactions and the developments of engineered devices are the three main areas explored to overcome the shortcomings of the traditional nanofabrication approaches. This review proposes a focus on the current techniques used in nanogel design, highlighting the upgrades in physico-chemical methodologies, microfluidics and 3D printing. Polymers and biomolecules can be combined to produce ad hoc nanonetworks according to the final curative aims, preserving the criteria of biocompatibility and biodegradability. Controlled polymerization, interfacial reactions, sol-gel transition, manipulation of the fluids at the nanoscale, lab-on-a-chip technology and 3D printing are the leading strategies to lean on in the next future and offer new solutions to the critical healthcare scenarios.

Keywords: nanogels; colloids; chemical crosslinking; physical crosslinking; microfluidics; lab-on-a-chip; 3D printing

1. Introduction

Nowadays, the design of nanocarriers plays a leading role in the formulation of advanced therapeutic treatments for several acute and chronic diseases, ranging from tumors to neurodegenerative scenarios. The manipulation of matter at the atomic and molecular levels offers different configurations of colloidal nanocarriers: organic (such as solid lipid nanoparticles, liposomes, dendrimers, polymeric nanoparticles and micelles), inorganic (carbon nanotubes, metallic and silica-derived nanostructures) and hybrid systems (combination of organic and inorganic materials) [1–3]. All of them have been investigated as efficient tools for the delivery of bioactive species, designing the so-called 'controlled drug delivery' strategy, where the pivotal aim is the tunable release of the payload within the therapeutic concentration range, hence minimizing the ineffectiveness or the potential side effects of the treatment [4]. However, the relevance of the nanocarriers is not limited to this task: they also serve as platforms for diagnosis, monitoring or theranostics, in order to promote the administration of a single nanosystem for a combinatorial and multifunctional therapy [5,6]. This approach involves different routes to obtain nanovehicles with high selectivity towards target cells, and the consequent internalization and nanostability within the cytosol. Many strategies have been proposed to tune the physical, chemical and mechanical features of the nanocarriers. In particular, nanostructured gels (nanogels) have gained considerable interest and have become the subject of several studies focused on novel production methods and new areas of application.

Nanogels (NGs) can be defined as submicron-sized hydrogels, formed by physically or chemically crosslinked polymeric chains which give rise to a three-dimensional (3D) tunable porous network with a high capacity to absorb water, without actually dissolving into an aqueous medium [7]. Typically, they are characterized by spherical shape, but other configurations can occur depending on the fabrication methods: for example, micromolding techniques and photolithography allow nanogel size and shape to be controlled by modulating the surface energy and the chemical interactions among the polymers [8–11]. Moreover, NGs can be designed to have a crosslinked core-shell, a core-shell-corona or a bulk (similar to a 'ball of wool') structure.

The distinctive feature of these nanomaterials is their swelling behavior. The polymers used in NG synthesis absorb water giving a mostly hydrophilic nature to the final nanonetwork, capable of incorporating a great amount of water or biological fluids while maintaining its structural integrity. The phenomenon is led by the contact between the nanoscaffold and the solvent molecules: the latter interact with the polymeric surface *via* dipole-dipole interactions, London dispersion forces or hydrogen bonds, and penetrate the network. The polymer chains start to elongate, expanding the nanostructure, until the elastic retroactive force counterbalances the deformation of the system. The resulting condition is an equilibrium of stretching–shrinking, which allows NGs to uptake more than 90% w/w of aqueous solution [12]. The ability to retain water enhances the diffusion and the exchange of ions, metabolites, and biomolecules to and from the biological compartments (i.e., tissue fluids and organs), in order to maintain the biological-chemical balance with the surrounding environment: this shows good synergy between NGs and biological applications [13,14], making them highly promising biocompatible candidates. Moreover, NG swelling can be engineered to be sensitive to external stimuli, such as pH and temperature variations, ionic strength, and solvent affinity: this helps to tune NG porosity, stiffness and size, so as to achieve a controlled, triggered response at the target site [5,15,16].

Therefore, in drug delivery, NGs not only protect the payload from undesired degradation and early release, but also actively participate to the delivery process.

The smart choice of the polymers and the resulting architectural versatility enable NGs to incorporate a plethora of hydrophilic and lipophilic molecules, ranging from inorganic compounds to biomolecules such as amino acids, proteins and nucleic acids (DNA, RNA), without compromising their gel-like behavior.

NG multifunctionality is a key property which is difficult to find in other types of nanoparticulate systems and is the result of considerable research efforts to improve and optimize NG formulations.

This review aims to give an overview on the synthetic routes commonly used to fabricate NGs for drug delivery and targeted therapies. The discussion will be focused on the description of the traditional and advanced methodologies for NG synthesis, highlighting the physico-chemical principles, the main advantages and limitations, and the most common precursors in NG design. For each synthetic strategy, applicative examples will be provided. Moreover, a summary of the main chemical reactions involved in the modification of NGs will be addressed. Indeed, the functionalization of nanogels represents a complementary approach to face the challenge of developing nanoscaffolds for selective targeting and encapsulating entities with very different physical properties within the same carrier.

2. Fundamental Criteria in NG Synthesis

Beyond the swelling behavior, which can be classified as a 'superior' property of the NGs, the main features to be considered in the synthesis of these nanomaterials are: high biocompatibility, biodegradability, colloidal stability, high surface area, high loading capacity ensuring a sustained and targeted drug delivery, and active/passive drug release thanks to the particle size and the surface properties [11]. In addition, other features that can be tuned by carefully controlling the NG synthetic routes include:

- Release of both water-soluble and oil-soluble bioactive compounds;
- Versatility in administration route (i.e., mucosal or parenteral pathway);

- Low immunogenicity and reduced NG elimination by the mononuclear phagocytic system (MPS);
- Optimization of NG permeability;
- Enhancement of the solubility of low-molecular-weight drugs;
- Reduction of the drug payload compared to standard drug administration.

NGs are typically composed of natural or synthetic polymers, or a combination thereof. However, their smart formulation can encompass the inclusion of inorganic components or the grafting of specific bio-moieties on the polymeric backbone. In the first case, NGs can work as imaging probes, incorporating a wide variety of diagnostic and contrast agents for different types of biomedical applications. These systems are usually defined as 'nanohybrid nanogels' and are aimed at increasing the circulation half-lives of small molecules, serving as a highly convenient platform for combined delivery of therapeutic molecules [17]. In the second case, the conjugation of targeting ligands, antibodies, or peptides encourages the mechanism of NG active/passive targeting to the site of interest and the controlled release of the therapeutic payload.

Different strategies have been developed to address multiple applicative scenarios; however, they can all be traced back to the fundamental principles of chemistry and physics: interactions among the reactive groups of different molecules and physical parameters—such as viscosity, density and rheology—represent the basis and the key knowledge of NG design.

3. Traditional NG Synthesis

The methods for NG synthesis can be divided into chemical and physical ones, as shown in Figure 1. Generally, the former gives rise to nanonetworks characterized by strong covalent bonds that improve the colloidal stability under in vitro and in vivo conditions, essential for limiting the leakage of the payload induced by unwanted dissociation of the gel network. These bonds can be distinguished in:

- Cleavable linkers under specific external stimuli (pH and temperature variations);
- Stable bonds which provide the gel with the ability to retain its shape under physico-chemical stress [18,19].

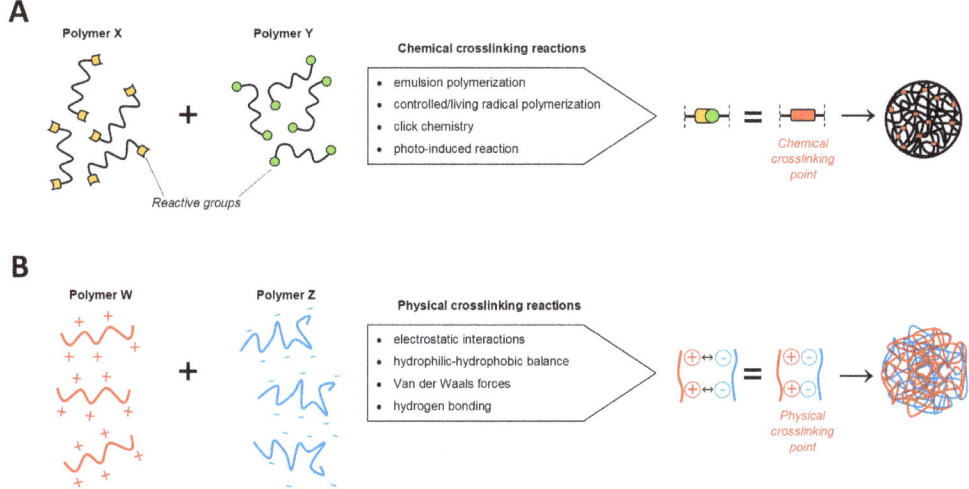

Figure 1. Summary of the chemical and physical crosslinking methods in nanogels (NG) design. (**A**) The formation of covalent bonds (crosslinking points, in red) between the reactive moieties (yellow and green) of polymers X and Y can be addressed through different chemical reactions exploiting the nature of the reacting groups. (**B**) The physical interactions between polymers W and Z enables the formation of a self-assembled nanoscaffold.

Chemical crosslinking is the most developed and the most versatile strategy for NG production. On the other side, physical assembling of NGs involves a controlled aggregation mechanism led by reversible non-covalent connections (hydrophilic/hydrophobic, electrostatic, hydrogen bonding, Van der Waals, or host-guest interactions). Despite the relatively weak nanostructure due to the physical nature of the crosslinking, this process is more flexible because chemical reactions are not involved, and it occurs at mild conditions in aqueous media [20].

3.1. Chemical Routes

Chemical crosslinking includes the following main NG synthetic routes: emulsion polymerization, controlled/living radical polymerization, click chemistry and photo-induced crosslinking. The starting materials are low-molecular-weight monomers, polymer precursors or polymers with specific terminal or pendular reactive groups.

3.1.1. Emulsion Polymerization

Emulsion-based polymerization works through the formation of monodisperse kinetically stable droplets in a continuous phase. The rationale underlying this process is to keep the polymerization in a confined space (the droplets), whose size would affect the dimension of the final product. The dispersion of organic droplets containing the reactive monomers/polymers in aqueous solution (oil-in-water, O/W emulsion) is usually indicated as direct emulsion polymerization; whereas the aqueous droplets dispersed in an organic medium (water-in-oil, W/O emulsion) is known as inverse emulsification polymerization. NG formulation involves the use of monomers, initiators, catalysts, and crosslinking agents. Generally, the process occurs in three steps: nucleation, precursor nanoparticle growth and polymerization [21]. Two main approaches can be acknowledged. In the first one, all the reagents are dissolved in the dispersed phase and photo-initiators are preferred to activate the reaction mechanism via homolytic degradation. In the second approach, different monomers can be dissolved in the dispersed and continuous phases, respectively. A catalyst and a crosslinker are typical components of the droplets, while an initiator is added to the continuous phase. Here, thermal-initiators and reactives that are degradable via water radiolysis are commonly chosen. The resulting first radicals react with the monomers generating monomer radicals which enter the droplets via diffusion to react with the other components and form the nanonetwork: this pathway is promoted by the use of surfactants, which reduce the interface energy between the organic and the aqueous phases, or by the hydrophilic-lipophilic balance of the chemical structure of the radical itself. Raghupathi et al. [22] proposed the design of redox-responsive NGs through inverse emulsification method using hydroxylethylacrylamide and cysteine diacrylamide with the surfactant Brij-L4; the resulting nanoscaffolds were intended to encapsulate proteins as a hydrophilic cargo and to tune their release under specific stimulus triggers (Figure 2). Peres and coworkers [23] prepared glutamic acid-based NGs using methylenebis(acrylamide) as a crosslinker and sodium dodecyl sulfate (SDS) as a surfactant, and they demonstrated the effect of NG swelling behavior under pH variations in a drug delivery application.

Surfactants are used to improve the formation and stability of nanodroplets, modulating their dimension and giving nanogels with diameters less than 150 nm [24–26]. The resulting nano-emulsion has a high surface area and allows active components to penetrate easily and faster in the dispersed phase. The concentration of surfactant can also affect the polymerization reaction: when it exceeds the critical micellar concentration (CMC), the arrangement of surfactant at the organic-water interphase results in the formation of micelles encapsulating monomers. The consequent addition of the initiator to the emulsion gives rise to the polymerization process only after diffusing within the micelles, which therefore act as nanoreactors. Both hydrophobic and hydrophilic precursors can be used to form NGs, since the emulsion system guarantees their solubility and enhances their interfacial permeability. However, limitations of this method are related to the high amount of surfactant required and to the difficulties in achieving the complete purification of the obtained NGs. Indeed, the removal of surfactant may result in additional waste-water treatments, increasing the process costs, or in undesired contaminants in the final product. In particular, the use of non-amphiphilic polymers as nanogel precursors requires the addition of surfactants to form the nanoparticulate, and their subsequent removal during the process [27].

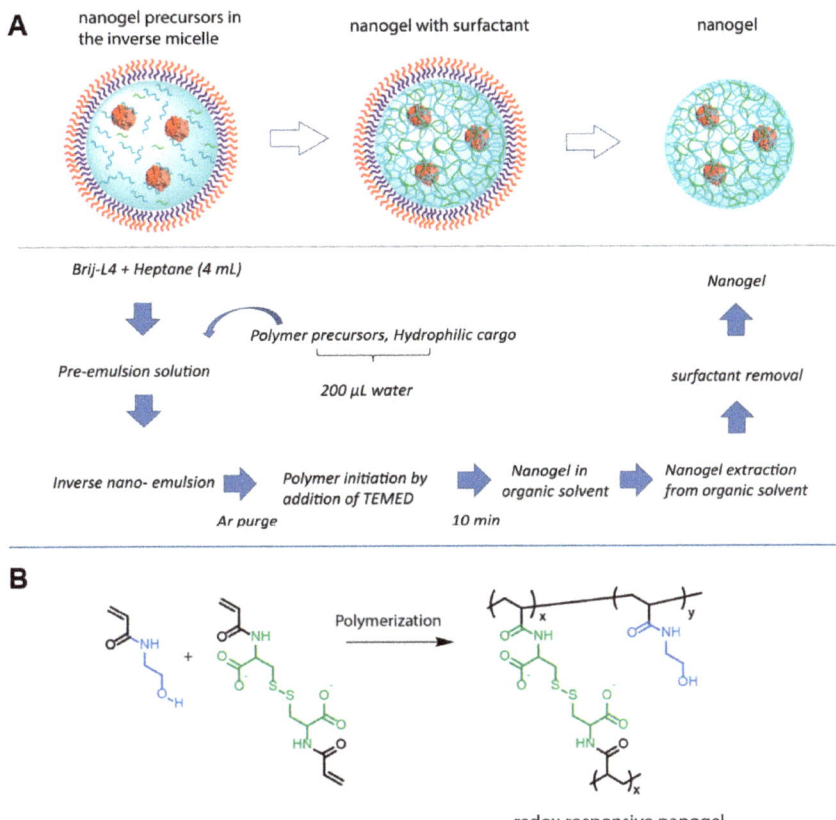

Figure 2. NG synthesis by inverse emulsification in the presence of Brij-L4 surfactant. (**A**) Protocol used in NG formation; (**B**) polymerization between hydroxylethylacrylamide and cystine diacrylamide giving rise to the nanonetwork. Reprinted with permission from Raghupathi et al. [22]. Copyright (2017) American Chemical Society.

For these reasons, in the last years, many attempts have been devoted to developing surfactant-free emulsion polymerization (SFEP) methods. Some strategies are based on W/O emulsions, modulating the stability of the system through the volumetric ratio of the two phases. For example, Cheng and coworkers [28] have synthetized redox-responsive nanogels for drug delivery using an organic solution of PEGylated poly(amido amine) functionalized with disulfide bonds and an aqueous solution containing the drug: their mixing (through ultrasonication, shaking, or homogenization) generated a spontaneous stable W/O emulsion, where the polymer reduced the interfacial tension and filled up the water droplet. The crosslinked polymer network was generated via the intermolecular disulfide exchange reaction in the aqueous phase. Otherwise, the organic phase could be composed by the monomers: Ashrafizadeh et al. [29] proposed the design of amphiphilic pH-responsive NGs using a mixture of acrylate monomers and a dimethacrylate cross-linker combined with an aqueous solution of initiatior to activate the emulsion polymerization. Furthermore, another innovative approach has been discussed by Wang and collaborators [30], who have prepared NGs by emulsion-free photopolymerization: they combined the self-emulsification of poly(ethylene glycol) diacrylate monomer and the small irradiation region of a low-cost semiconductor laser to achieve a spatiotemporally controllable photopolymerization.

Another technique, which represents a milestone in NG design, is the emulsification-solvent evaporation [31,32]. In this case, the approach is preferentially based on the covalent crosslinking of preformed polymer chains, instead of monomers polymerization, to provide excellent opportunities for producing nanogels with tunable pore size [33,34]. Vinogradov, who share with Akiyoshi's group the development of the first nanogels, focused on this method to formulate a wide range of nanonetworks. In this procedure, an activated polymer is dissolved in an organic phase and the other one in water; the addition, under vigorous stirring or sonication, of the former to the latter gives rise to a O/W emulsion, followed by the evaporation in vacuum of the organic solvent and the progressive maturation of the NGs in the aqueous phase. The formation of covalent bonds between the polymers starts at the interface and continues in water. In this scenario, the presence of reactive groups is essential to generate the nanoparticles: these moieties can be existing in the polymer backbone or can be grafted to it by post-polymerization functionalization, and they shall maintain their reactivity in the reaction system. Equally, the choice of the organic solvent is important: it should be poorly soluble in the continuous phase, have high volatility and low boiling point, be able to dissolve the polymer and preferably hold reduced toxicity. The main solvents used are: dichloromethane, ethyl acetate and ethyl formate [31].

Literature reports several application examples of this technique. Vinogradov and coworkers synthetized cationic NGs composed of imidazole-activated PEG and branched PEI for antiviral drug delivery against HIV infection in the brain [35–37] and amphiphilic cationic NGs based on cholesterol-ε-polylysine to deliver therapeutic nucleoside reverse transcriptase inhibitors (NRTI) in the central nervous system, limiting the neurotoxicity [38]. A similar protocol was used by Mauri and coworkers to design NGs made of PEG and linear PEI for drug delivery in astrocytes [39] and microglia [40] as treatments for spinal cord injury. Li and coworkers used carboxymethyl chitosan in the fabrication of NGs for doxorubicin release in tumor-like multicellular spheroids [41].

3.1.2. Controlled/Living Radical Polymerization

Another polymerization technique of major interest is the controlled/living radical polymerization (CLRP). Since 1990s, CLRP has been explored to synthetize either crosslinked particles or gels with well-defined polymer molecular weight through the addition of crosslinking agents [42]. It can be conducted under simple polymerization conditions and a wide range of chemical reactive groups and solvents, including protic media such as water, can be used to achieve a high control on the polymerization process, almost as good as that of living anionic polymerization, and to maintain the characteristic tolerance and flexibility of a free radical process [43]. The use of a wide range of vinylic monomers allows the production of nanogels with different composition, dimensions, and architectures, including core–shell and hollow configurations. Moreover, functional initiators or macroinitiators ensure the grafting of specific moieties in the NG inner core or surface, facilitating multivalent bioconjugation [7,44]. Similar to conventional radical polymerization, CLRP mechanism is based on four elementary steps: initiation, propagation, transfer chain reaction and termination. The essential differences lie in:

- The initiation phase, which is faster (compared to the corresponding propagation and termination reactions) than in standard radical polymerization reactions;
- The generation of a dynamic equilibrium between a low concentration of radicals and a large amount of dormant reactivatable species in the propagation phase;
- A considerably slower global kinetics than conventional radical polymerization.

As discussed in detail by Sanson et al. [44], these features promote an ideal condition of an almost constant number of chains throughout the polymerization, which are initiated nearly at the same time and having the same growth rate: this ensures an inner control over molar mass distribution and architecture. Furthermore, the slow CLRP mechanism allows a gradual reactivation of the dormant chains, permitting them to diffuse and be homogeneously distributed. As a result, the crosslinking points forming the NGs are more evenly distributed within the nanonetwork, which features a locally branched polymeric structure, more regular than the standard radical approaches (where the formation of dense/nodular crosslinking domains occurs, producing a heterogeneous nanostructure) [45]. The success of CLRP in NG design is related to the control of the network formation: the nanoscaffolds can be modulated in size and molar mass. Depending on the concentration of monomers and crosslinking agent, the NG building blocks can have different ranges of molar mass and branched structure, tuning the final morphology and structural composition (such as the porosity) of the nanogels and thus achieving a variety of architectures [46–48]. As reported in Table 1, the pivotal strategies of CLRP are: nitroxide-mediated polymerization (NMP) [49], atom transfer radical polymerization (ATRP) [50], reversible addition–fragmentation chain transfer polymerization (RAFT) [51], iodine-mediated polymerization (RITP) [52], and polymerization/macromolecular design via the interchange of xanthates (MADIX) [53].

Table 1. Controlled/living radical polymerization (CLRP) techniques used in polymerization and NG design.

CLRP Method	Main Features	References
NMP	<ul><li>Suitable for surfactant-free nanoscaffold formation in aqueous dispersion.</li><li>Reversible termination mechanism between the growing-propagating radicals and the nitroxide, which acts as a control agent.</li><li>Well-defined structures of the polymeric backbone with the opportunity to add functional moieties before NG formation.</li></ul>	[54–56]
ATRP	<ul><li>Fast dynamic equilibrium between radicals and dormant alkyl halides.</li><li>Catalytic process to activate the dormant alkyl halide, generating a complex with higher oxidation state and a macro-radical 'living' polymer. The latter can propagate or is deactivated back to 'dormant' polymer, preventing bimolecular termination.</li><li>Variations vs the traditional use of metal catalysts: activation by electron transfer (ARGET-ATRP), electrochemically mediated transfer (eATRP), zerovalent metals in supplemental activators and reducing agents (SARA-ATRP), initiators for continuous activator regeneration (ICAR-ATRP), photochemical reduction (photoATRP), and metal-free ATRP.</li></ul>	[42,57–60]
RAFT	<ul><li>Metal-free polymerization chemistry.</li><li>Use of fast reversible transfer of chain transfer agents (CTA, typically a thio-carbonyl-thio group) to control the propagation of radicals and the formation of dormant species. It ensures a prolonged lifetime of growing chains.</li><li>Minimal to null undesirable background polymerization.</li></ul>	[61–65]
RITP	<ul><li>Reversible deactivation of radicals mediated by iodine.</li><li>Synthesis of well-defined polymers with programmable compositions and architectures, narrow molecular weight distributions.</li><li>Simple reaction system, various types of monomers, mild reaction conditions and tolerance to the reactant purity.</li></ul>	[66,67]
MADIX	<ul><li>The use of the xanthate promotes the design of NGs with high-density of crosslinking points and polymer chains without formation of macrostructures.</li><li>Possibility of generating core-shell star nanosystems by chain extension from branched polymeric precursors.</li><li>It can be applied in combination with RAFT.</li></ul>	[53,68,69]

Recently, Dinari and coworkers [58] have applied ATRP in fabricating dual responsive lignin-based nanogels for controlled release of curcumin, whereas in the work of Lou et al. [59] a core-shell NGs with thermo- and redox-sensitive properties was developed via ATRP using poly-N-isopropylacrylamide and zwitterionic copolymer blocks containing poly(sulfobetaine methacrylate) and a lactose motif of poly(2-lactobionamidoethyl methacrylamide). The resulting NGs demonstrated selective targeting towards hepatoma receptors and were used for doxorubicin delivery in hepatic cells (HepG2). On the other side, as shown in Figure 3, Piogé and coworkers [61] have proposed the design of Low Critical Solution Temperature (LCST)-type thermosensitive NGs made of poly[poly(ethylene glycol) methyl ether acrylate] and poly-N-isopropylacrylamide (PPEGA-b-PNIPAM), combining RAFT mechanism and high-frequency ultrasound. Poly et al. [68] synthetized poly(vinyl acetate) NGs by radical crosslinking controlled by xanthate, which ensured a high yield and a homogeneous structure in the nanonetwork.

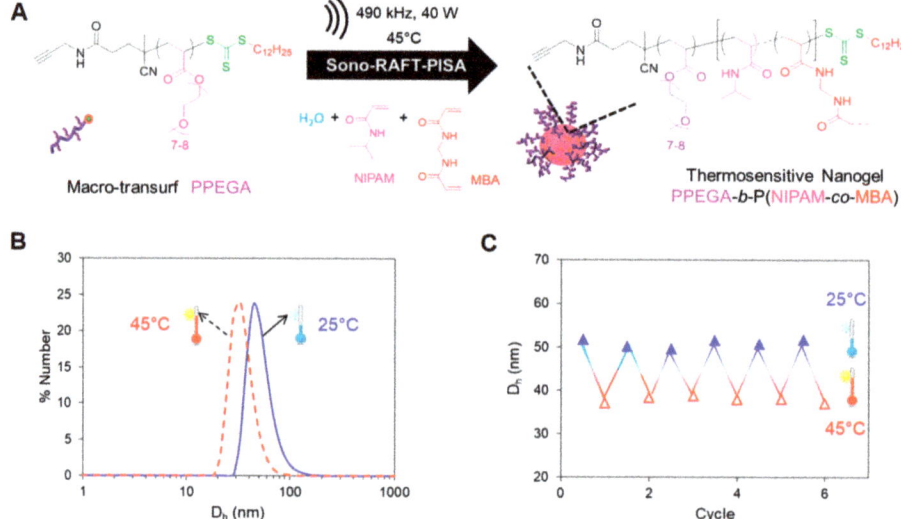

Figure 3. Sonochemically induced reversible addition–fragmentation chain transfer polymerization (RAFT) polymerization to design thermosensitive NGs. (**A**) Synthesis of (PPEGA-b-PNIPAM) NGs; (**B**,**C**) Dynamic Light Scattering analysis (DLS) of the obtained NGs: the size (hydrodynamic diameter, D_h) of the nanonetwork differs in heating (45 °C) and cooling (25 °C) experiments, demonstrating the thermosensitive behavior. Adapted with permission from Piogè et al. [61]. Copyright (2018) American Chemical Society.

3.1.3. Click Chemistry

Emerging NG designs have also taken advantage of click chemistry approaches [70,71]. This strategy ensures short reaction time, high yield and purity, regiospecificity, versatility and aqueous reaction conditions. The mechanism involves the presence of azide and alkyne groups on the building blocks to form a stable conjugate (triazole, Figure 4). Copper-catalyzed and copper-free strain-promoted azide−alkyne cycloadditions are the reference reactions.

As discussed by Sharpless [72], the reaction is generally not affected by steric factors and substituted primary, secondary, tertiary, aromatic azides can readily participate in this transformation with alkyne-derived components. A potential limitation of this approach is related to starting materials missing of the reactive groups: however, azide and triple bond can be easily grafted on polymeric backbones (for example, via nucleophilic substitution [73,74]) and the resulting compounds are extremely stable in standard conditions [75]. Indeed, high tolerance to oxygen, aqueous and common organic solvents, synthesis conditions, biological molecules, and pH has been demonstrated [76]. Zhang and coworkers [77] synthetized crosslinked prodrug nanogels using polyethylene glycol (PEG) modified with polypropargyl glutamate and doxorubicin functionalized azide: the click reaction led to 3D nanoscaffolds feasible for selective intracellular drug delivery in tumors, such as human breast adenocarcinoma and cervical cancer. Additionally, the work of Ding et al. [78] showed the conjugation between dibenzocyclooctyl-modified DNA and azide-modified polycaprolactone (PCL) by a copper-free click reaction: the result was a nucleic acid-based NG, designed for gene delivery and antitumor therapy.

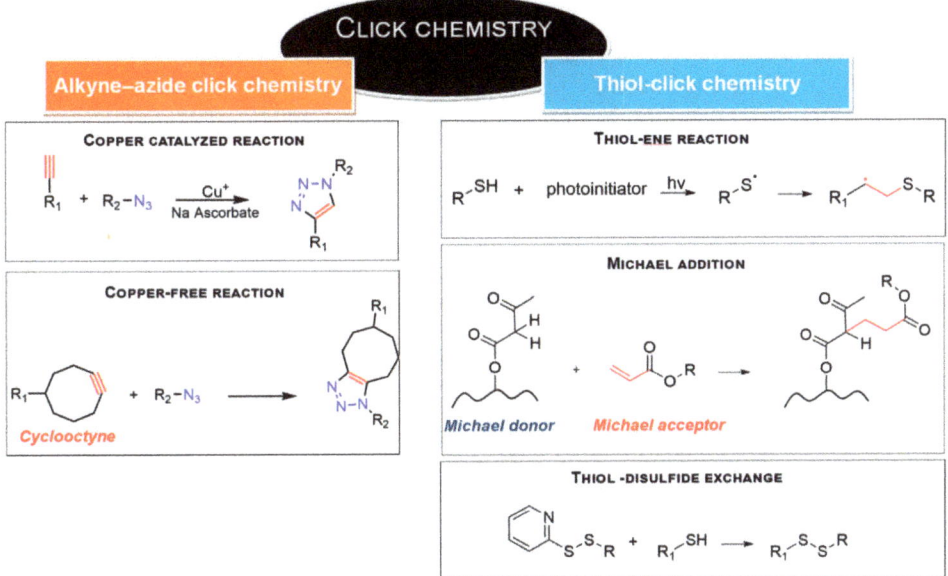

Figure 4. Click chemistry reactions involved in NG synthesis.

Another reaction that might be classified as a 'click chemistry' approach is the thiol-click chemistry [79,80] (Figure 4). Mainly, it includes thiol–alkene and thiol–alkyne reactions, nucleophilic Michael addition and disulfide exchange; pH-sensitive NGs can be produced with this strategy: for example, the polymerization between methoxy polyethyleneglycol acrylate (mPEGA), pentaerythritol tetra(3-mercaptopropionate) (PT) and ortho ester diacrylamide (OEAM) was investigated by Wang and collaborators [81] to obtain NGs characterized by an acid-cleavable network enable to perform pH-triggered drug release in intracellular acid environment. Zhang and coworkers [82] used thiol-alkene reaction to develop highly structured dendritic NGs through the divergent growth approach from bifunctional or monofunctional PEG and propionic acid-derivative monomers, for the successful delivery of chemotherapeutics in 3D pancreatic spheroids tumor model.

3.1.4. Photo-Induced Crosslinking

Photo-crosslinking represents an alternative approach to the above-described polymerization-based design of NGs. It is defined as a 'clean method', because no crosslinking agents and/or catalysts are necessary, and no by-products are formed [83,84]. Polymers functionalized with photo-activatable or dimerizable groups are used to give rise to a stable NG architecture, produced through the formation of covalent bonds. The reaction involves the presence of a photoinitiator, which is converted into reactive radical species via photolysis or light-induced cleavage. The rate of formation of the initial radicals affects the spatial distribution of the covalent crosslinking and, consequently, the microscopic and macroscopic characteristics of the formed nanonetwork (i.e., swelling behavior, stability and surface area). Together with the polymer concentration, the intensity of the incident light, the type and concentration of the photoinitiator, the quantum yield and the number of radicals generated per photolysis event are the main parameters affecting photo-crosslinking reaction [85]. However, although the photo-induced crosslinking is highly efficient and characterized by short reaction time, the initiator may induce cytotoxicity in the produced nanoscaffolds: for this reason, the choice of photoinitiator is intended to cause the minimal toxicity [18]. For example, Irgacure 2959 is well tolerated by many cell types and suitable for optimal NG design [86].

In the last years, Chen and collaborators [87] synthetized pH-degradable polyvinyl alcohol (PVA)-based NGs through photo-crosslinking of thermal-preinduced nanoaggregates. The synthesis was conducted in aqueous solution, using Irgacure as initiator and under UV exposure in an inert gas atmosphere. Paclitaxel was encapsulated and its release kinetics was investigated in acidic and physiological conditions: the responsive behavior of NGs ensured a pH-triggered drug release under intracellular acidic conditions, highlighting this system as a promising nanoplatform for delivery of active molecules in tumoral scenarios. Kim and coworkers [88] obtained gelatin methacrylate NGs by UV-induced nanonetwork formation: the resulting system was applied as nanocarrier for transdermal delivery of hydrophilic macromolecules.

Moreover, photo-activation principles are also applied in the photo-mediated redox crosslinking. The nanonetwork is obtained by polymers carrying phenol moieties, that can be activated, in the presence of a photosensitizer, via photooxidation, generating a subsequent radical coupling between the reactive groups and the final crosslinking [89]. Regarding the photosensitizer molecules, their choice falls on dyes or additives that absorb light and promote a transition towards an excited state, which ensures their capability of oxidizing the reactive groups of interest. They also have high quantum yield and sufficient stability to catalyze the photooxidation [90].

3.2. Physical Routes

The formation of 3D nanoscaffolds can be also reached by polymer-polymer and polymer-biomolecule physical interactions. In this case, supramolecular crosslinking occurs, due to the generation of nanoaggregates via self-assembly. Ionic, hydrophilic-hydrophobic, Van der Waals and hydrogen bonds are the driving forces, without the addition of crosslinking agents that might cause undesired interactions during NG formation and the encapsulation of bioactives, affecting the performance of drug loading. Compared to chemical strategies, NG engineering through physical crosslinking involves mild synthesis conditions, mostly in water, limited adverse toxic effects, and NG size can be tuned by regulating the polymers concentration and the experimental conditions, such as ionic strength, temperature, and pH.

Temperature-responsive and pH-responsive monomers or polymers can be used to address the self-assembly of nanonetworks and to control the drug release performance. These starting materials can also modulate the NG structure, shape and size according to the external stimuli. Thermo-sensitive compounds undergo a reversible phase transition in aqueous media in response to temperature variations; in particular, thermo-sensitive polymers exhibit a critical solution behavior according to the temperature-dependent polymer-polymer and polymer-solvent interactions. This defines the critical solution temperature (CST) as the key parameter that manages the nanonetwork assembly: it represents the value at which the polymer, dissolved in the solvent, generates a second separated phase where the chains tend to form a compacted coil state [91]. NGs can be formed by materials characterized by a low critical solution temperature (LCST) or an upper critical solution temperature (UCST): in the first case, the generation of the nanoscaffolds occurs at values above LCST; whereas in the second case, NGs are synthetized at temperatures below UCST. In addition, both these temperatures can be tuned through the functionalization of the polymer backbones with specific moieties: LCST can decrease by grafting pendular hydrophobic moieties [6], whereas the UCST phase transition can be affected by changing the polarity of the polymers and their capability of engaging in hydrogen and electrostatic interactions [92]. Such tunability allows the synthesis of NGs in the desired temperature window and the generation of NGs with specific operative temperatures for applications including hyperthermia and theranostics. Thermo-sensitive polymers can bear amine, polyether, vinyl ether, acrylate, and hydrophobic groups. For example, Paradossi's group has widely discussed the use of poly(N-isopropylacrylamide) (PNIPAM) to develop nanoscaffolds with temperature-responsive properties, which enable the active targeting of tumor cells and sustained drug release [93,94]. Sliwa et al. [95] have synthetized a thermo-responsive nanogel by PNIPAM and vinylimidazole, characterized by size variations with changing temperature or pH, which modulates the release of the dye Orange II. Instead, Ohshio et al. [96] exploited the UCST of a random copolymer with lateral ureido groups and primary amines combined with poly(2-methacryloyloxyethyl phosphorylcholine (PMPC) to fabricate NGs, which encapsulated hydrophobic fluorescence probes and bovine serum albumin below the UCST and released them, in a controlled manner, above the UCST.

On the other hand, pH-sensitive components possess ionizable groups which undergo protonation/deprotonation upon external pH changes. As a result, new local interactions occur among the monomers and polymers encouraging their spatial reorganization and assembling into an ordered pattern. This mechanism is driven by weak acidic or basic moieties that either accept or release protons in response to a change in the environmental pH. Polymers having carboxyl, amine, pyridine, sulfonic or phosphate groups are typically described as pH-responsive polymers because the ionization of these moieties affects the surface activity, the chain conformation and the solubility, and results in a different structural architecture of the final NGs [97]. Acrylate and methacrylate derivatives are commonly used to design pH-responsive nanoscaffolds. For example, methacrylic acid (MAA) and methyl acrylate (MA) were reported to form NGs [98] in which the pH sensitivity is correlated to their carboxyl groups: at basic pH, these groups are deprotonated and the mutual Coulombic repulsions induce NG swelling; on the other hand, at lower pH levels, the carboxyl moieties are not ionized, leading to the shrinking of the nanonetwork. Additionally, natural polymers, including chitosan, alginate, hyaluronic acid, carboxymethyl cellulose and gelatin, are used to produce NGs with swelling-deswelling variation and designed for precisely controlled drug release with the external stimulus. Indeed, exploiting this behavior, it is possible to regulate the delivery and release of drugs in a wide range of tumoral scenarios according to the pH gradients between the tumor microenvironment and the normal physiological environment [99].

However, physically crosslinked nanosystems possess lower mechanical strength and are more easily degradable than their covalently crosslinked counterparts: as a result, they are more sensitive to the sol−gel transition caused by environmental stimuli changes [20] and their application in harsh in vivo conditions, such as in blood circulation, is questioned [18,100]. Candidate materials for this approach are chemical structures presenting a hydrophilic framework and several grafted hydrophobic moieties, or protonatable groups: examples are gelatin, cholesterol, polysaccharides and pollulans [101–104]. For example, Nakai and coworkers [105] designed anionic NGs for protein delivery exploiting the self-assembly of cholesteryl groups grafted on hyaluronic acid (HA, Figure 5). Moshe et al. [106] modified PVA with hydrophobic moieties (isopropylacrylamide), obtaining an amphiphilic polymer with self-assembling properties, which gave rise to 3D nanonetworks. The further functionalization with boric acid domains improved NG stability, promoting their use in a sprayable form for mucosal tissue applications.

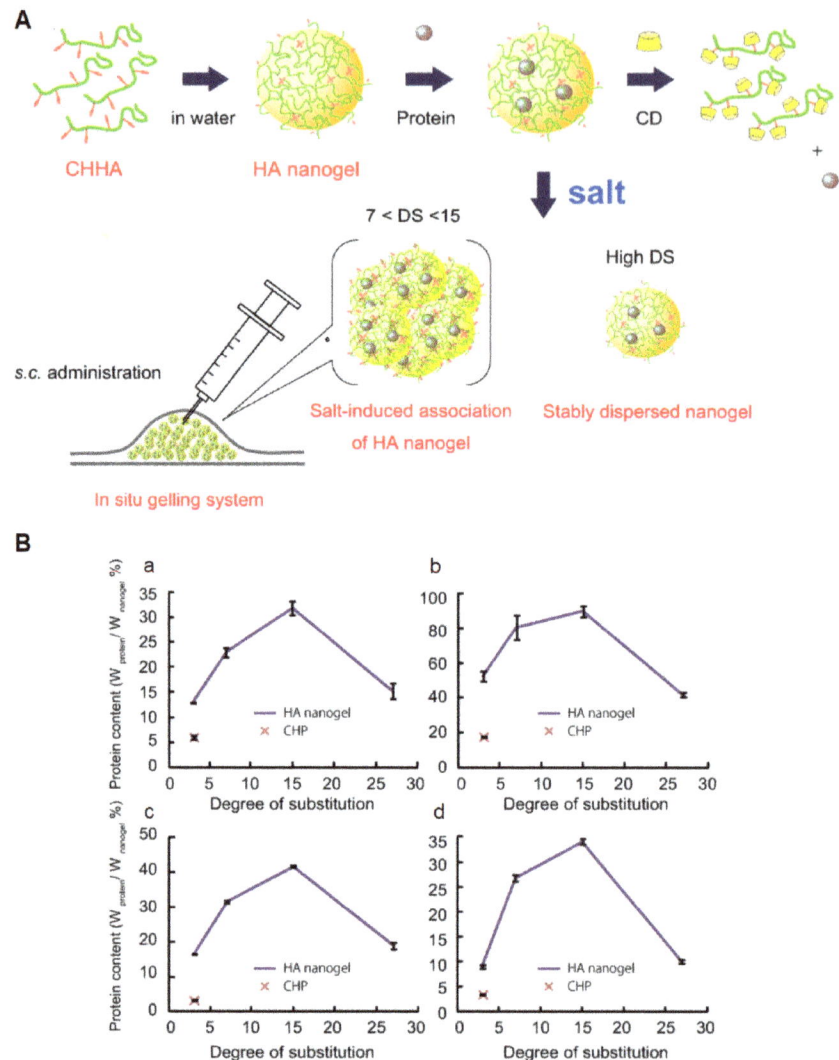

Figure 5. Self-assembling of hyaluronic acid (HA) NGs functionalized with cholesteryl groups. (**A**) Scheme of nanoscaffold design, protein loading and administration. (**B**) Protein-hosting capacity in the proposed NGs, using different payloads: rhGH (**a**), EPO (**b**), lysozyme (**c**) and exendin-4 (**d**). Adapted with permission from Nakai et al. [105]. Copyright (2012) WILEY-VCH Verlag GmbH and Co. KGaA.

3.3. Other Functionalization Strategies

To enhance the performance of NGs in drug delivery and to elicit specific therapeutic responses, the grafting of specific groups, including biomolecules, such as peptides, proteins, and growth factors represents a key approach. It can be performed through the orthogonal functionalization of the starting materials using different techniques, which require the presence of specific chemical moieties. In particular, the functionalization is usually addressed in the following ways:

- Click chemistry. The same chemical groups discussed in the previous section for NG formation can be applied to link bio-functionalities to the nanocarriers.

- Activation of esters to form amide bonds, under mild conditions. Amides are versatile linkages, characterized by unique stability towards extreme chemical environments.
- Isocyanate-based chemistry, also through modifications with alcohols, amines, and thiols. The methodology ensures high yields and stability; however, some limitations occur in the biological applications due to toxicity issues and to the sensitivity of the isocyanate to moisture.
- Imine and oxime linkages. They ensure bond reversibility due to the imine equilibrium and potential oxime hydrolysis under aqueous acidic conditions; this approach can be used to design NG systems where the release of biomolecules is tuned by hydrolysis activation.
- Ring-opening reaction. It represents a very versatile method to graft desired heteroatoms on the polymer backbone.

A thorough description of these reactions is reported by Blasco et al. [107] and Mauri et al. [12].

4. Advanced Fabrication Technologies

Associated to the need of having specific functional moieties for tunable drug release and targeted therapy, one major challenge is the control on NG size distribution coupled to process scalability. Conventional fabrication methods can result in extensive NG polydispersity and practical difficulties to modulate the physicochemical properties of the nanocarriers, unless the polymeric material is changed. In this context, microfluidics offers high-controllable and large-scale production yields, opening new scenarios to advanced NG design [108]. Moreover, the integration with 3D printing technologies can offer additional benefits to the production of 3D structured NGs for biomedical and pharmaceutical applications [109].

4.1. Microfluidic-Assisted Fabrication

Microfluidics has emerged as an innovative and advantageous approach to manipulate small amounts of reagents with accurate control on mixing and physical processes at the microscale. The advantages offered by microfluidics include miniaturization, minimized reagent consumption, decreased reaction time and enhanced process accuracy and efficiency [110]. In this framework, microfluidic technology has allowed the synthesis of a wide range of micro- and nanoproducts, including NGs, with a superior control over the product yield and the requested physico-chemical properties [110].

Indeed, by controlling the microfluidic conditions (including chip design, fluid rheology, and flow rates), it is possible to customize the nanosystem size, polydispersity, surface properties, payload delivery, and release profile [111]. However, due to the complexity of fluid dynamic processes, costly iterative approaches are often required to achieve the desired performance. Numerical investigations have been carried out and experimentally validated on several microfluidic devices to gain a predictive understanding of the most influential design parameters [112,113]. A significant contribution to the field has been made by Lashkaripour et al. [114], who developed an open-source tool (DAFD: Design Automation of Fluid Dynamics) that leverages machine learning to predict the parameters (e.g., droplet diameter and production rate) of flow-focusing droplet generators. Although the initial version supports a limited set of fluids, this tool has the potential to enable application-specific design optimization facilitating the adoption of functional microfluidic platforms without demanding expertise and resources.

Furthermore, since microfluidic devices often run with continuous flows, they supply high-throughput production of nanonetworks with the same quality over time, paving the way to industrial scale up [115,116]. All these benefits currently promote microfluidics as a robust approach for the synthesis of size-controlled NGs as well as for the encapsulation of antibodies, cells, or proteins with potential outcomes in the landscape of drug delivery and cell therapies [117].

4.1.1. Materials for Microfluidic Devices

Besides structural and operating parameters (e.g., channel dimensions, number of inlets), the choice of the material for the fabrication of the microfluidic device is a fundamental crossroad [118]. Indeed, each material has its own features regarding processability, cost, capacity to withstand high pressures and flow rates, and compatibility with organic solvents.

Polydimethylsiloxane (PDMS) and poly(methyl methacrylate) (PMMA) are among the most commonly used polymers for microfluidic devices. PDMS devices are usually obtained by soft-lithography [119]. In a standard soft-lithography process, PDMS replicas are obtained from lithographic masters. Although 3D printing can be alternatively used to manufacture thermoplastic masters for PDMS devices, the printing resolution is still not enough to justify a shift from traditional lithographic techniques [120,121]. The PDMS replica is then bonded to a glass substrate to seal the channels of the microfluidic device (Figure 6A). Alternatively, the obtained replica can be used as a stamp for the hot embossing of fluoropolymers [122]. Although PDMS can be easily tailored to a variety of geometries, ranging from 3D channel structures to multicompartment systems [123,124], its main disadvantage is the susceptibility to swelling when exposed to many common organic solvents, such as acetone [125]. Furthermore, the low elastic modulus of PDMS represents a shortcoming for applications requiring high pressure and high flow rate, because these parameters significantly alter the channel geometry.

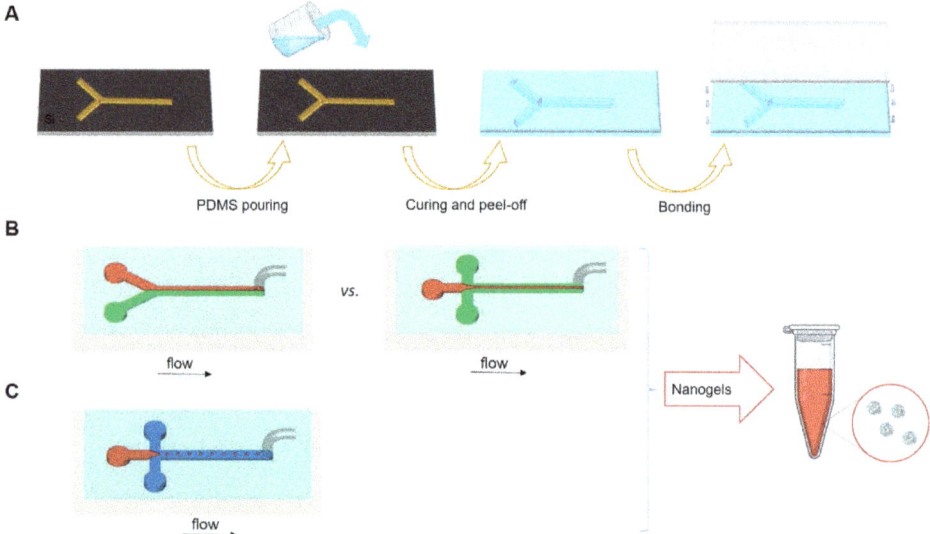

Figure 6. Microfluidics-assisted NG production. (**A**) Polydimethylsiloxane (PDMS) replica molding; (**B**,**C**) microfabricated microfluidic platforms: schematic of a Y-shaped microfluidic mixer (**B**, left), cross-shaped planar flow focusing mixer (**B**, right) and droplet microfluidic platform (**C**).

Alternative polymers for microfluidic applications include polytetrafluoroethylene (PTFE) and cyclic olefin copolymer (COC), which are microstructured by hot-embossing techniques [126,127]. In addition, microfluidic devices can be also prepared by laser direct-writing of thermoplastic materials (for example, polystyrene [128] and polycarbonate [129]) at low cost and fast speed.

Although less used, glass capillary microfluidic devices have also received considerable attention. Glass capillaries can tolerate a wider range of solvents compared to their lithographically fabricated PDMS counterparts and have excellent resistance to high pressures and flow rates. Moreover, their wettability can be easily modified into hydrophobic or hydrophilic by simple chemical reactions with an appropriate surface modifier [130]. However, a laborious and technically challenging fabrication procedure often hampers the use of these devices [131].

4.1.2. Microfluidic Mixing

The main applications of microfluidics for NG synthesis are focused on the nanoprecipitation and self-assembly principles. These approaches are based on the fast mixing of a solution containing the NG building blocks (polymers and biomolecules) with a miscible non-solvent of the polymers, and their use is encouraged by the straightforward procedures and their data reproducibility [132]. Several microfluidic platforms have been developed to produce nanosystems with precisely controlled diameters: the precipitation or the quick assembly of nanonetwork precursors result in smaller particle size and narrower size distribution, in comparison to the conventional methods [133,134].

In its simpler version, passive mixing has been implemented using Y- and T-shaped microfluidic schemes, which consist of two inlets and one outlet (Figure 6B). In this approach, the mixing primarily occurs at the interface of the two parallel streams flowing alongside and strongly depends on the diffusion rate [116]. To further increase the mixing, a variety of barriers to generate a chaotic flow of the molecules in the mixing channel have been added, as shown in the herringbone mixer [135] and Tesla structured channel [136].

In addition to these setups, hydrodynamic flow focusing (HFF) systems are among the most recognized continuous flow mixing techniques [137]. A 2D HFF mixer usually consists of three inlet microchannels and one central outlet channel organized as such that two outer fluid flows of non-solvent (continuous phases) horizontally compress the central fluid (dispersed phase) containing the nanoparticle precursors (Figure 6B). The squeezing of the core solution by the side sheath flows results in shorter mixing time leading to rapid nanoparticles production [111]. Besides the 2D-HFF devices, where the central flow is only focused on the horizontal plane, 3D-HFF have been developed to further improve the size uniformity of the synthesized nanomaterials [138].

Alginate NGs were successfully synthesized by Bazban-Shotorbani et al. [108] using HFF procedure in a PDMS microfluidic platform. Alginate aqueous solution was used as the core flow and was hydrodynamically focused by the lateral $CaCl_2$ streams into a narrowly focused stream; hence, the polymer chains were crosslinked by calcium ions to form the alginate-based nanonetwork. Considering the anionic polyelectrolytic nature of the alginate chains, NGs were formed through a controlled diffusion-mediated mass transfer of Ca^{2+} ions into the focused polymer solution stream. The key parameter in determining NG size is the flow ratio of inlet sheath and core flows, which in turn affects the mixing time. The fine tuning of process parameters has led the production of smaller and more monodisperse nanostructures than the corresponding counterparts obtainable by the standard bulk synthesis. Furthermore, a correlation between on-chip mixing time and the average pore size of the synthesized alginate NGs was also demonstrated. In particular, a higher flow ratio between core and sheath fluid resulted in smaller and more compact nanonetwork which in turn led to higher encapsulation efficiency and slower and more controlled drug release. Indeed, NGs synthesized by the conventional bulk methods usually have a burst release and low encapsulation efficiency due to their large pore size. On the base of these findings, Mahmoudi and coworkers [139] investigated the effect of TGF-β3-loaded alginate NGs obtained through a co-flow microfluidic system on the differentiation of mesenchymal stem cells (MSCs) demonstrating their superior performance in terms of TGF-β3 release and better chondrogenic differentiation respect to the bulk synthesized nanoscaffolds.

Although alginate has gained far more attention given its unique features such as ease of gelation and biocompatibility, to date, various kinds of polymeric NGs have been obtained through microfluidic-based approaches. Among the most notable examples, Agnello and collaborators [140] employed hyaluronic acid (HA) derivatives, functionalized with octadecyl and ethylenediamine moieties to produce nanostructures by modulating its self-assembly into a split-and-recombine micromixer. The peculiar property of the obtained HA-based specimens lies in the responsive behavior to the ionic strength, which can be exploited to produce a controlled coacervation of the polymer when mixed with aqueous salt solutions. The resulting NG size can be finely controlled in the range 150–400 nm by regulating the flow ratio. Furthermore, thanks to the very mild conditions used for their production, these nanomaterials have been used for the encapsulation of Imatinib, a FDA-approved drug to treat cell proliferation in leukemia [141]. Antiangiogenic potential of Imatinib-loaded NGs was also evaluated in vitro by using human retinal pigment epithelial cells and human umbilical vein endothelial cells.

The key concepts of controlled drug release and targeted cellular uptake have recently been addressed by the work of Elkassih et al. [142], where a fully degradable disulfide crosslinked nanogel was synthesized by oxidative radical polymerization of 2,2′-(ethylenedioxy)diethanethiol (EDDET) with different crosslinking agents. A commercial microfluidic mixing platform with herringbone rapid mixing features was used in the presence of a non-ionic Pluronic surfactant to ensure NG stability. Considering the poly(EDDET) backbone and the crosslinking junctions entirely characterized by disulfide bonds, these NGs were able to degrade intracellularly in response to redox potential through thiol-disulfide exchange reactions. Other examples of redox-responsive nanocarriers have been recently reviewed by Mi et al. [143] and successfully applied to target drugs in specific human pathological conditions, such as cancer cells characterized by remarkable higher levels of glutathione.

4.1.3. Droplet-Based Microfluidics

Droplet-based microfluidics is an important subcategory of the microfluidic technologies, which manipulates immiscible/partially miscible phases to generate droplets. The droplet phase is termed as dispersed phase, while the medium phase is identified as continuous phase. Droplet generation occurs when the discrete phase intersects with an immiscible continuous phase which forces it to break into the bulk fluid stream [125] (Figure 6C). The generated droplets can then be subjected to an external trigger, such as UV light or ionic environment, to induce their crosslinking. This droplet-based microfluidic method can also allow the simultaneous loading of cells, drugs or growth factors by simply adding them into the pre-gel solution, before injection into the inner stream channel for droplet formation [144].

Conventional droplet-based microfluidics has been widely used to generate monodisperse W/O and O/W microemulsions with extremely high controllability. The most popular geometries of droplet generators applied in single emulsion preparation are co-flow, flow-focusing, and T-junction, working under a dripping or jetting regimen [118,137]. The droplet size can be tuned by adjusting the flow rates and ratios of the two phases; however, it is also affected by the viscosity of the dispersed phase, the channel and orifice diameter, and the flow regimes. By a fine tuning of process parameters using low concentrations of nanoparticle precursors, the generation of monodisperse nano-emulsions can be theoretically obtained. Indeed, if the dispersed phase is partially miscible with the continuous one, the formed micro-droplets begin to shrink after the initial creation of an interface leading to the formation of nanoscale systems. Rondeau and Cooper-White [145] fabricated crosslinked alginate-based NGs with an average size ranging from 10 to 300 nm through W/O droplet microfluidics using a low-concentrated aqueous alginate solution and dimethyl carbonate as the dispersed and continuous phase, respectively. Since water and the continuous phase are partially miscible, water diffuses from the polymeric droplets into the transport fluid causing the shrinkage of the drops and the condensation of the polymer phase. However, due to the low concentrations of particle building material and the slow formation of droplets (one by one), the production efficiency of NGs by droplet microfluidics is usually limited and much lower than that of micron-sized particles [137].

To address this issue, controlled generation of nanosized mono-dispersed droplets was recently achieved through a novel droplet microfluidics/nanofluidics strategy [146].

Monodisperse nanosized W/O emulsions with controllable sizes were formed from three different protein solutions (i.e., reconstituted silk fibroin, β-lactoglobulin, and lysozyme), which were used as the dispersed phase in a continuous immiscible oil phase (a fluorinated oil containing 2% w/w fluorosurfactant). These nano-emulsions were then incubated to promote protein self-assembly resulting in the formation of protein nanoscaffolds stabilized by supramolecular fibrils from the three different proteins. The capability of these NGs to penetrate through mammalian cell membrane and deliver their cargo intracellularly was demonstrated, showing their potential as nano-vehicles for drug delivery in biomedical applications. Table 2 enlists the most recent applications of NGs synthesized through microfluidics-assisted technologies.

Table 2. Microfluidic approaches used in NG design (n.a. = not available drug release studies).

Microfluidic Technique	Materials	Nanogel Size (nm)	Payload	Application Field	References
Microfluidic mixer	Alginate	68–138	Bovine serum albumin (BSA)	Protein delivery	[108]
		45–125	TGF-β3	Growth factors delivery and tissue engineering	[139]
	Hyaluronic acid derivatives	150–400	n.a.	Drug and peptide delivery	[140]
	Hyaluronic acid-cyRGDC derivative	193.2–242.9	Imatinib	Antiangiogenic effect	[141]
	Hyaluronic acid, octenyl and succinic anhydride	115–321	Antimicrobial-peptidomimetic	Antibacterial activity	[147]
		174–194	Anti-biofilm peptide DJK-5	Pseudomonas aeruginosa LESB58 high bacterial density infection	[148]
	2,2′(ethylenedioxy) diethanethiol (EDDET)	60–70	Rhodamine B	Tumor therapy	[142]
Droplet generator	Alginate	10–300	n.a.	Multipurpose	[145]
	Proteins (reconstituted silk fibroin, β-lactoglobulin, lysozyme)	50–2500	Fluorescent marker	Intracellular delivery	[146]
	Hyaluronic acid	80–160	Proteins	Cancer therapy	[149]

4.2. The Challenge of 3D printing

3D printing represents a milestone in the development of advanced biomaterials for tissue engineering and regenerative medicine approach. In particular, it has revolutionized the traditional design of hydrogel systems improving their properties as excellent cell carriers. This technique has also been implemented for the generation of 3D structured nanogels, where they are 3D-printed in a configuration of drug/photoinitiator-loaded nanoparticles, liposomes, or nano-emulsions suspended in hydrogel matrices [109]. NGs are generally incorporated in hydrogel without affecting the rheological properties of 3D-printed systems and can also carry photoinitiators to improve gelation process or mechanical strength of the printed scaffolds [150]. For example, Liu and coworkers [151] incorporated Pluronic-based NGs encapsulating simvastatin into a 3D-printed porous titanium alloys for orthopedic applications; the combination of them has allowed overcoming the main drawbacks related to the poor compatibility of titanium with bone ingrowth and the limited mechanical properties of NGs, resulting in a composite biomaterial able to promote osteogenesis and neovascularization. Otherwise, NGs intended as building units of a hydrogel network are discussed in the work of Muller et al. [152]: they mixed pure and acrylate-modified Pluronic and, after the printing process and UV exposure, inter-linked nanostructures were formed. By subsequent elution of the unmodified Pluronic, nanostructured hydrogel represented the final product suitable as a bioink and for cell encapsulation.

However, 3D printing of NGs mandates for further research in the academic and industrial sector, and it represents one of the current challenges in the design of these smart nano-vehicles.

5. Conclusions

NGs are promising nanosystems to treat a wide range of acute and chronic healthcare scenarios. Different techniques have been developed over the years to meet the criteria of tunable size, morphology, physico-chemical properties, controlled drug delivery and selective cellular uptake. Chemical crosslinking methods, such as polymerization and interfacial reaction, have been increasingly improved to overcome the shortcomings of available formulations with a particular attention to biocompatibility issues. On the other side, physical approaches have been refined to produce NGs showing higher degree of response to biochemical stimuli. Microfluidics and 3D printing represent innovative strategies to produce customized nanoscaffolds for a wide range of applications, thanks to their high efficiency, low cost, and scalability. At present, no conclusive strategy can be identified for NG production: arguably, the smart combination of physico-chemical design principles with advanced manufacturing platforms (e.g., microfluidics and additive manufacturing) will be the driver for innovation in the field of NGs in the near future.

Author Contributions: Conceptualization, E.M. and S.M.G.; writing—original draft preparation, E.M., S.M.G. and A.R.; writing—review and editing, E.M., S.M.G. and A.R.; visualization, M.T. and A.R.; supervision, A.R. All authors have read and agreed to the published version of the manuscript.

Funding: This research received no external funding.

Conflicts of Interest: The authors declare no conflict of interest.

References

1. Su, S.; Kang, P.M. Systemic Review of Biodegradable Nanomaterials in Nanomedicine. *Nanomaterials* **2020**, *10*, 656. [CrossRef] [PubMed]
2. Lombardo, D.; Kiselev, M.A.; Caccamo, M.T. Smart Nanoparticles for Drug Delivery Application: Development of Versatile Nanocarrier Platforms in Biotechnology and Nanomedicine. *J. Nanomater.* **2019**, *2019*, 3702518. [CrossRef]
3. Saldanha, P.L.; Lesnyak, V.; Manna, L. Large scale syntheses of colloidal nanomaterials. *Nano Today* **2017**, *12*, 46–63. [CrossRef]
4. Chamundeeswari, M.; Jeslin, J.; Verma, M.L. Nanocarriers for drug delivery applications. *Environ. Chem. Lett.* **2019**, *17*, 849–865. [CrossRef]
5. Vicario-de-la-Torre, M.; Forcada, J. The Potential of Stimuli-Responsive Nanogels in Drug and Active Molecule Delivery for Targeted Therapy. *Gels* **2017**, *3*, 16. [CrossRef]
6. Ghaeini-Hesaroeiye, S.; Razmi Bagtash, H.R.; Boddohi, S.; Vasheghani-Farahani, E.; Jabbari, E. Thermoresponsive Nanogels Based on Different Polymeric Moieties for Biomedical Applications. *Gels* **2020**, *6*, 20. [CrossRef]
7. Soni, K.S.; Desale, S.S.; Bronich, T.K. Nanogels: An overview of properties, biomedical applications and obstacles to clinical translation. *J. Control. Release* **2016**, *240*, 109–126. [CrossRef]
8. Gratton, S.E.A.; Pohlhaus, P.D.; Lee, J.; Guo, J.; Cho, M.J.; DeSimone, J.M. Nanofabricated particles for engineered drug therapies: A preliminary biodistribution study of PRINT™ nanoparticles. *J. Control. Release* **2007**, *121*, 10–18. [CrossRef]
9. Lockhart, J.N.; Hmelo, A.B.; Harth, E. Electron beam lithography of poly(glycidol) nanogels for immobilization of a three-enzyme cascade. *Polym. Chem.* **2018**, *9*, 637–645. [CrossRef]
10. Lima, C.S.A.d.; Balogh, T.S.; Varca, J.P.R.O.; Varca, G.H.C.; Lugão, A.B.; A Camacho-Cruz, L.; Bucio, E.; Kadlubowski, S.S. An Updated Review of Macro, Micro, and Nanostructured Hydrogels for Biomedical and Pharmaceutical Applications. *Pharmaceutics* **2020**, *12*, 970. [CrossRef] [PubMed]
11. Kaur, M.; Sudhakar, K.; Mishra, V. Fabrication and biomedical potential of nanogels: An overview. *Int. J. Polym. Mater. Polym. Biomater.* **2019**, *68*, 287–296. [CrossRef]
12. Mauri, E.; Perale, G.; Rossi, F. Nanogel Functionalization: A Versatile Approach To Meet the Challenges of Drug and Gene Delivery. *ACS Appl. Nano Mater.* **2018**, *1*, 6525–6541. [CrossRef]
13. Qureshi, M.A.; Khatoon, F. Different types of smart nanogel for targeted delivery. *J. Sci. Adv. Mater. Dev.* **2019**, *4*, 201–212. [CrossRef]
14. Yin, Y.; Hu, B.; Yuan, X.; Cai, L.; Gao, H.; Yang, Q. Nanogel: A Versatile Nano-Delivery System for Biomedical Applications. *Pharmaceutics* **2020**, *12*, 290. [CrossRef]

15. Torchilin, V.P. Multifunctional, stimuli-sensitive nanoparticulate systems for drug delivery. *Nat. Rev. Drug Discov.* **2014**, *13*, 813–827. [CrossRef] [PubMed]
16. Drozdov, A.D.; Sanporean, C.G.; deClaville Christiansen, J. Modeling the effect of ionic strength on swelling of pH-sensitive macro- and nanogels. *Mater. Today Commun.* **2016**, *6*, 92–101. [CrossRef]
17. Eslami, P.; Rossi, F.; Fedeli, S. Hybrid Nanogels: Stealth and Biocompatible Structures for Drug Delivery Applications. *Pharmaceutics* **2019**, *11*, 71. [CrossRef]
18. Zhang, X.; Malhotra, S.; Molina, M.; Haag, R. Micro- and nanogels with labile crosslinks—From synthesis to biomedical applications. *Chem. Soc. Rev.* **2015**, *44*, 1948–1973. [CrossRef]
19. Hamzah, Y.B.; Hashim, S.; Rahman, W.A.W.A. Synthesis of polymeric nano/microgels: A review. *J. Polym. Res.* **2017**, *24*, 134. [CrossRef]
20. Sasaki, Y.; Akiyoshi, K. Nanogel engineering for new nanobiomaterials: From chaperoning engineering to biomedical applications. *Chem. Rec.* **2010**, *10*, 366–376. [CrossRef] [PubMed]
21. Lovell, P.A.; Schork, F.J. Fundamentals of Emulsion Polymerization. *Biomacromolecules* **2020**, *21*, 4396–4441. [CrossRef] [PubMed]
22. Raghupathi, K.; Eron, S.J.; Anson, F.; Hardy, J.A.; Thayumanavan, S. Utilizing Inverse Emulsion Polymerization To Generate Responsive Nanogels for Cytosolic Protein Delivery. *Mol. Pharm.* **2017**, *14*, 4515–4524. [CrossRef]
23. Peres, L.B.; dos Anjos, R.S.; Tappertzhofen, L.C.; Feuser, P.E.; de Araújo, P.H.H.; Landfester, K.; Sayer, C.; Muñoz-Espí, R. pH-responsive physically and chemically cross-linked glutamic-acid-based hydrogels and nanogels. *Eur. Polym. J.* **2018**, *101*, 341–349. [CrossRef]
24. Gonzalez-Ayon, M.A.; Cortez-Lemus, N.A.; Zizumbo-Lopez, A.; Licea-Claverie, A. Nanogels of Poly(N-Vinylcaprolactam) Core and Polyethyleneglycol Shell by Surfactant Free Emulsion Polymerization. *Soft Mater.* **2014**, *12*, 315–325. [CrossRef]
25. Gupta, A.; Eral, H.B.; Hatton, T.A.; Doyle, P.S. Nanoemulsions: Formation, properties and applications. *Soft Matter* **2016**, *12*, 2826–2841. [CrossRef]
26. Wik, J.; Bansal, K.K.; Assmuth, T.; Rosling, A.; Rosenholm, J.M. Facile methodology of nanoemulsion preparation using oily polymer for the delivery of poorly soluble drugs. *Drug Deliv. Transl. Res.* **2020**, *10*, 1228–1240. [CrossRef] [PubMed]
27. Urakami, H.; Hentschel, J.; Seetho, K.; Zeng, H.; Chawla, K.; Guan, Z. Surfactant-Free Synthesis of Biodegradable, Biocompatible, and Stimuli-Responsive Cationic Nanogel Particles. *Biomacromolecules* **2013**, *14*, 3682–3688. [CrossRef]
28. Cheng, W.; Wang, G.; Kumar, J.N.; Liu, Y. Surfactant-Free Emulsion-Based Preparation of Redox-Responsive Nanogels. *Macromol. Rapid Commun.* **2015**, *36*, 2102–2106. [CrossRef]
29. Ashrafizadeh, M.; Tam, K.C.; Javadi, A.; Abdollahi, M.; Sadeghnejad, S.; Bahramian, A. Synthesis and physicochemical properties of dual-responsive acrylic acid/butyl acrylate cross-linked nanogel systems. *J. Colloid Interface Sci.* **2019**, *556*, 313–323. [CrossRef]
30. Wang, X.; Peng, Y.; Peña, J.; Xing, J. Preparation of ultrasmall nanogels by facile emulsion-free photopolymerization at 532 nm. *J. Colloid Interface Sci.* **2021**, *582*, 711–719. [CrossRef]
31. Li, M.; Rouaud, O.; Poncelet, D. Microencapsulation by solvent evaporation: State of the art for process engineering approaches. *Int. J. Pharm* **2008**, *363*, 26–39. [CrossRef] [PubMed]
32. Watts, P.J.; Davies, M.C.; Melia, C.D. Microencapsulation using emulsification/solvent evaporation: An overview of techniques and applications. *Crit. Rev. Ther. Drug* **1990**, *7*, 235–259.
33. Hennink, W.E.; van Nostrum, C.F. Novel crosslinking methods to design hydrogels. *Adv. Drug Deliv. Rev.* **2002**, *54*, 13–36. [CrossRef]
34. Kabanov, A.V.; Vinogradov, S.V. Nanogels as pharmaceutical carriers: Finite networks of infinite capabilities. *Angew. Chem. Int. Ed.* **2009**, *48*, 5418–5429. [CrossRef]
35. Gerson, T.; Makarov, E.; Senanayake, T.H.; Gorantla, S.; Poluektova, L.Y.; Vinogradov, S.V. Nano-NRTIs demonstrate low neurotoxicity and high antiviral activity against HIV infection in the brain. *Nanomedicine* **2014**, *10*, 177–185. [CrossRef]
36. Vinogradov, S.; Batrakova, E.; Kabanov, A. Poly(ethylene glycol)-polyethyleneimine NanoGel (TM) particles: Novel drug delivery systems for antisense oligonucleotides. *Colloid Surf. B* **1999**, *16*, 291–304. [CrossRef]
37. Senanayake, T.H.; Gorantla, S.; Makarov, E.; Lu, Y.; Warren, G.; Vinogradov, S.V. Nanogel-Conjugated Reverse Transcriptase Inhibitors and Their Combinations as Novel Antiviral Agents with Increased Efficacy against HIV-1 Infection. *Mol. Pharm.* **2015**, *12*, 4226–4236. [CrossRef]
38. Warren, G.; Makarov, E.; Lu, Y.; Senanayake, T.; Rivera, K.; Gorantla, S.; Poluektova, L.Y.; Vinogradov, S.V. Amphiphilic Cationic Nanogels as Brain-Targeted Carriers for Activated Nucleoside Reverse Transcriptase Inhibitors. *J. Neuroimmune Pharmacol.* **2015**, *10*, 88–101. [CrossRef]
39. Vismara, I.; Papa, S.; Veneruso, V.; Mauri, E.; Mariani, A.; De Paola, M.; Affatato, R.; Rossetti, A.; Sponchioni, M.; Moscatelli, D.; et al. Selective Modulation of A1 Astrocytes by Drug-Loaded Nano-Structured Gel in Spinal Cord Injury. *ACS Nano* **2020**, *14*, 360–371. [CrossRef]
40. Mauri, E.; Veglianese, P.; Papa, S.; Mariani, A.; De Paola, M.; Rigamonti, R.; Chincarini, G.M.F.; Rimondo, S.; Sacchetti, A.; Rossi, F. Chemoselective functionalization of nanogels for microglia treatment. *Eur. Polym. J.* **2017**, *94*, 143–151. [CrossRef]
41. Li, S.; Hu, L.; Li, D.; Wang, X.; Zhang, P.; Wang, J.; Yan, G.; Tang, R. Carboxymethyl chitosan-based nanogels via acid-labile ortho ester linkages mediated enhanced drug delivery. *Int. J. Biol. Macromol.* **2019**, *129*, 477–487. [CrossRef]

42. Oh, J.K.; Siegwart, D.J.; Lee, H.-i.; Sherwood, G.; Peteanu, L.; Hollinger, J.O.; Kataoka, K.; Matyjaszewski, K. Biodegradable Nanogels Prepared by Atom Transfer Radical Polymerization as Potential Drug Delivery Carriers: Synthesis, Biodegradation, in Vitro Release, and Bioconjugation. *J. Am. Chem. Soc.* **2007**, *129*, 5939–5945. [CrossRef] [PubMed]
43. Zetterlund, P.B.; Kagawa, Y.; Okubo, M. Controlled/Living Radical Polymerization in Dispersed Systems. *Chem. Rev.* **2008**, *108*, 3747–3794. [CrossRef] [PubMed]
44. Sanson, N.; Rieger, J. Synthesis of nanogels/microgels by conventional and controlled radical crosslinking copolymerization. *Polym. Chem.* **2010**, *1*, 965–977. [CrossRef]
45. Gao, H.; Matyjaszewski, K. Synthesis of functional polymers with controlled architecture by CRP of monomers in the presence of cross-linkers: From stars to gels. *Prog. Polym. Sci.* **2009**, *34*, 317–350. [CrossRef]
46. Gao, H.; Li, W.; Matyjaszewski, K. Synthesis of Polyacrylate Networks by ATRP: Parameters Influencing Experimental Gel Points. *Macromolecules* **2008**, *41*, 2335–2340. [CrossRef]
47. Kim, S.; Sikes, H.D. Radical polymerization reactions for amplified biodetection signals. *Polym. Chem.* **2020**, *11*, 1424–1444. [CrossRef]
48. Wang, C.E.; Stayton, P.S.; Pun, S.H.; Convertine, A.J. Polymer nanostructures synthesized by controlled living polymerization for tumor-targeted drug delivery. *J. Control. Release* **2015**, *219*, 345–354. [CrossRef]
49. Nicolas, J.; Guillaneuf, Y.; Lefay, C.; Bertin, D.; Gigmes, D.; Charleux, B. Nitroxide-mediated polymerization. *Prog. Polym. Sci.* **2013**, *38*, 63–235. [CrossRef]
50. Rodrigues, P.R.; Vieira, R.P. Advances in atom-transfer radical polymerization for drug delivery applications. *Eur. Polym. J.* **2019**, *115*, 45–58. [CrossRef]
51. Perrier, S. 50th Anniversary Perspective: RAFT Polymerization—A User Guide. *Macromolecules* **2017**, *50*, 7433–7447. [CrossRef]
52. Ni, Y.; Zhang, L.; Cheng, Z.; Zhu, X. Iodine-mediated reversible-deactivation radical polymerization: A powerful strategy for polymer synthesis. *Polym. Chem.* **2019**, *10*, 2504–2515. [CrossRef]
53. Taton, D.; Baussard, J.-F.; Dupayage, L.; Poly, J.; Gnanou, Y.; Ponsinet, V.; Destarac, M.; Mignaud, C.; Pitois, C. Water soluble polymeric nanogels by xanthate-mediated radical crosslinking copolymerisation. *Chem. Commun.* **2006**, 1953–1955. [CrossRef]
54. Delaittre, G.; Save, M.; Charleux, B. Nitroxide-Mediated Aqueous Dispersion Polymerization: From Water-Soluble Macroalkoxyamine to Thermosensitive Nanogels. *Macromol. Rapid Commun.* **2007**, *28*, 1528–1533. [CrossRef]
55. Amamoto, Y.; Kikuchi, M.; Otsuka, H.; Takahara, A. Arm-replaceable star-like nanogels: Arm detachment and arm exchange reactions by dynamic covalent exchanges of alkoxyamine units. *Polym. J.* **2010**, *42*, 860–867. [CrossRef]
56. Zetterlund, P.B.; Alam, N.; Okubo, M. Effects of the oil–water interface on network formation in nanogel synthesis using nitroxide-mediated radical copolymerization of styrene/divinylbenzene in miniemulsion. *Polymer* **2009**, *50*, 5661–5667. [CrossRef]
57. Kumar, P.; Wasim, L.; Chopra, M.; Chhikara, A. Co-delivery of Vorinostat and Etoposide via Disulfide Cross-Linked Biodegradable Polymeric Nanogels: Synthesis, Characterization, Biodegradation, and Anticancer Activity. *AAPS PharmSciTech* **2018**, *19*, 634–647. [CrossRef]
58. Dinari, A.; Abdollahi, M.; Sadeghizadeh, M. Design and fabrication of dual responsive lignin-based nanogel via "grafting from" atom transfer radical polymerization for curcumin loading and release. *Sci. Rep.* **2021**, *11*, 1962. [CrossRef]
59. Lou, S.; Zhang, X.; Zhang, M.; Ji, S.; Wang, W.; Zhang, J.; Li, C.; Kong, D. Preparation of a dual cored hepatoma-specific star glycopolymer nanogel via arm-first ATRP approach. *Int. J. Nanomed.* **2017**, *12*, 3653–3664. [CrossRef]
60. Bencherif, S.A.; Washburn, N.R.; Matyjaszewski, K. Synthesis by AGET ATRP of Degradable Nanogel Precursors for In Situ Formation of Nanostructured Hyaluronic Acid Hydrogel. *Biomacromolecules* **2009**, *10*, 2499–2507. [CrossRef]
61. Piogé, S.; Tran, T.N.; McKenzie, T.G.; Pascual, S.; Ashokkumar, M.; Fontaine, L.; Qiao, G. Sono-RAFT Polymerization-Induced Self-Assembly in Aqueous Dispersion: Synthesis of LCST-type Thermosensitive Nanogels. *Macromolecules* **2018**, *51*, 8862–8869. [CrossRef]
62. Bhuchar, N.; Sunasee, R.; Ishihara, K.; Thundat, T.; Narain, R. Degradable Thermoresponsive Nanogels for Protein Encapsulation and Controlled Release. *Bioconjugate Chem.* **2012**, *23*, 75–83. [CrossRef] [PubMed]
63. Liu, J.; Stansbury, J.W. RAFT-mediated control of nanogel structure and reactivity: Chemical, physical and mechanical properties of monomer-dispersed nanogel compositions. *Dent. Mater.* **2014**, *30*, 1252–1262. [CrossRef] [PubMed]
64. Fu, W.; Luo, C.; Morin, E.A.; He, W.; Li, Z.; Zhao, B. UCST-Type Thermosensitive Hairy Nanogels Synthesized by RAFT Polymerization-Induced Self-Assembly. *ACS Macro Lett.* **2017**, *6*, 127–133. [CrossRef]
65. Zuo, Y.; Guo, N.; Jiao, Z.; Song, P.; Liu, X.; Wang, R.; Xiong, Y. Novel reversible thermoresponsive nanogel based on poly(ionic liquid)s prepared via RAFT crosslinking copolymerization. *J. Polym. Sci. A Polym. Chem.* **2016**, *54*, 169–178. [CrossRef]
66. Matyjaszewski, K.; Gaynor, S.; Wang, J.-S. Controlled Radical Polymerizations: The Use of Alkyl Iodides in Degenerative Transfer. *Macromolecules* **1995**, *28*, 2093–2095. [CrossRef]
67. Maiti, S.; Samanta, P.; Biswas, G.; Dhara, D. Arm-First Approach toward Cross-Linked Polymers with Hydrophobic Domains via Hypervalent Iodine-Mediated Click Chemistry. *ACS Omega* **2018**, *3*, 562–575. [CrossRef]
68. Poly, J.; Wilson, D.J.; Destarac, M.; Taton, D. Synthesis of Poly(vinyl acetate) Nanogels by Xanthate-Mediated Radical Crosslinking Copolymerization. *Macromol. Rapid Commun.* **2008**, *29*, 1965–1972. [CrossRef]
69. Etchenausia, L.; Khoukh, A.; Deniau Lejeune, E.; Save, M. RAFT/MADIX emulsion copolymerization of vinyl acetate and N-vinylcaprolactam: Towards waterborne physically crosslinked thermoresponsive particles. *Polym. Chem.* **2017**, *8*, 2244–2256. [CrossRef]

70. Hein, C.D.; Liu, X.-M.; Wang, D. Click Chemistry, A Powerful Tool for Pharmaceutical Sciences. *Pharm Res.* 2008, *25*, 2216–2230. [CrossRef]
71. Mauri, E.; Rossi, F. Click chemistry for improving properties of bioresorbable polymers for medical applications. In *Bioresorbable Polymers for Biomedical Applications*; Woodhead Publishing: Cambridge, UK, 2017; pp. 303–329. [CrossRef]
72. Rostovtsev, V.V.; Green, L.G.; Fokin, V.V.; Sharpless, K.B. A Stepwise Huisgen Cycloaddition Process: Copper(I)-Catalyzed Regioselective "Ligation" of Azides and Terminal Alkynes. *Angew. Chem. Int. Ed.* 2002, *41*, 2596–2599. [CrossRef]
73. Mauri, E.; Moroni, I.; Magagnin, L.; Masi, M.; Sacchetti, A.; Rossi, F. Comparison between two different click strategies to synthesize fluorescent nanogels for therapeutic applications. *React. Funct. Polym.* 2016, *105*, 35–44. [CrossRef]
74. Obhi, N.K.; Peda, D.M.; Kynaston, E.L.; Seferos, D.S. Exploring the Graft-To Synthesis of All-Conjugated Comb Copolymers Using Azide–Alkyne Click Chemistry. *Macromolecules* 2018, *51*, 2969–2978. [CrossRef]
75. Kolb, H.C.; Sharpless, K.B. The growing impact of click chemistry on drug discovery. *Drug Discov. Today* 2003, *8*, 1128–1137. [CrossRef]
76. Bock, V.D.; Hiemstra, H.; van Maarseveen, J.H. CuI-Catalyzed Alkyne–Azide "Click" Cycloadditions from a Mechanistic and Synthetic Perspective. *Eur. J. Org. Chem.* 2006, *2006*, 51–68. [CrossRef]
77. Zhang, Y.; Ding, J.; Li, M.; Chen, X.; Xiao, C.; Zhuang, X.; Huang, Y.; Chen, X. One-Step "Click Chemistry"-Synthesized Cross-Linked Prodrug Nanogel for Highly Selective Intracellular Drug Delivery and Upregulated Antitumor Efficacy. *ACS Appl. Mater. Inter.* 2016, *8*, 10673–10682. [CrossRef] [PubMed]
78. Ding, F.; Mou, Q.; Ma, Y.; Pan, G.; Guo, Y.; Tong, G.; Choi, C.H.J.; Zhu, X.; Zhang, C. A Crosslinked Nucleic Acid Nanogel for Effective siRNA Delivery and Antitumor Therapy. *Angew. Chem. Int. Ed.* 2018, *57*, 3064–3068. [CrossRef] [PubMed]
79. Hoyle, C.E.; Lowe, A.B.; Bowman, C.N. Thiol-click chemistry: A multifaceted toolbox for small molecule and polymer synthesis. *Chem. Soc. Rev.* 2010, *39*, 1355–1387. [CrossRef]
80. Pinelli, F.; Perale, G.; Rossi, F. Coating and Functionalization Strategies for Nanogels and Nanoparticles for Selective Drug Delivery. *Gels* 2020, *6*, 6. [CrossRef]
81. Wang, J.; Wang, X.; Yan, G.; Fu, S.; Tang, R. pH-sensitive nanogels with ortho ester linkages prepared via thiol-ene click chemistry for efficient intracellular drug release. *J. Colloid Interf Sci.* 2017, *508*, 282–290. [CrossRef]
82. Zhang, Y.; Andrén, O.C.J.; Nordström, R.; Fan, Y.; Malmsten, M.; Mongkhontreerat, S.; Malkoch, M. Off-Stoichiometric Thiol-Ene Chemistry to Dendritic Nanogel Therapeutics. *Adv. Funct. Mater.* 2019, *29*, 1806693. [CrossRef]
83. Neamtu, I.; Rusu, A.G.; Diaconu, A.; Nita, L.E.; Chiriac, A.P. Basic concepts and recent advances in nanogels as carriers for medical applications. *Drug Deliv.* 2017, *24*, 539–557. [CrossRef] [PubMed]
84. Kim, K.; Choi, H.; Choi, E.S.; Park, M.-H.; Ryu, J.-H. Hyaluronic Acid-Coated Nanomedicine for Targeted Cancer Therapy. *Pharmaceutics* 2019, *11*, 301. [CrossRef]
85. Lim, K.S.; Galarraga, J.H.; Cui, X.; Lindberg, G.C.J.; Burdick, J.A.; Woodfield, T.B.F. Fundamentals and Applications of Photo-Cross-Linking in Bioprinting. *Chem. Rev.* 2020, *120*, 10662–10694. [CrossRef]
86. Williams, C.G.; Malik, A.N.; Kim, T.K.; Manson, P.N.; Elisseeff, J.H. Variable cytocompatibility of six cell lines with photoinitiators used for polymerizing hydrogels and cell encapsulation. *Biomaterials* 2005, *26*, 1211–1218. [CrossRef]
87. Chen, W.; Hou, Y.; Tu, Z.; Gao, L.; Haag, R. pH-degradable PVA-based nanogels via photo-crosslinking of thermo-preinduced nanoaggregates for controlled drug delivery. *J. Control. Release* 2017, *259*, 160–167. [CrossRef] [PubMed]
88. Kim, J.; Gauvin, R.; Yoon, H.J.; Kim, J.-H.; Kwon, S.-M.; Park, H.J.; Baek, S.H.; Cha, J.M.; Bae, H. Skin penetration-inducing gelatin methacryloyl nanogels for transdermal macromolecule delivery. *Macromol. Res.* 2016, *24*, 1115–1125. [CrossRef]
89. Cui, X.; Soliman, B.G.; Alcala-Orozco, C.R.; Li, J.; Vis, M.A.M.; Santos, M.; Wise, S.G.; Levato, R.; Malda, J.; Woodfield, T.B.F.; et al. Rapid Photocrosslinking of Silk Hydrogels with High Cell Density and Enhanced Shape Fidelity. *Adv. Healthc. Mater.* 2020, *9*, 1901667. [CrossRef]
90. DeRosa, M.C.; Crutchley, R.J. Photosensitized singlet oxygen and its applications. *Coordin. Chem. Rev.* 2002, *233–234*, 351–371. [CrossRef]
91. Sánchez-Moreno, P.; de Vicente, J.; Nardecchia, S.; Marchal, J.A.; Boulaiz, H. Thermo-Sensitive Nanomaterials: Recent Advance in Synthesis and Biomedical Applications. *Nanomaterials* 2018, *8*, 935. [CrossRef]
92. Niskanen, J.; Tenhu, H. How to manipulate the upper critical solution temperature (UCST)? *Polym. Chem.* 2017, *8*, 220–232. [CrossRef]
93. Ruscito, A.; Chiessi, E.; Toumia, Y.; Oddo, L.; Domenici, F.; Paradossi, G. Microgel Particles with Distinct Morphologies and Common Chemical Compositions: A Unified Description of the Responsivity to Temperature and Osmotic Stress. *Gels* 2020, *6*, 34. [CrossRef]
94. Cerroni, B.; Pasale, S.K.; Mateescu, A.; Domenici, F.; Oddo, L.; Bordi, F.; Paradossi, G. Temperature-Tunable Nanoparticles for Selective Biointerface. *Biomacromolecules* 2015, *16*, 1753–1760. [CrossRef]
95. Śliwa, T.; Jarzębski, M.; Andrzejewska, E.; Szafran, M.; Gapiński, J. Uptake and controlled release of a dye from thermo-sensitive polymer P(NIPAM-co-Vim). *React. Funct. Polym.* 2017, *115*, 102–108. [CrossRef]
96. Ohshio, M.; Ishihara, K.; Maruyama, A.; Shimada, N.; Yusa, S.-i. Synthesis and Properties of Upper Critical Solution Temperature Responsive Nanogels. *Langmuir* 2019, *35*, 7261–7267. [CrossRef]
97. Kocak, G.; Tuncer, C.; Bütün, V. pH-Responsive polymers. *Polym. Chem.* 2017, *8*, 144–176. [CrossRef]

98. Argentiere, S.; Blasi, L.; Morello, G.; Gigli, G. A Novel pH-Responsive Nanogel for the Controlled Uptake and Release of Hydrophobic and Cationic Solutes. *J. Phys. Chem. C* **2011**, *115*, 16347–16353. [CrossRef]
99. Li, Z.; Huang, J.; Wu, J. pH-Sensitive nanogels for drug delivery in cancer therapy. *Biomater. Sci.* **2021**, *9*, 574–589. [CrossRef] [PubMed]
100. Moreira Teixeira, L.S.; Feijen, J.; van Blitterswijk, C.A.; Dijkstra, P.J.; Karperien, M. Enzyme-catalyzed crosslinkable hydrogels: Emerging strategies for tissue engineering. *Biomaterials* **2012**, *33*, 1281–1290. [CrossRef] [PubMed]
101. Sawada, S.-i.; Yukawa, H.; Takeda, S.; Sasaki, Y.; Akiyoshi, K. Self-assembled nanogel of cholesterol-bearing xyloglucan as a drug delivery nanocarrier. *J. Biomater. Sci. Polym. Ed.* **2017**, *28*, 1183–1198. [CrossRef]
102. Morimoto, N.; Hirano, S.; Takahashi, H.; Loethen, S.; Thompson, D.H.; Akiyoshi, K. Self-Assembled pH-Sensitive Cholesteryl Pullulan Nanogel As a Protein Delivery Vehicle. *Biomacromolecules* **2013**, *14*, 56–63. [CrossRef]
103. Grimaudo, M.A.; Concheiro, A.; Alvarez-Lorenzo, C. Nanogels for regenerative medicine. *J. Control. Release* **2019**, *313*, 148–160. [CrossRef] [PubMed]
104. Kim, K.; Bae, B.; Kang, Y.J.; Nam, J.-M.; Kang, S.; Ryu, J.-H. Natural Polypeptide-Based Supramolecular Nanogels for Stable Noncovalent Encapsulation. *Biomacromolecules* **2013**, *14*, 3515–3522. [CrossRef] [PubMed]
105. Nakai, T.; Hirakura, T.; Sakurai, Y.; Shimoboji, T.; Ishigai, M.; Akiyoshi, K. Injectable Hydrogel for Sustained Protein Release by Salt-Induced Association of Hyaluronic Acid Nanogel. *Macromol. BioSci.* **2012**, *12*, 475–483. [CrossRef] [PubMed]
106. Moshe, H.; Davizon, Y.; Menaker Raskin, M.; Sosnik, A. Novel poly(vinyl alcohol)-based amphiphilic nanogels by non-covalent boric acid crosslinking of polymeric micelles. *Biomater. Sci.* **2017**, *5*, 2295–2309. [CrossRef] [PubMed]
107. Blasco, E.; Sims, M.B.; Goldmann, A.S.; Sumerlin, B.S.; Barner-Kowollik, C. 50th Anniversary Perspective: Polymer Functionalization. *Macromolecules* **2017**, *50*, 5215–5252. [CrossRef]
108. Bazban-Shotorbani, S.; Dashtimoghadam, E.; Karkhaneh, A.; Hasani-Sadrabadi, M.M.; Jacob, K.I. Microfluidic Directed Synthesis of Alginate Nanogels with Tunable Pore Size for Efficient Protein Delivery. *Langmuir* **2016**, *32*, 4996–5003. [CrossRef]
109. Cho, H.; Jammalamadaka, U.; Tappa, K. Nanogels for Pharmaceutical and Biomedical Applications and Their Fabrication Using 3D Printing Technologies. *Materials* **2018**, *11*, 302. [CrossRef]
110. Chou, W.-L.; Lee, P.-Y.; Yang, C.-L.; Huang, W.-Y.; Lin, Y.-S. Recent Advances in Applications of Droplet Microfluidics. *Micromachines* **2015**, *6*, 1249–1271. [CrossRef]
111. Zhang, Y.; Liu, D.; Zhang, H.; Santos, H.A. Chapter 7—Microfluidic mixing and devices for preparing nanoparticulate drug delivery systems. In *Microfluidics for Pharmaceutical Applications*; Santos, H.A., Liu, D., Zhang, H., Eds.; William Andrew Publishing: Norwich, NY, USA, 2019; pp. 155–177. [CrossRef]
112. Rahimi, M.; Shams Khorrami, A.; Rezai, P. Effect of device geometry on droplet size in co-axial flow-focusing microfluidic droplet generation devices. *Colloids Surf. A Physicochem. Eng. Asp.* **2019**, *570*, 510–517. [CrossRef]
113. Deng, C.; Wang, H.; Huang, W.; Cheng, S. Numerical and experimental study of oil-in-water (O/W) droplet formation in a co-flowing capillary device. *Colloids Surf. A Physicochem. Eng. Asp.* **2017**, *533*, 1–8. [CrossRef]
114. Lashkaripour, A.; Rodriguez, C.; Mehdipour, N.; Mardian, R.; McIntyre, D.; Ortiz, L.; Campbell, J.; Densmore, D. Machine learning enables design automation of microfluidic flow-focusing droplet generation. *Nat. Commun.* **2021**, *12*, 25. [CrossRef]
115. Ding, S.; Anton, N.; Vandamme, T.F.; Serra, C.A. Microfluidic nanoprecipitation systems for preparing pure drug or polymeric drug loaded nanoparticles: An overview. *Expert Opin. Drug Deliv.* **2016**, *13*, 1447–1460. [CrossRef]
116. Kim, Y.; Lee Chung, B.; Ma, M.; Mulder, W.J.M.; Fayad, Z.A.; Farokhzad, O.C.; Langer, R. Mass Production and Size Control of Lipid–Polymer Hybrid Nanoparticles through Controlled Microvortices. *Nano Lett.* **2012**, *12*, 3587–3591. [CrossRef] [PubMed]
117. Oh, J.K.; Drumright, R.; Siegwart, D.J.; Matyjaszewski, K. The development of microgels/nanogels for drug delivery applications. *Prog. Polym. Sci.* **2008**, *33*, 448–477. [CrossRef]
118. Damiati, S.; Kompella, U.B.; Damiati, S.A.; Kodzius, R. Microfluidic Devices for Drug Delivery Systems and Drug Screening. *Genes* **2018**, *9*, 103. [CrossRef] [PubMed]
119. Xia, Y.; Whitesides, G.M. SOFT LITHOGRAPHY. *Ann. Rev. Mater. Sci.* **1998**, *28*, 153–184. [CrossRef]
120. Amin, R.; Knowlton, S.; Hart, A.; Yenilmez, B.; Ghaderinezhad, F.; Katebifar, S.; Messina, M.; Khademhosseini, A.; Tasoglu, S. 3D-printed microfluidic devices. *Biofabrication* **2016**, *8*, 022001. [CrossRef] [PubMed]
121. McDonald, J.C.; Whitesides, G.M. Poly(dimethylsiloxane) as a Material for Fabricating Microfluidic Devices. *Acc. Chem. Res.* **2002**, *35*, 491–499. [CrossRef] [PubMed]
122. Begolo, S.; Colas, G.; Viovy, J.L.; Malaquin, L. New family of fluorinated polymer chips for droplet and organic solvent microfluidics. *Lab. Chip* **2011**, *11*, 508–512. [CrossRef]
123. Taylor, A.M.; Rhee, S.W.; Tu, C.H.; Cribbs, D.H.; Cotman, C.W.; Jeon, N.L. Microfluidic Multicompartment Device for Neuroscience Research. *Langmuir* **2003**, *19*, 1551–1556. [CrossRef]
124. Zhao, C.-X. Multiphase flow microfluidics for the production of single or multiple emulsions for drug delivery. *Adv. Drug Deliv. Rev.* **2013**, *65*, 1420–1446. [CrossRef] [PubMed]
125. Lee, W.; Walker, L.M.; Anna, S.L. Role of geometry and fluid properties in droplet and thread formation processes in planar flow focusing. *Phys. Fluids* **2009**, *21*, 032103. [CrossRef]
126. Nunes, P.S.; Ohlsson, P.D.; Ordeig, O.; Kutter, J.P. Cyclic olefin polymers: Emerging materials for lab-on-a-chip applications. *Microfluid. Nanofluidics* **2010**, *9*, 145–161. [CrossRef]

27. Ren, K.; Dai, W.; Zhou, J.; Su, J.; Wu, H. Whole-Teflon microfluidic chips. *Proc. Natl. Acad. Sci. USA* **2011**, *108*, 8162–8166. [CrossRef]
28. Li, H.; Fan, Y.; Kodzius, R.; Foulds, I.G. Fabrication of polystyrene microfluidic devices using a pulsed CO_2 laser system. *Microsyst Technol* **2012**, *18*, 373–379. [CrossRef]
29. Qi, H.; Chen, T.; Yao, L.; Zuo, T. Micromachining of microchannel on the polycarbonate substrate with CO_2 laser direct-writing ablation. *Opt. Lasers Eng* **2009**, *47*, 594–598. [CrossRef]
30. Shah, R.K.; Shum, H.C.; Rowat, A.C.; Lee, D.; Agresti, J.J.; Utada, A.S.; Chu, L.-Y.; Kim, J.-W.; Fernandez-Nieves, A.; Martinez, C.J.; et al. Designer emulsions using microfluidics. *Mater. Today* **2008**, *11*, 18–27. [CrossRef]
31. Benson, B.R.; Stone, H.A.; Prud'homme, R.K. An "off-the-shelf" capillary microfluidic device that enables tuning of the droplet breakup regime at constant flow rates. *Lab. Chip* **2013**, *13*, 4507–4511. [CrossRef]
32. Torino, E.; Russo, M.; Ponsiglione, A.M. Chapter 6—Lab-on-a-chip preparation routes for organic nanomaterials for drug delivery. In *Microfluidics for Pharmaceutical Applications*; Santos, H.A., Liu, D., Zhang, H., Eds.; William Andrew Publishing: Norwich, NY, USA, 2019; pp. 137–153. [CrossRef]
33. Feng, Q.; Sun, J.; Jiang, X. Microfluidics-mediated assembly of functional nanoparticles for cancer-related pharmaceutical applications. *Nanoscale* **2016**, *8*, 12430–12443. [CrossRef]
34. Wang, J.; Chen, W.; Sun, J.; Liu, C.; Yin, Q.; Zhang, L.; Xianyu, Y.; Shi, X.; Hu, G.; Jiang, X. A microfluidic tubing method and its application for controlled synthesis of polymeric nanoparticles. *Lab Chip* **2014**, *14*, 1673–1677. [CrossRef] [PubMed]
35. Stroock, A.D.; Dertinger, S.K.W.; Ajdari, A.; Mezić, I.; Stone, H.A.; Whitesides, G.M. Chaotic Mixer for Microchannels. *Science* **2002**, *295*, 647–651. [CrossRef] [PubMed]
36. Hong, C.C.; Choi, J.W.; Ahn, C.H. A novel in-plane passive microfluidic mixer with modified Tesla structures. *Lab Chip* **2004**, *4*, 109–113. [CrossRef]
37. Liu, D.; Zhang, H.; Fontana, F.; Hirvonen, J.T.; Santos, H.A. Microfluidic-assisted fabrication of carriers for controlled drug delivery. *Lab Chip* **2017**, *17*, 1856–1883. [CrossRef] [PubMed]
38. Lu, M.; Ho, Y.-P.; Grigsby, C.L.; Nawaz, A.A.; Leong, K.W.; Huang, T.J. Three-Dimensional Hydrodynamic Focusing Method for Polyplex Synthesis. *ACS Nano* **2014**, *8*, 332–339. [CrossRef]
39. Mahmoudi, Z.; Mohammadnejad, J.; Razavi Bazaz, S.; Abouei Mehrizi, A.; Saidijam, M.; Dinarvand, R.; Ebrahimi Warkiani, M.; Soleimani, M. Promoted chondrogenesis of hMCSs with controlled release of TGF-β3 via microfluidics synthesized alginate nanogels. *Carbohydr. Polym.* **2020**, *229*, 115551. [CrossRef]
40. Agnello, S.; Bongiovì, F.; Fiorica, C.; Pitarresi, G.; Palumbo, F.S.; Di Bella, M.A.; Giammona, G. Microfluidic Fabrication of Physically Assembled Nanogels and Micrometric Fibers by Using a Hyaluronic Acid Derivative. *Macromol. Mater. Eng.* **2017**, *302*, 1700265. [CrossRef]
41. Bongiovì, F.; Fiorica, C.; Palumbo, F.S.; Pitarresi, G.; Giammona, G. Hyaluronic acid based nanohydrogels fabricated by microfluidics for the potential targeted release of Imatinib: Characterization and preliminary evaluation of the antiangiogenic effect. *Int. J. Pharm.* **2020**, *573*, 118851. [CrossRef]
42. Elkassih, S.A.; Kos, P.; Xiong, H.; Siegwart, D.J. Degradable redox-responsive disulfide-based nanogel drug carriers via dithiol oxidation polymerization. *Biomater. Sci.* **2019**, *7*, 607–617. [CrossRef]
43. Mi, P. Stimuli-responsive nanocarriers for drug delivery, tumor imaging, therapy and theranostics. *Theranostics* **2020**, *10*, 4557–4588. [CrossRef]
44. Wang, H.; Chen, Q.; Zhou, S. Carbon-based hybrid nanogels: A synergistic nanoplatform for combined biosensing, bioimaging, and responsive drug delivery. *Chem. Soc. Rev.* **2018**, *47*, 4198–4232. [CrossRef]
45. Rondeau, E.; Cooper-White, J.J. Biopolymer Microparticle and Nanoparticle Formation within a Microfluidic Device. *Langmuir* **2008**, *24*, 6937–6945. [CrossRef]
46. Toprakcioglu, Z.; Challa, P.K.; Morse, D.B.; Knowles, T. Attoliter protein nanogels from droplet nanofluidics for intracellular delivery. *Sci. Adv.* **2020**, *6*, eaay7952. [CrossRef]
47. Kłodzińska, S.N.; Molchanova, N.; Franzyk, H.; Hansen, P.R.; Damborg, P.; Nielsen, H.M. Biopolymer nanogels improve antibacterial activity and safety profile of a novel lysine-based α-peptide/β-peptoid peptidomimetic. *Eur. J. Pharm. Biopharm.* **2018**, *128*, 1–9. [CrossRef] [PubMed]
48. Kłodzińska, S.N.; Pletzer, D.; Rahanjam, N.; Rades, T.; Hancock, R.E.W.; Nielsen, H.M. Hyaluronic acid-based nanogels improve in vivo compatibility of the anti-biofilm peptide DJK-5. *Nanomedicine* **2019**, *20*, 102022. [CrossRef]
49. Huang, K.; He, Y.; Zhu, Z.; Guo, J.; Wang, G.; Deng, C.; Zhong, Z. Small, Traceable, Endosome-Disrupting, and Bioresponsive Click Nanogels Fabricated via Microfluidics for CD44-Targeted Cytoplasmic Delivery of Therapeutic Proteins. *ACS Appl. Mater. Inter.* **2019**, *11*, 22171–22180. [CrossRef] [PubMed]
50. Pawar, A.A.; Saada, G.; Cooperstein, I.; Larush, L.; Jackman, J.A.; Tabaei, S.R.; Cho, N.-J.; Magdassi, S. High-performance 3D printing of hydrogels by water-dispersible photoinitiator nanoparticles. *Sci. Adv.* **2016**, *2*, e1501381. [CrossRef] [PubMed]
51. Liu, H.; Li, W.; Liu, C.; Tan, J.; Wang, H.; Hai, B.; Cai, H.; Leng, H.J.; Liu, Z.J.; Song, C.L. Incorporating simvastatin/poloxamer 407 hydrogel into 3D-printed porous Ti_6Al_4V scaffolds for the promotion of angiogenesis, osseointegration and bone ingrowth. *Biofabrication* **2016**, *8*, 045012. [CrossRef]
52. Müller, M.; Becher, J.; Schnabelrauch, M.; Zenobi-Wong, M. Nanostructured Pluronic hydrogels as bioinks for 3D bioprinting. *Biofabrication* **2015**, *7*, 035006. [CrossRef]

Review

Nanogels as a Versatile Drug Delivery System for Brain Cancer

Brielle Stawicki, Tyler Schacher and Hyunah Cho *

School of Pharmacy and Health Sciences, Fairleigh Dickinson University, 230 Park Ave., Florham Park, NJ 07932, USA; brielled@student.fdu.edu (B.S.); tschach@student.fdu.edu (T.S.)
* Correspondence: hyunahc@fdu.edu; Tel.: +1-973-443-8234

Abstract: Chemotherapy and radiation remain as mainstays in the treatment of a variety of cancers globally, yet some therapies exhibit limited specificity and result in harsh side effects in patients. Brain tissue differs from other tissue due to restrictions from the blood–brain barrier, thus systemic treatment options are limited. The focus of this review is on nanogels as local and systemic drug delivery systems in the treatment of brain cancer. Nanogels are a unique local or systemic drug delivery system that is tailorable and consists of a three-dimensional polymeric network formed via physical or chemical assembly. For example, thermosensitive nanogels show promise in their ability to incorporate therapeutic agents in nano-structured matrices, be applied in the forms of sprays or sols to the area from which a tumor has been removed, form adhesive gels to fill the cavity and deliver treatment locally. Their usage does come with complications, such as handling, storage, chemical stability, and degradation. Despite these limitations, the current ongoing development of nanogels allows patient-centered treatment that can be considered as a promising tool for the management of brain cancer.

Keywords: nanogel; drug delivery; brain cancer

1. Introduction

With the continuous input of scientific medical research, cancer treatments have improved substantially over the past decade. Brain cancer proposes a unique situation, having similarities to other forms of cancer in the body, yet major differences in the diversity of intracranial neoplasms, genetic heterogeneity, complexity of the organ in which it resides and physiological features of the cranial cavity that limit treatment options [1]. A multitude of brain tumor types exist that are categorized based on their location of origin and malignancy properties, and surgical removal and chemotherapy serve as vital options for the majority of types that are deemed treatable. In 2020, the Central Brain Tumor Registry of the United States reported an overall primary malignant tumor incidence rate of 7.08 per 100,000, an estimated 123,484 cases, and a non-malignant tumor incidence rate of 16.71 per 100,000, 291,927 cases [2–5]. In 2021, it is projected that approximately 84,170 new cases will be diagnosed in the United States, highlighting the necessity of ever-expanding efficacious treatment options for patients. Brain tumors can be classified based on their location within the cranial cavity, presumptive origin, and microscopic similarities [6]. Internal malignant tumors include the common tumor type, gliomas, which presumptively derive from glial tissues [7]. They are further categorized into astrocytomas and oligodendrogliomas and subdivided into grades based on tumor pathological characteristics. These grades dictate treatment options and responsiveness. Other internal malignant tumors include ependymomas, affecting the ependymal cells of the four ventricles of the brain and the spinal cord canal, and gangliogliomas. Extrinsic malignant tumors, such as meningiomas and schwannomas arise from dura matter and Schwann cells, respectively.

Patients may experience both general and localized symptoms prior to diagnosis and radiographic visualization of their brain tumors. These symptoms include headaches, nausea, seizures, and vomiting due to increased intracranial pressure [7]. The specific symptoms and signs produced by brain tumors vary with the location of the tumor. For example, patients with the tumors located in or subjacent to cortical regions may present with language dysfunction, visual field abnormalities, or focal seizures [8]. Tumors arising in the brain stem may cause rapidly progressing cranial neuropathies as well as motor and sensory deficits [8]. Despite the location of cranial infliction, surgical debulking of the tumor remains a competent first-line treatment option and is used in conjunction with radiotherapy and/or systemic chemotherapy.

A vital consideration that must be taken when beginning systemic chemotherapy for brain malignancies is the blood–brain barrier (BBB) [9]. This barrier consists of a microvascular system that supplies nutrients to the central nervous system. The blood vessels possess unique properties that allow them to vigorously regulate molecules, ions, and cells moving from the blood to the CNS tissues, resulting in CNS homeostasis and the prevention of entrance of toxins and pathogens [9]. Physically, it is composed of continuous capillaries with endothelial cells attached via tight junctions that are able to restrict paracellular diffusion of solutes. P-glycoprotein efflux transporters limit lipophilic solute entrance to the brain. These gatekeeping properties thus also prevent pharmacological substances, such as chemotherapy, from entering and working in brain tissues. The BBB is seen as an obstacle that must be overcome to treat brain metastasis; thus, the developed therapeutic agents are specifically engineered with this in consideration. Generally, therapeutic agents with a molecular mass of <400–600 Da that are lipid soluble have greater BBB penetration [10]. Although a variety of chemotherapy agents are available, not all have the ability to cross the blood–brain barrier, thus limiting treatment options for patients presenting with malignant gliomas. Temozolomide, an alkylating agent, is a viable option. Its mechanism of action consists of the transfer of its alkyl group at the O6 and N7 guanine positions causing DNA double strand breaks and apoptosis inside the nucleus of cancer cells [11]. Historically, the nitrosoureas (e.g., carmustine, lomustine) and vincristine have been the most widely used class of chemotherapy agents due to their physicochemical properties that enable them to penetrate the BBB and exert therapeutic effects [12–14]. Nitrosoureas undergo biotransformation via non-enzymatic decomposition to active metabolites with a mechanism of action similar to alkylating agents. However, these agents are limited by serious nausea, vomiting, and an increased risk of secondary malignancies, due to their overall carcinogenic nature, myelosuppression, infertility, and mucositis [15].

Radiation oncology is based on the principles of x-ray machines and directs harmful ionizing radiation to kill cancer cells [16]. Ionizing radiation deposits energy in cancer cells which can directly cause death or result in detrimental genetic mutations. These genetic mutations alter DNA, causing both single and double strand breaks, preventing further tumor growth. As a result of genomic instability, the cells die via apoptosis, necrosis, mitotic catastrophe, autophagy, and other mechanisms [16]. When used in conjunction with neurosurgery, radiation can be used prior to shrink the tumor size, or after to remove the cancerous cells that remain in the area [16]. However, it does not come without its limitations. Radiation commonly results in acute radiation central nervous system toxicity, characterized by nausea, drowsiness, and ataxia. Late effects can be seen 9 months to 10 years after therapy and include focal necrosis, CNS neurological dysfunction, MRI visible white matter alterations, and necrotizing encephalopathy [17]. Cranial radiation may affect other body systems, causing endocrine abnormalities due to a disruption of normal pituitary/hypothalamic axis function and leading to a need for increasing monitoring of anterior and posterior pituitary hormone levels [18].

2. Gliadel Wafers for Postsurgical Brain Cancer Treatment

The realization of less-than-ideal characteristics of available radiation and systemic chemotherapy treatment options has led to the search for novel adjunctive "implanted" therapies. BCNU (Gliadel) wafer therapy, approved for use in 1996 by the FDA's Oncology Drug Advisory Committee, was the first implantable drug delivery system used to deliver carmustine, directly to the site of a surgically resected tumor [19]. Upon tumor removal, the wafers are implanted, providing direct treatment and limiting systemic side effects. The wafer is dime-sized and consists of polifeprosan 20, a biodegradable polyanhydride copolymer. Through slow erosion of the polymer matrix, polifeprosan 20 releases carmustine gradually for an extended period of time, approximately 2–3 weeks [20]. Generally, up to 8 wafers are placed in the tumor cavity, each with 7.7 mg of carmustine, for a total dose of 61.6 mg [21]. Westphal et al. demonstrated that the median survival time after gliadel wafer implantation was 13.8 months compared to 11.6 months in the placebo-group patients [22]. A decrease in mortality by 30% in those treated with the gliadel wafers was also reported [22]. Further studies compared the efficacy of the wafer versus classical chemotherapy agents, such as temozolomide, and showed a difference in peak survival. The absolute percentage gain of survival over placebo with gliadel wafers showed peaks at 12 months versus 18 months with temozolomide [23]. Thus, the sequential use of the two agents was proposed and decreed the "Stupp protocol" [24]. Clinical trials have shown an increased median survival time, including Stupp et al., which compared a combination gliadel wafer implantation and temozolomide to temozolomide treatment alone [24]. In a 5 year follow-up, 97% of patients treated with solely temozolomide died, compared to 89% of patients who received combination therapy [24]. The overall survival was 14.6 months in the wafer plus temozolomide group, and 12.1 months for wafer alone [24]. Post-implantation of gliadel wafers, patients should be monitored for adverse effects and complications, including hemotoxicity [25]. There has been a concern associated with an increased risk of intracranial infections, brain abscess, and cerebrospinal fluid leaks. Other side effects include headaches, cerebral oedema, drowsiness, and seizures [26]. When following the Stupp protocol, multiple studies have demonstrated that there were no unexpected adverse effects or increased incidence reported [27,28].

The gliadel wafer is a rigid, disc-shaped wafer compressed from the mixture of spray-dried polyanhydride and carmustine [19]. Though proven modestly effective, gliadel wafer therapy has well-recognized drawbacks, including limited drug loading capability for a single hydrophilic drug, uneven drug release due to the rigid structure incapable of intimate contacting with surrounding tissues and cumbersome to maximally cover the resection cavity (requires cutting/overlapping the wafers).

Nanogel-based local delivery of chemotherapy has shown great promise in overcoming the weaknesses of gliadel wafer therapy. A thermosensitive nanogel formulation, OncoGel, is a triblock copolymer comprised of poly(D,L-lactide-co-glycolide), (PLGA) and poly(ethylene glycol) (PEG) with the basic structure of PLGA-b-PEG-b-PLGA. OncoGel contains paclitaxel at 6 mg/mL and makes a sol-to-gel transition at 37 °C. OngoGel entered a phase II clinic trial for treating esophageal cancer [29]. Although found to have low toxicity and reduce tumor burden, OncoGel failed in this clinical study due to insufficient efficacy in esophageal cancer. Tyler et al. proved that OncoGel containing 6.3 mg/mL paclitaxel was safe for intracranial injection in 18 Fischer-344 rats bearing glioma and most effective when administered in combination with radiation therapy [30]. Torres et al. used computational mass transport simulations to investigate the effectiveness of paclitaxel delivery from OncoGel [31]. The effective therapeutic concentrations were maintained for > 30 days using OngoGel whereas those were maintained for ca. 4 days for carmustine released from gliadel wafers. This result was simulated due to the controlled release of paclitaxel within the degradation lifetime of the OncoGel matrix. Nanogels are bioadhesive, thus does not require additives to secure it against the cavity surface after brain surgery. Unlike gliadel wafers that require cutting and overlapping wafers to properly cover cavities of different sizes and shapes, the dose of the drugs loaded in nanogels can be easily controlled and adjusted using a syringe to offer patient-centered treatment (considering the size/shape of the tumor resection cavity). In this short review paper, we summarized the desirable properties of nanogels and possible obstacles with their development and use, and highlighted the application of biocompatible nanogels as a drug delivery system in brain cancer.

3. Nanogel

3.1. Overview and Preparation of Nanogels

Nanogels have a three-dimensional "nanoscopic" structure composed of a variety of natural polymers, synthetic polymers or a combination of both. Nanogels are formed via physical or chemical assemblies of polymers that carry amphiphilic macromolecular chains (Table 1) [32]. Physical assembly relies on the physical interactions/entanglements via hydrogen bonds, electrostatic, van der Waals and hydrophobic interactions, and chemical cross-linking utilizes covalent bonds [32]. Aforementioned nanogels carrying paclitaxel, OncoGel, is one of the examples formed via physical self-assembly [31]. In water, below the critical gelation temperature (CGT), PLGA-b-PEG-b-PLGA (ABA type) copolymers create loops sharing a hydrophobic PLGA center and form nanoscopic micelles (Figure 1A) [33]. Paclitaxel (logP 3.0) is physically entrapped in the core of the micelles. Gelation of PLGA-b-PEG-b-PLGA occurs above the CGT. As the temperature increases, hydrophobic interactions among PLGA segments become stronger, leading to aggregation of micelles, decrease in the mobility of water, and increase in the viscosity (gelation occurs). Akiyoshi et al. physically-assembled nanogels using hydrophobized cholesterol-bearing pullulan that deliver insulin [34]. Nanogels were ca. 20–30 nm in diameter and self-assembled into nanogels with insulin (Figure 1C) [34]. Another way of physically assembling nanogels is to suspend/immobilize nanoparticles in a hydrogel matrix. Giovannini et al. formed silica nanoparticle (ca. 100 nm in diameter) and gold nanoparticles (ca. 80 nm in diameter) and suspended them in Fmoc-galactosamine-based hydrogels [35]. Nanogels carrying silica and gold nanoparticles decreased the premature drug release of loaded drugs as the hydrogels restricted the movement of the nanoparticles and retarded the aggregation of nanoparticles. Overall, the introduction of hydrogels improved the stability of nanoparticles.

Nanogels can be formed from polymer precursors vial chemical cross-linking that utilizes some functional groups such as disulfide, amine and imine or via photo-induced cross-linking (Table 1) [32]. Ryu et al. developed nanogels composed of polymers that carry an oligoethyleneglycol unit for hydrophilicity and a pyridyl disulfide (PDS)-derived thioethylmethacrylate for the cross-linkability [36]. Disulfide bonds impart structural stability for hydrophobic payloads. Disulfide can reversibly reduce to thiol, as a function of thiol concentrations of the environment. As thiols are highly reduced in cells, by using thiol-disulfide exchange, disulfide bonds are rapidly degraded, releasing the payloads selectively under the reduced condition in cells. Elkassih et al. developed fully degradable disulfide cross-linked nanogels using oxidative radical polymerization of 2,2'-(ethylenedioxy)diethanethiol (EDDET) as a monomer with different cross-linkers, including pentaerythritol tetramercaptoacetate (PETMA) [37]. As the poly(EDDET) backbone repeated structures and cross-linking junctions were composed entirely of disulfide bonds, nanogels were able to degrade into thiols intracellularly in response to the reducing agent glutathione present inside of cells. Amine groups are widely used in the preparation of nanogels because of their high reactivity with carboxylic acids, activated esters, isocyanates and iodides [32]. This technique provides an opportunity to introduce various stimuli-response properties into nanogels by incorporating a diamine cross-linker. Kockelmann et al. developed nanogels carrying imidazoquinolinen using active-ester-containing amphiphilic poly(methacrylate) block copolymers [38]. The amphiphilic reactive ester block copolymers self-assembled into precursor micelles, whose cores were functionalized by mono-amine-bearing entities, cross-linked with pH-degradable bisamines, and finally converted into fully hydrophilic nanogels. The authors stated that these nanogels provided safe and controllable drug delivery strategies in immunotherapy for cancers, considering their slightly acidic environment. Liao et al. developed functionalized polymeric nanogels with pH-responsive benzoic-imine cross-linkages [39]. The polymer was synthesized by one-step cross-linking of the branched poly(ethylenimine)-g-methoxy poly(ethylene glycol) (PEI-g-mPEG) copolymer with hydrophobic terephthalaldehyde (TPA) molecules. The functionalized polymeric nanogels were comprised of multiple hydrophobic benzoic-imine-rich spherical domains covered by positively-charged PEI networks. The external PEI network and mPEG attracted large amounts of water whereas the TPA created the colloidal core more hydrophobic and compact. The nanogels exhibited acid-triggered drug release (indocyanine green incorporated in hydrophobic core via pi-pi stacking) by the cleavage of benzoic-imine bonds in response to pH reduction from 7.8 to 6.4. He et al. developed photoresponsive nanogels, which utilize light to reversibly change the cross-linking density of nanogel particles [40]. Poly(ethylene oxide) and poly[2-(2-methoxyethoxy)ethylmethacrylate-co-4-methyl-[7-(methacryloyl)oxyethyloxy]coumarin] (PEO-b-P(MEOMA-co-CMA)) were synthesized. The reversible photo-cross-linking reaction was provided by the photodimerization of coumarin side groups under irradiation at λ >310 nm and the photocleavage of cyclobutane bridges under irradiation at λ <260 nm, reducing the degree of cross-linking and accelerating the rate of drug release.

Table 1. Main nanogel assembly techniques.

Assembly	Reactions	Properties	References
Physical	Micellar	Self-assembly using triblock copolymers or branched polymers	[31,34]
	Hybrid (nanoparticles suspended in hydrogels)	Nanoparticles immobilized in hydrogels	[35,41–45]
Cross-linking	Disulfide	Cross-linked via thiol-disulfide exchange reaction, Cleaved in response to glutathione	[36,37,46]
	Amide	High reactivity with carboxylic acids, activated esters, isocyanates and iodides	[38,47]
	Imine	Stable under physiological conditions and labile at acidic pH	[39,48]
	Photo-induced	Photo-induced cross-linking or cleavage	[40,49]

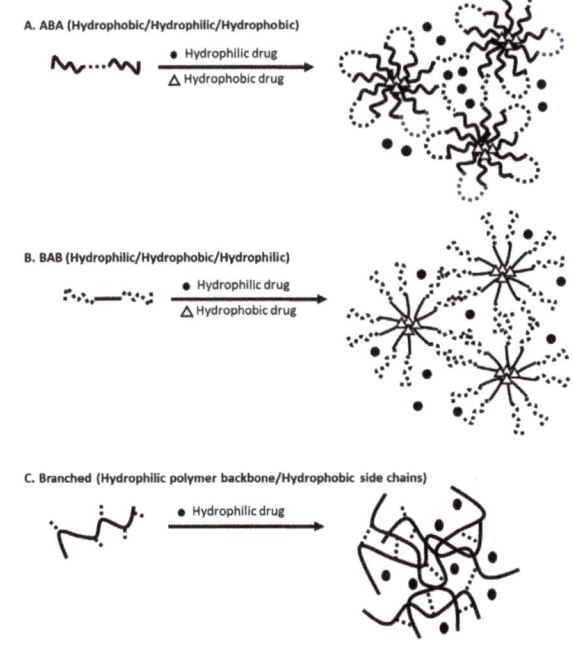

Figure 1. Illustrations of physically self-assembled nanogels loaded with hydrophilic (solid circle) and hydrophobic (empty triangle) drugs using (A) ABA, (B) BAB, and (C) branched polymers (modified from [50]).

3.2. Desired Properties of Nanogels for Drug Delivery

An option to consider for the future treatment of brain tumors is a drug delivery system known as the nanogel. A nanogel is a three-dimensional hydrogel that is formed by connection of nano-scopic micelles dispersed within an aqueous medium ("nano-in-hydrogels") having an inherent capability to incorporate hydrophobic molecules in the core of the micelles while maintaining a hydrophilic exterior [50]. Similar to hydrogels, nanogels are (i) mostly hydrophilic in nature, soft, resembling the texture of soft tissues, bioadhesive, biocompatible, and biodegradable [51]. One of the most widely reported biocompatible nanogels are chitosan-based nanogels. Pereira et al. performed a thorough study of biocompatibility of a glycol chitosan nanogel, one of the highly biocompatible chitosan derivatives, in vitro [52]. Glycol chitosan nanogels did not induce noticeable metabolic activities and did not affect the cell membrane integrity in 3T3 fibroblast, HMEC human microvascular endothelial and RAW mouse leukemia monocyte macrophage cell lines. Glycol chitosan nanogels were poorly internalized by murine macrophages. Blood compatibility of glycol chitosan nanogels was confirmed by hemolysis and whole blood clotting time assays.

A number of nanogels demonstrate (ii) stimuli-responsive behaviors to release drugs in response to external stimuli [53]. One of the widely explored external stimuli is temperature. Some nanogels are designed to make a sol-to-gel transition at 37 °C, body temperature. Below 37 °C, nanogels are in a sol form. At 37 °C, nanogels begin to increase the viscosity, forming a semisolid gel. At ambient temperature, the viscosity of the sol is low allowing the formulation to pass through the syringe/needle. When injected, at 37 °C, nanogels are formed conforming to a shape of body cavity. Poly(ethylene oxide)-block-poly(propylene oxide)-block-poly(ethylene oxide) (PEO-b-PPO-b-PEO), also known as poloxamer, Pluronics, or Kolliphor, has been explored widely to create thermos-responsive nanogels [50,51,54]. Poloxamer 407 is a triblock copolymer with a hydrophobic central PPO core and two hydrophilic PEO side chains. At the concentrations of PEO-b-PPO-b-PEO of 20–30% w/w, the copolymers reach the critical micelle concentration; it is at this point where the micelles reorder themselves into a lattice [54]. Upon the elevation of the environmental temperature (at 37 °C), the hydrophilic chains are desolvated as the hydrogen bonds between the solvent and chains begin to break leading to chain entanglement. The resultant product is a gel that allows for the gradual release of hydrophobic drugs and the more rapid release of hydrophilic drugs in the insertion site. Thermo-responsive behavior of nanogels can also serve an ideal dermal drug delivery system. Gerecke et al. reported that thermo-responsive nanogels synthesized from dendritic polyglycerol with poly(glycidyl methyl ether-co-ethyl glycidyl ether) were capable of enhancing penetration through biological barriers such as the stratum corneum and were taken up by keratinocytes of human skin without cytotoxic or genotoxic effect [55].

Another desirable behavior of nanogels is pH sensitivity. Nanogels are designed to undergo the cleavage of the polymer networks/linkages under acidic conditions (mimicking tumor environment) and degrade completely, utilizing various cross-linkers, but are stable in physiological environment. Yang et al. developed a pH-triggered hyaluronic acid nanogel system by copolymerization between methacrylate hyaluronic acid and a cross linker containing ortho ester groups [56]. This nanogel system carrying doxorubicin demonstrated excellent tumor homing and selective tumor cell uptake, resulting in superior anticancer efficacy in HepG2 human liver cancer cell spheroids. The rapid release of doxorubicin was observed under endo/lysosomal conditions due to the pH-triggered cleavage of ortho ester linkages. Kang et al. developed a pH-responsive, chemically cross-linked nanogel using dopamine hydrochloride-conjugated carbonized hyaluronic acid [57]. Release of doxorubicin from this nanogel system was pH-dependent, resulting in 80% of doxorubicin released in 30 h at pH 5.0. Less than 20% of doxorubicin was released from nanogels at pH 6 and 7.4. This pH-dependency was caused by the cleavage of boronate ester bond between catechol and boronic acid under acidic conditions.

Multi-stimuli responsive nanogels have been shown to exhibit greater "fine-tuning" effect compared to their singular-stimuli responsive counterparts. Salehi et al. dual-stimuli responsive nanogels were composed of poly(N-isopropylacrylamide-dimethylaminoethyl methacrylatequaternary ammonium alkyl halide-methacrylic acid) and poly(N-isopropylacrylamide-dimethylami-noethyl methacrylate quaternary ammonium alkyl halide-methacrylic acid-hydroxyethyl methacrylate) [58]. This nanogel system showcased the capability of stimuli-triggered-controlled release behaviors mediated by temperature and pH values and were administered for the simultaneous delivery of doxorubicin and methotrexate. The release of both drugs were accelerated at pH 4 and 5.5 but arrested at pH 7.4. The release of both drugs was more rapid at 40 °C than at 37 °C. The authors highlighted that the prolonged and constant drug release pattern along with the dual-stimuli responsive behaviors offer a cancer treatment option without the frequent administration of multi-drugs. Pan et al. developed multi-stimuli responsive nanogels using the tailored modified sugarcane bagasse cellulose [59]. In the presence of a disulfide crosslinking agent, cystamine bisacrylamide, the in situ free radical copolymerization of methacrylated monocarboxylic sugarcane bagasse cellulose and N-isopropylacrylamide was processed, thus leading to redox (in the presence of reducing agent), pH (below 5.8) and temperature (above 32 °C)-responsive nanogels.

Nanogels are one of the excellent drug delivery systems with (iii) the capability of incorporating and delivering a wide range of drugs by immobilizing them through covalent or non-covalent interactions. Notably, nanogels demonstrate (iv) controlled the release of multi-drugs with the primary release mechanism being diffusion of drug followed by the degradation of polymeric matrix. Cho et al. investigated theragnostic effects of thermosensitive PLGA-b-PEG-b-PLGA nanogels carrying paclitaxel, rapamycin, an NIR imaging agent in ovarian cancer-bearing mice [60]. Nanogels made a sol-to-gel transition at 37 °C and slowly released drugs at a simultaneous release rate in response to the physical degradation of nanogels. Nanogels released ca. 26% of the payloads within 48 h whereas a micellar liquid formulation released the payloads at more rapid rate, reaching 68–70% content release within 48 h. This thermos-responsive nanogel system enabled loco-regional delivery of multi-payloads by forming a gel-depot in the peritoneal cavity of ovarian cancer-bearing mice. In the control, without treatment, animals bearing ES-2- human ovarian cancer increased tumor burden significantly from 100% to 3480 ± 445% within 3 days. A single intraperitoneal (IP) injection of nanogels remarkably decreased tumor burden from 100% down to 7 ± 1% on day 3. A single intravenous (IV) or IP injection of micelles containing the same drugs did not show the therapeutic effectiveness, demonstrating increase of tumor burden from 100% to 110 ± 21% for IP micelles and 100% to 471 ± 236% for IV micelles. Cho et al. also developed 3D printed nanogel discs constructed of PEO-PPO-PEO and therapeutic payloads, paclitaxel and rapamycin [54]. The authors emphasized the convenience in use, proposing that in clinical settings, healthcare providers could place the disc into the peritoneal cavity post-surgical ovarian tumor removal without concerns of unsuccessful delivery of medications nor detrimental effects on patient recovery known as peritoneal adhesion.

Modified nanogels can (v) target the specific receptors or tissues. Su et al. synthesized thermo- and pH-responsive poly (N-isopropyl acrylamide-co-acrylic acid) nanogels [61]. Fluorescent bovine serum albumin (BSA) encapsulated gold nanoclusters were conjugated onto the surface of nanogels, followed by functionalization of tumor targeting peptide iRGD onto the BSA for tumor targeting. Nanoparticles carrying doxorubicin were ca. 182 nm in diameter. The targeted nanogel system enhanced intracellular uptake of the payload, doxorubicin, into the vein endothelial (HUVECs) and the extravascular tumor (B16) cells. Chen et al. designed and constructed a smart nanogel platform integrating both receptor-mediated targeting (RMT) and environment-mediated targeting (EMT) strategies to heighten the tumor accumulation and cellular uptake of drugs [62]. A phenylboronic acid (PBA) and morpholine (MP) dual-modified polypeptide nanogel (PMNG) were prepared. The PBA ligand selectively targeted sialyl (SA) overly expressed on the highly metastatic tumor cells. The MP ligand favored extracellular pH condition (ca. pH 6.5) and facilitated the internalization of drugs into cells. In vivo, this smart, targeted nanogel system carrying doxorubicin exhibited the outstanding efficacy of the inhibition of metastatic nodules in C57/BL mice bearing subcutaneous melanomas.

The aforementioned properties of nanogels make them outstanding candidates for pharmaceutical/biomedical applications as a drug delivery system, specifically involving brain cancer (Table 2).

Table 2. Comparison of nanogels, wafers, and liquid dosage forms for their properties as drug delivery systems for brain cancer.

Properties	Nanogels	Wafers	Nanoparticle-Based Liquid Dosage Forms
Route of administration	IV, implant, intratumoral, nasal	Implant	IV, intratumoral, nasal
Multi-drug delivery	Yes	Maybe (not known)	Yes
Delivery of hydrophobic drugs	Yes	Maybe (not known)	Yes
Form of drugs	Encapsulated in nanoparticles	Free form	Encapsulated in nanoparticles
Dose adjustment	Yes (via syringe)	Yes but manipulation needed (e.g., cutting, inserting multiple wafers)	Yes (via syringe)
Surface modification for targeted drug delivery	Yes	No	Yes
Long residence time	Yes	Yes	No
Controllable drug release	Yes (stimuli-responsive, diffusion followed by physical degradation)	Yes (physical erosion)	Yes (stimuli-responsive, diffusion followed by physical degradation)
Suitable for intratumoral injection	Yes	No	Yes
Available as a spray delivery system	Yes	No	Yes
Biocompatible and biodegradable	Yes	Yes	Yes
Convenience in handling	+	+++	++
Conforming to the shape/size of the resection cavity post-surgery	Yes (intimate contacting with surrounding tissues)	No (stiff)	No (easily washed away by the interstitial fluid)

4. Nanogels That Deliver Drugs to Brain

4.1. Nanogels That Cross the BBB

The ability to transport of compounds across the BBB is a fundamental requirement to treat and diagnose various brain diseases [63]. The BBB prevents compounds from reaching therapeutic concentrations in the brain, thereby hampering the therapeutic/diagnostic efficacy. Many studies have elucidated a few factors for compounds, especially nanoparticles, to penetrate the BBB and reach the brain [64]. Nanoscopic drug delivery systems can cross the BBB in a variety of ways, including endocytosis, receptor-mediated transcytosis, or the enhanced permeability and retention (EPR) effect [65]. The EPR effect exploits the leaky vasculature of solid tumors where the nanoparticles extravasate locally into the tumor tissues, resulting in slow release of encapsulated drugs into the brain tumor tissue. Intravenous injection of nanogels could be a potential method for drug delivery for brain tumors relying on the EPR effect.

The molecular and particle "sizes" and "hydrophilicity" of the compounds are considered to be predominantly important factors enabling their migration across these barriers [66–69]. Small, hydrophilic molecules may cross the BBB via paracellular transportation, but it may be limited by the regulation of the transient relaxation of tight junctions between the endothelial cells [70]. Small, lipophilic molecules enter the brain tissues via transcellular diffusion [70]. However, the route of transcellular diffusion involves the traversing of the luminal membrane and cytosol prior to reaching the brain tissue, which represents a challenge due to the tendency of lipophilic substances to remain within the cell membrane. Kimura et al. prepared ultra-fine hydrophilic nanogels (average particle size between ca. 5 and 21 nm) carrying Gd-DOTA for brain imaging. In mice, it was confirmed that intravenously injected nanogels helped Gd-DOTA enter brain parenchyma through the BBB.

Ribovski et al. added one more advantage of nanogels in delivering compounds across the BBB, which is "low stiffness" [63]. The authors investigated the effect of nanogel stiffness on nanogel transport across the in vitro BBB model and calculated the fraction of internalized nanogels that reached the basolateral compartment of the BBB model. The softer nanogels showed a 2-fold higher secretion at the basal side of the BBB model compared to the stiff nanogels. The authors hypothesized that low nanogel stiffness promoted intracellular trafficking and transcytosis.

She et al. demonstrated that the "biocompatibility" of nanogels that mimic the cellular membrane was the key factor for effective drug delivery across the BBB [67]. The authors synthesized an azobenzene-based cross-linker to construct hypoxia-degradable zwitterionic phosphorylcholine nanogels. This nanogel was degradable in hypoxic environment, leading to the collapse of nanogels and rapid release of drugs in hypoxic tissue of glioblastoma. Nanogels were able to pass through the BBB and exhibited the high accumulation of the payload in glioblastoma tissue due to the phosphorylcholine mimicking cellular membrane. Nanogels were able to deliver doxorubicin effectively to the brain of mice and demonstrated the superior therapeutic behaviors in treating glioblastoma.

4.2. Nanogel Use in Brain Cancers

Nanogels could be used during surgery; the removal of tumor tissue could be followed by the insertion of a nanogel, which would then harden and provide a protective layer or a filler of the resection cavity where the tumor was removed. In addition, nanogels could be loaded with multi-therapeutic agents in order to keep the tumor at bay for the foreseeable future (Figure 2). Unlike the wafers, the dose of loaded drugs can be easily controlled in a syringe based on the size/shape of resection cavity created by removing tumor tissues (Table 2). For nanogels, cutting or overlapping multiple units is not required. In addition to the loco-regional therapeutic benefits, some reports highlighted that diluted nanogels administered intravenously improved selective accumulation of nanogels in brain tumor tissues in vivo. Nanogels could also deliver therapeutic agents intravenously or intranasally to target brain tumor tissues prior to or after a surgical procedure.

Lin et al. prepared MRI traceable, rapidly gelating hydrogels by blending negatively charged carboxymethyl cellulose-grafted poly(N-isopropylacrylamide-co-methacrylic acid) and positively charged gadopentetic acid/branched polyethylenimine [71]. Hydrogels carried hydrophilic epirubicin and hydrophobic paclitaxel (PTX) incorporated in bovine serum albumin nanoparticles (BSA/PTX NPs: average diamer of 181.7 ± 3.9 nm). Hydrogels exhibited free-flowing sol phase at 4°C and made a transition to non-flowing gel phase at 37 °C. In vivo hydrogels carrying epirubicin and BSA/PTX NPs were implanted to the residual tumor tissues after surgical tumor resection in humam glioma U87 tumor-bearing mice. Hydrogels carrying epirubicin and BSA/PTX NPs showed remarkable tumor growth inhibition with the medium survival of 69 days. Notably, the average survival spans for animals in control group (receiving no treatment after surgical tumor removal) was ca. 27 days. The authors presumed that nanogels facilitated "tumor-priming effect" by releasing epirubicin rapidly at the first stage to prime tumor tissues to maximue therapeutic efficacy of paclitaxel released later.

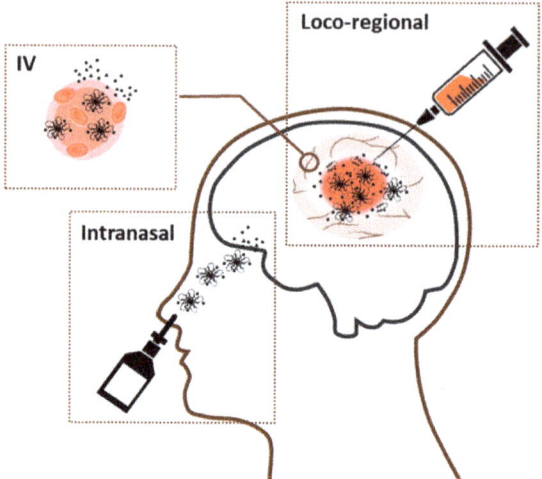

Figure 2. Administrations of nanogels that target brain.

McCrorie et al. designed a spray delivery system consisting of a bioadhesive hydrogel (pectin) and poly(ethylene glycol)-block-polylactic acid (PEG-b-PLA) micelles carrying etoposide and olaparib (the average diameter of 70 nm) [72]. The release was rapid with a burst release of 5% for olaparib in the first 30 min followed by 85% after 24 h. For etoposide, there was a similar initial release of 9% in the first 30 min followed by 83% after 24 h. For both drugs, 100% of the drug was released after 48 h. A pectin-based hydrogel showed the potential to adhere to brain tissues due to the bioadhesive forces, instead of being washed away by the interstitial fluid. Following insertion of nanogels, pectin degrades slowly over 14 days within the brain. There was no neurotoxicity observed in mice. The authors also simulated surgical brain tumor removal followed by spray-delivering nanogels. While under general anesthesia, a small incision was made through the skin along the midline of the skull [72]. A larger drill bit was used to enlarge the burr hole to approximately 1–2 mm. Some brain tissues were removed and nanogels carrying drugs were sprayed into the linings of a surgical pseudo-resection cavity. Sequential biopsies were taken from below this cavity to determine successful delivery of the system and assess depth of penetration. The burr hole was then plugged with bone wax and the skin sutured shut. The authors observed the presence of nanoparticles in the surround tissues up to 1.5 cm away from the cavity. It was evident that sprayable hydrogels containing nanoparticles could be a great loco-regional treatment modality post-surgical brain tumor resection.

Picone et al. developed poly(N-vinyl pyrrolidone)-co-acrylic acid nanogels conjugated with insulin for intranasal delivery of insulin [73]. The average particle size of nanogels was ca. 70 nm. Free insulin or the nanogel system carrying insulin was administered intranasally in mice, and the localization of insulin in the different parts of brain were analyzed. Nanogels increased the levels of insulin in the olfactory bulb, hippocampus and cerebral cortex at statistically greater levels at 30 and 60 min from intranasal administration. These results imply that the mucoadhesive properties of nanogels increased the retention time of insulin, facilitating muco-penetration of insulin. It also appears that insulin conjugated to nanogels was more resistant to the action of proteolytic enzymes in the nasal cavity. The literature has suggested that via intranasal application, drugs (low and large molecular weight drugs) can be transported into the brain via the olfactory and the trigeminal nerve pathways [73]. Drugs transported via the olfactory pathway enter the rostral area of the brain, whereas the trigeminal nerve route facilitates drug delivery to the caudal area of the brain. Although this nanogel system was not specifically used to treat brain cancer, it clearly presented that mucoadhesive nanogels were capable of augmenting the level of drugs in the olfactory bulb, hippocampus and cerebral cortex, benefiting the effective delivery of therapeutic agents across the BBB.

Shatsberg et al. designed disulfide crosslinked nanogels based on the polyglycerol-scaffold to deliver microRNA for glioblastoma therapy [74]. The primary amine groups with higher pKa in this nanogel system were protonated at neutral pH, imparting the positive charge to bind microRNA and facilitating cellular uptake via endocytosis. The secondary amines with lower pKa were protonated at endosomal pH, providing this nanogel system with endosomal escape capacities (via proton sponge effect). This nanogel system was designed to be cleaved under the intracellular reductive conditions. The polyplex formed between nanogels and microRNA were ca., 140–170 nm in hydrodynamic diameter. Nanogels enabled internalization of microRNA into U-87 MG glioblastoma multiforme cells whereas intracellular localization of naked microRNA was hardly observed. U-87 MG glioblastoma multiforme-bearing mice received nanogels carrying microRNA intratumorally on days 0, 3, 7, and 10. Nanogels carrying microRNA helped restore the tumor suppressor role of miR-34a in the xenograft mice, resulting in remarkable inhibition of tumor growth. The authors stated that nanogels were chosen as microRNA delivery carrier for glioblastoma therapy due to their controllable size, shape, functionality, good mechanical properties, presence of the voids allowing encapsulation of multi-drugs, and tunability of drug release profiles resulting extended drug circulation time in the blood.

Azadi et al. developed nanogels using chitosan and polyanionic pentasodium triphosphate [75]. The nanogels carrying methotrexate was ca. 118.54 ± 15.93 nm in diameter. The plasma and brain concentrations of methotrexate at different time points following intravenous administration of the diluted nanogels versus free drug provided the evidence that nanogels improved the efficacy of drug delivery in the brain. A 2.4-fold increase in drug plasma concentration and a 10–15-fold increase of drug concentration in the brain were obtained in male Sprague–Dawley rats as a result of the intravenous administration of methotrexate-loaded nanogels. The authors called this the "Trojan Horse" effect. The authors explained that this effect was presented due to the longer retention time of nanogels in the brain which, in turn, compensate for the drug efflux from the brain to the circulation.

Jiang et al. developed pH/temperature-sensitive poly(N-isopropylacrylamide-co-acrylic acid) nanogels carrying citric acid-coated Fe_3O_4 nanoparticles [76]. After conjugated with Cy 5.5-labeled lactoferrin, the resultant nanogels serve as bifunctional contrast agent for both MRI and intraoperative optical imaging for glioma. The nanogels had a mean hydrodynamic diameter of 95.5 ± 6.2 nm. In vivo application of IV nanogels on glioma detection with MRI and fluorescence imaging were evaluated in rats bearing C6 glioma. Nanogels appeared to selectively accumulate in the tumor tissues and could be used for the pre-operative MRI diagnosis of the glioma. The optical imaging ability of the nanogels was verified by acquiring ex vivo fluorescence images. A significant fluorescence signal was observed only in the brain tumor region of the rat receiving nanogels. The nanogels were proven to be biocompatible with no noticeable toxic effects detected in important biological functions and major organs.

5. Challenges and Prospects for Nanogel-Based Drug Delivery to Brain

Despite their possible novel uses, it is also important to address some setbacks of nanogel use. As aforementioned, a great property of nanogels is their ability to release drugs stimulated by external stimuli (e.g., temperature, pH, enzyme), but this property also has a downside. Should the nanogel arrive an environment where it can degrade to release drugs before it reaches the target site of action, significant problems can arise with delivery of a drug to the off-target, leading to adverse reactions. Considering the unque feature of nanogels releasing drugs slowly and gradually and their prolonged residence time in the body, the off-terget effect may exacerbate the adverse reaction.

Other limitations to nanogels include the particle size and polydispersity of nanocarriers and issues associated with polymer degradation. The impermeable characteristics of the BBB have been considered to be the main reason for the failure to achieve therapeutic drug concentrations in the brain tissue [77]. Especially, the BBB prevents many large molecules, including peptides and medicinal macromolecules, from entering the brain and the rest of the central nervous system. It is primarily because brain capillary endothelial cells are closely connected to each other by tight intercellular junctions and zonulae occludentes. Dehghankhold et al. indicated that the particle sizes of long-circulating drug delivery systems should range between 50 and 200 nm to deposit the systems in the brain [77]. Successful nanogel formulation requires the preparation of homogenous (polydispersity index PDI <0.7) nanocarriers of the average size of 50–200 nm, noting that very small particles (<0 nm) are rapidly cleared via the renal system. Nanogels are commonly formulated with synthetic polymers and organic/inorganic solvents. It is especially crucial to investigate the toxicity of the degraded polymers in the brain and make sure to remove toxic solvent completely from the formulation. It is important to modify the polymers to maximize their bioadhesiveness, biocompatibility and biodegradabiliy. It is also ideal to design the nanogels wisely to minimize any toxic effects caused by fragmented/cleaved polymers after degradation of nanogels.

One of the last major challenges is the inconvenience related to handling and storage. This applies specifically to temperature-responsive nanogels with the gelation temperature lower than 37 °C. This may cause several issues including the premature gelation in the syringe/needle, instability problems affecting the product shelf-life and difficulty in handling.

Considering experimental results from the published articles and experts' opinions/reviews, we listed a few key desirable properties of nanogels carrying drugs that can help seamlessly incoporation of nanogel systems in brain cancer treatment regimen (Table 3). These include the tissue-like properties of gels, particle sizes of 50–200 nm and polydispersity index of nanoparticles below 0.7, the capability of loading multiple agents and releasing agents when desired at controlled and gradual manners, desired rheological patterns, prolonged retention of nanogels in the patient's body, and the storage stability.

Table 3. List of key desirable properties of nanogels desired for brain cancer therapy.

Properties	Nanogels
Gels	Bioadhesive, biocompatible, biodegradable, soft "tissue-like" texture, able to conform to the shape/size of the resection cavity
Nanoparticle size	50–200 nm with PDI <0.7
Payloads	Multiple (hydrophilic and hydrophobic) agents (therapeutics and/or diagnostics)
Drug release	• Demonstrate controlled and gradual release of drugs only when exposed to stimuli (e.g., pH, enzyme) • Demonstrate simultaneous or sequential release of multi-drugs • Minimize premature drug release
Rheology	• Maintain the viscosity under shear stress and at storage/handling • Design the system to increase the viscosity only when exposed to specific stimuli (e.g., temperature)
Modification	• Conjugate targeting moiety and/or imaging agents • Include polymers that maximize biodegradability and biocompatibility
Gelation	Make a sol-to-gel transition rapidly by responding to stimuli
Retention time	Retain extended period of time to increase drug concentrations in plasma and brain tissues
Degradation	• Degrade rapidly when no longer needed • Leave no residual polymers • Does not produce toxic byproduct/degraded polymer fragments
Administration	Exhibit the versatility in routes of administration (e.g., loco-regional, intravenous, and intranasal)
Storage	• Maintain product stability at storage • Does not require special storage conditions (e.g., freezer)

6. Conclusions

Nanogel technology presents an opportunity for viable, lucrative, and efficient future treatments of brain cancer. Brain cancer treatment's critical obstacles are the BBB, diversity of intracranial neoplasms, and the complexity of the organ in which it resides, limiting the treatment options. Nanogels provide local or systemic treatment options that respect the BBB and the physiological feature of the cranial cavity while limiting adverse effects.

Author Contributions: Writing—original draft preparation, B.S., T.S. and H.C.; writing—review and editing, B.S., T.S. and H.C.; visualization, B.S. and H.C.; supervision, H.C. All authors have read and agreed to the published version of the manuscript.

Funding: This research received no external funding.

Conflicts of Interest: The authors declare no conflict of interest.

References

1. Woodworth, G.F.; Dunn, G.P.; Nance, E.A.; Hanes, J.; Brem, H. Emerging Insights into Barriers to Effective Brain Tumor Therapeutics. *Front. Oncol.* **2014**, *4*, 126. [CrossRef]
2. Ferlay, J.; Colombet, M.; Soerjomataram, I.; Parkin, D.M.; Pineros, M.; Znaor, A.; Bray, F. Cancer statistics for the year 2020: An overview. *Int. J. Cancer* **2021**, 1–12. [CrossRef]
3. Mathur, P.; Sathishkumar, K.; Chaturvedi, M.; Das, P.; Sudarshan, K.L.; Santhappan, S.; Nallasamy, V.; John, A.; Narasimhan, S.; Roselind, F.S.; et al. Cancer Statistics, 2020: Report From National Cancer Registry Programme, India. *JCO Glob. Oncol.* **2020**, *6*, 1063–1075. [CrossRef]
4. Ostrom, Q.T.; Patil, N.; Cioffi, G.; Waite, K.; Kruchko, C.; Barnholtz-Sloan, J.S. CBTRUS Statistical Report: Primary Brain and Other Central Nervous System Tumors Diagnosed in the United States in 2013–2017. *Neuro-Oncology* **2020**, *22*, iv1–iv96. [CrossRef]
5. Siegel, R.L.; Miller, K.D.; Jemal, A. Cancer statistics, 2020. *CA A Cancer J. Clin.* **2020**, *70*, 7–30. [CrossRef]
6. Louis, D.N.; Perry, A.; Reifenberger, G.; von Deimling, A.; Figarella-Branger, D.; Cavenee, W.K.; Ohgaki, H.; Wiestler, O.D.; Kleihues, P.; Ellison, D.W. The 2016 World Health Organization Classification of Tumors of the Central Nervous System: A summary. *Acta Neuropathol.* **2016**, *131*, 803–820. [CrossRef] [PubMed]
7. DeAngelis, L.M. Brain tumors. *N. Engl. J. Med.* **2001**, *344*, 114–123. [CrossRef]
8. Ahmed, R.; Oborski, M.J.; Hwang, M.; Lieberman, F.S.; Mountz, J.M. Malignant gliomas: Current perspectives in diagnosis, treatment, and early response assessment using advanced quantitative imaging methods. *Cancer Manag. Res.* **2014**, *6*, 149–170. [CrossRef]
9. Daneman, R.; Prat, A. The blood-brain barrier. *Cold Spring Harbor Perspect. Biol.* **2015**, *7*, a020412. [CrossRef]
10. Dong, X. Current Strategies for Brain Drug Delivery. *Theranostics* **2018**, *8*, 1481–1493. [CrossRef]
11. Friedman, H.S.; Kerby, T.; Calvert, H. Temozolomide and treatment of malignant glioma. *Clin. Cancer Res. Off. J. Am. Assoc. Cancer Res.* **2000**, *6*, 2585–2597.
12. Wick, W.; Winkler, F. Regimen of procarbazine, lomustine, and vincristine versus temozolomide for gliomas. *Cancer* **2018**, *124*, 2674–2676. [CrossRef]
13. Lassman, A.B. Procarbazine, lomustine and vincristine or temozolomide: Which is the better regimen? *CNS Oncol.* **2015**, *4*, 341–346. [CrossRef]
14. Brada, M.; Stenning, S.; Gabe, R.; Thompson, L.C.; Levy, D.; Rampling, R.; Erridge, S.; Saran, F.; Gattamaneni, R.; Hopkins, K.; et al. Temozolomide versus procarbazine, lomustine, and vincristine in recurrent high-grade glioma. *J. Clin. Oncol. Off. J. Am. Soc. Clin. Oncol.* **2010**, *28*, 4601–4608. [CrossRef]
15. Stone, J.B.; DeAngelis, L.M. Cancer-treatment-induced neurotoxicity–focus on newer treatments. *Nat. Rev. Clin. Oncol.* **2016**, *13*, 92–105. [CrossRef]
16. Baskar, R.; Lee, K.A.; Yeo, R.; Yeoh, K.W. Cancer and radiation therapy: Current advances and future directions. *Int. J. Med Sci.* **2012**, *9*, 193–199. [CrossRef]
17. Smart, D. Radiation Toxicity in the Central Nervous System: Mechanisms and Strategies for Injury Reduction. *Semin. Radiat. Oncol.* **2017**, *27*, 332–339. [CrossRef] [PubMed]
18. Pekic, S.; Miljic, D.; Popovic, V. Hypopituitarism Following Cranial Radiotherapy. In *Endotext*; Feingold, K.R., Anawalt, B., Boyce, A., Chrousos, G., de Herder, W.W., Dhatariya, K., Dungan, K., Grossman, A., Hershman, J.M., Hofland, J., et al., Eds.; Endotext. MDText.com, Inc.: South Dartmouth, MA, USA, 2000.
19. Perry, J.; Chambers, A.; Spithoff, K.; Laperriere, N. Gliadel wafers in the treatment of malignant glioma: A systematic review. *Curr. Oncol.* **2007**, *14*, 189–194. [CrossRef]
20. Wenbin Dang, T.D.; Peter, Y.; Yong, Z.; David, N.; Charles, S.C.; Betty, T.; Henry, B. Effects of GLIADEL®wafer initial molecular weight on the erosion of wafer and release of BCNU. *J. Control. Release* **1996**, *42*, 83–92. [CrossRef]
21. Fisher, J.P.; Adamson, D.C. Current FDA-Approved Therapies for High-Grade Malignant Gliomas. *Biomedicines* **2021**, *9*, 324. [CrossRef]

22. Westphal, M.; Hilt, D.C.; Bortey, E.; Delavault, P.; Olivares, R.; Warnke, P.C.; Whittle, I.R.; Jaaskelainen, J.; Ram, Z. A phase 3 trial of local chemotherapy with biodegradable carmustine (BCNU) wafers (Gliadel wafers) in patients with primary malignant glioma. *Neuro-Oncology* **2003**, *5*, 79–88. [CrossRef]
23. Spiegel, B.M.; Esrailian, E.; Laine, L.; Chamberlain, M.C. Clinical impact of adjuvant chemotherapy in glioblastoma multiforme: A meta-analysis. *CNS Drugs* **2007**, *21*, 775–787. [CrossRef]
24. Stupp, R.; Hegi, M.E.; Mason, W.P.; van den Bent, M.J.; Taphoorn, M.J.; Janzer, R.C.; Ludwin, S.K.; Allgeier, A.; Fisher, B.; Belanger, K.; et al. Effects of radiotherapy with concomitant and adjuvant temozolomide versus radiotherapy alone on survival in glioblastoma in a randomised phase III study: 5-year analysis of the EORTC-NCIC trial. *Lancet Oncol.* **2009**, *10*, 459–466. [CrossRef]
25. Miglierini, P.; Bouchekoua, M.; Rousseau, B.; Hieu, P.D.; Malhaire, J.P.; Pradier, O. Impact of the per-operatory application of GLIADEL wafers (BCNU, carmustine) in combination with temozolomide and radiotherapy in patients with glioblastoma multiforme: Efficacy and toxicity. *Clin. Neurol. Neurosurg.* **2012**, *114*, 1222–1225. [CrossRef]
26. Weber, E.L.; Goebel, E.A. Cerebral edema associated with Gliadel wafers: Two case studies. *Neuro-Oncology* **2005**, *7*, 84–89. [CrossRef] [PubMed]
27. Larocca, R.V.; Vitaz, T.W.; Morassutti, D.J.; Doyle, M.J.; Glisson, S.D.; Hargis, J.B.; Goldsmith, G.H.; Cervera, A.; Stribinskiene, L.; New, P. A phase II study of radiation with concomitant and then sequential temozolomide (TMZ) in patients (pts) with newly diagnosed supratentorial high grade malignant glioma (MG) who have undergone surgery with carmustine (BCNU) wafer insertion. *J. Clin. Oncol.* **2005**, *23*, 1547. [CrossRef]
28. Salmaggi, A.; Milanesi, I.; Silvani, A.; Gaviani, P.; Marchetti, M.; Fariselli, L.; Solero, C.L.; Maccagnano, C.; Casali, C.; Guzzetti, S.; et al. Prospective study of carmustine wafers in combination with 6-month metronomic temozolomide and radiation therapy in newly diagnosed glioblastoma: Preliminary results. *J. Neurosurg.* **2013**, *118*, 821–829. [CrossRef]
29. Elstad, N.L.; Fowers, K.D. OncoGel (ReGel/paclitaxel)–clinical applications for a novel paclitaxel delivery system. *Adv. Drug Deliv. Rev.* **2009**, *61*, 785–794. [CrossRef]
30. Tyler, B.; Fowers, K.D.; Li, K.W.; Recinos, V.R.; Caplan, J.M.; Hdeib, A.; Grossman, R.; Basaldella, L.; Bekelis, K.; Pradilla, G.; et al. A thermal gel depot for local delivery of paclitaxel to treat experimental brain tumors in rats. *J. Neurosurg.* **2010**, *113*, 210–217. [CrossRef]
31. Torres, A.J.; Zhu, C.; Shuler, M.L.; Pannullo, S. Paclitaxel delivery to brain tumors from hydrogels: A computational study. *Biotechnol. Prog.* **2011**, *27*, 1478–1487. [CrossRef]
32. Neamtu, I.; Rusu, A.G.; Diaconu, A.; Nita, L.E.; Chiriac, A.P. Basic concepts and recent advances in nanogels as carriers for medical applications. *Drug Deliv.* **2017**, *24*, 539–557. [CrossRef]
33. McKenzie, M.; Betts, D.; Suh, A.; Bui, K.; Kim, L.D.; Cho, H. Hydrogel-Based Drug Delivery Systems for Poorly Water-Soluble Drugs. *Molecules* **2015**, *20*, 20397–20408. [CrossRef]
34. Akiyoshi, K.; Kobayashi, S.; Shichibe, S.; Mix, D.; Baudys, M.; Kim, S.W.; Sunamoto, J. Self-assembled hydrogel nanoparticle of cholesterol-bearing pullulan as a carrier of protein drugs: Complexation and stabilization of insulin. *J. Control. Release Off. J. Control. Release Soc.* **1998**, *54*, 313–320. [CrossRef]
35. Giovannini, G.; Kunc, F.; Piras, C.C.; Stranik, O.; Edwards, A.A.; Hall, A.J.; Gubala, V. Stabilizing silica nanoparticles in hydrogels: Impact on storage and polydispersity. *RSC Adv.* **2017**, *7*, 19924–19933. [CrossRef]
36. Ryu, J.H.; Jiwpanich, S.; Chacko, R.; Bickerton, S.; Thayumanavan, S. Surface-functionalizable polymer nanogels with facile hydrophobic guest encapsulation capabilities. *J. Am. Chem. Soc.* **2010**, *132*, 8246–8247. [CrossRef] [PubMed]
37. Elkassih, S.A.; Kos, P.; Xiong, H.; Siegwart, D.J. Degradable redox-responsive disulfide-based nanogel drug carriers via dithiol oxidation polymerization. *Biomater. Sci.* **2019**, *7*, 607–617. [CrossRef]
38. Kockelmann, J.; Stickdorn, J.; Kasmi, S.; De Vrieze, J.; Pieszka, M.; Ng, D.Y.W.; David, S.A.; De Geest, B.G.; Nuhn, L. Control over Imidazoquinoline Immune Stimulation by pH-Degradable Poly(norbornene) Nanogels. *Biomacromolecules* **2020**, *21*, 2246–2257. [CrossRef]
39. Liao, S.C.; Ting, C.W.; Chiang, W.H. Functionalized polymeric nanogels with pH-sensitive benzoic-imine cross-linkages designed as vehicles for indocyanine green delivery. *J. Colloid Interface Sci.* **2020**, *561*, 11–22. [CrossRef]
40. He, J.; Tong, X.; Zhao, Y. Photoresponsive Nanogels Based on Photocontrollable Cross-Links. *Macromolecules* **2009**, *42*, 4845–4852. [CrossRef]
41. Dannert, C.; Stokke, B.T.; Dias, R.S. Nanoparticle-Hydrogel Composites: From Molecular Interactions to Macroscopic Behavior. *Polymers* **2019**, *11*, 275. [CrossRef]
42. Levin, M.; Sonn-Segev, A.; Roichman, Y. Structural changes in nanoparticle-hydrogel composites at very low filler concentrations. *J. Chem. Phys.* **2019**, *150*, 064908. [CrossRef]
43. Ayyub, O.B.; Kofinas, P. Enzyme Induced Stiffening of Nanoparticle-Hydrogel Composites with Structural Color. *ACS Nano* **2015**, *9*, 8004–8011. [CrossRef]
44. Baek, K.; Liang, J.; Lim, W.T.; Zhao, H.; Kim, D.H.; Kong, H. In situ assembly of antifouling/bacterial silver nanoparticle-hydrogel composites with controlled particle release and matrix softening. *ACS Appl. Mater. Interfaces* **2015**, *7*, 15359–15367. [CrossRef] [PubMed]
45. Thoniyot, P.; Tan, M.J.; Karim, A.A.; Young, D.J.; Loh, X.J. Nanoparticle-Hydrogel Composites: Concept, Design, and Applications of These Promising, Multi-Functional Materials. *Adv. Sci.* **2015**, *2*, 1400010. [CrossRef] [PubMed]

46. Ohya, Y.; Takahashi, A.; Kuzuya, A. Preparation of Biodegradable Oligo(lactide)s-Grafted Dextran Nanogels for Efficient Drug Delivery by Controlling Intracellular Traffic. *Int. J. Mol. Sci.* **2018**, *19*, 1606. [CrossRef] [PubMed]
47. Tan, J.P.; Tan, M.B.; Tam, M.K. Application of nanogel systems in the administration of local anesthetics. *Local Reg. Anesth.* **2010**, *3*, 93–100. [CrossRef]
48. Tan, H.; Jin, H.; Mei, H.; Zhu, L.; Wei, W.; Wang, Q.; Liang, F.; Zhang, C.; Li, J.; Qu, X.; et al. PEG-urokinase nanogels with enhanced stability and controllable bioactivity. *Soft Matter* **2012**, *8*, 2644–2650. [CrossRef]
49. Escobedo, H.D.; Stansbury, J.W.; Nair, D.P. Photoreactive nanogels as versatile polymer networks with tunable in situ drug release kinetics. *J. Mech. Behav. Biomed. Mater.* **2020**, *108*, 103755. [CrossRef]
50. Cho, H.; Jammalamadaka, U.; Tappa, K. Nanogels for Pharmaceutical and Biomedical Applications and Their Fabrication Using 3D Printing Technologies. *Materials* **2018**, *11*. [CrossRef]
51. Soni, K.S.; Desale, S.S.; Bronich, T.K. Nanogels: An overview of properties, biomedical applications and obstacles to clinical translation. *J. Control. Release Off. J. Control. Release Soc.* **2016**, *240*, 109–126. [CrossRef]
52. Pereira, P.; Pedrosa, S.S.; Correia, A.; Lima, C.F.; Olmedo, M.P.; Gonzalez-Fernandez, A.; Vilanova, M.; Gama, F.M. Biocompatibility of a self-assembled glycol chitosan nanogel. *Toxicol. In Vitro* **2015**, *29*, 638–646. [CrossRef]
53. Oishi, M.; Nagasaki, Y. Stimuli-responsive smart nanogels for cancer diagnostics and therapy. *Nanomedicine* **2010**, *5*, 451–468. [CrossRef] [PubMed]
54. Cho, H.; Jammalamadaka, U.; Tappa, K.; Egbulefu, C.; Prior, J.; Tang, R.; Achilefu, S. 3D Printing of Poloxamer 407 Nanogel Discs and Their Applications in Adjuvant Ovarian Cancer Therapy. *Mol. Pharm.* **2019**, *16*, 552–560. [CrossRef] [PubMed]
55. Gerecke, C.; Edlich, A.; Giulbudagian, M.; Schumacher, F.; Zhang, N.; Said, A.; Yealland, G.; Lohan, S.B.; Neumann, F.; Meinke, M.C.; et al. Biocompatibility and characterization of polyglycerol-based thermoresponsive nanogels designed as novel drug-delivery systems and their intracellular localization in keratinocytes. *Nanotoxicology* **2017**, *11*, 267–277. [CrossRef] [PubMed]
56. Yang, G.; Fu, S.; Yao, W.; Wang, X.; Zha, Q.; Tang, R. Hyaluronic acid nanogels prepared via ortho ester linkages show pH-triggered behavior, enhanced penetration and antitumor efficacy in 3-D tumor spheroids. *J. Colloid Interface Sci.* **2017**, *504*, 25–38. [CrossRef] [PubMed]
57. Kang, E.B.; Lee, G.B.; In, I.; Park, S.Y. pH-sensitive fluorescent hyaluronic acid nanogels for tumor-targeting and controlled delivery of doxorubicin and nitric oxide. *Eur. Polym. J.* **2018**, *101*, 96–104. [CrossRef]
58. Salehi, R.; Rasouli, S.; Hamishehkar, H. Smart thermo/pH responsive magnetic nanogels for the simultaneous delivery of doxorubicin and methotrexate. *Int. J. Pharm.* **2015**, *487*, 274–284. [CrossRef]
59. Pan, Y.; Liu, J.; Yang, K.; Cai, P.; Xiao, H. Novel multi-responsive and sugarcane bagasse cellulose-based nanogels for controllable release of doxorubicin hydrochloride. *Mater. Sci. Eng. C Mater. Biol. Appl.* **2021**, *118*, 111357. [CrossRef]
60. McKenzie, M.; Betts, D.; Suh, A.; Bui, K.; Tang, R.; Liang, K.; Achilefu, S.; Kwon, G.S.; Cho, H. Proof-of-Concept of Polymeric Sol-Gels in Multi-Drug Delivery and Intraoperative Image-Guided Surgery for Peritoneal Ovarian Cancer. *Pharm. Res.* **2016**, *33*, 2298–2306. [CrossRef]
61. Su, S.; Wang, H.; Liu, X.; Wu, Y.; Nie, G. iRGD-coupled responsive fluorescent nanogel for targeted drug delivery. *Biomaterials* **2013**, *34*, 3523–3533. [CrossRef]
62. Chen, J.; Ding, J.; Xu, W.; Sun, T.; Xiao, H.; Zhuang, X.; Chen, X. Receptor and Microenvironment Dual-Recognizable Nanogel for Targeted Chemotherapy of Highly Metastatic Malignancy. *Nano Lett.* **2017**, *17*, 4526–4533. [CrossRef] [PubMed]
63. Ribovski, L.; de Jong, E.; Mergel, O.; Zu, G.; Keskin, D.; van Rijn, P.; Zuhorn, I.S. Low nanogel stiffness favors nanogel transcytosis across an in vitro blood–brain barrier. *Nanomed. Nanotechnol. Biol. Med.* **2021**, *34*, 102377. [CrossRef] [PubMed]
64. Cardoso, F.L.; Brites, D.; Brito, M.A. Looking at the blood–brain barrier: Molecular anatomy and possible investigation approaches. *Brain Res. Rev.* **2010**, *64*, 328–363. [CrossRef]
65. Haumann, R.; Videira, J.C.; Kaspers, G.J.L.; van Vuurden, D.G.; Hulleman, E. Overview of Current Drug Delivery Methods Across the Blood–Brain Barrier for the Treatment of Primary Brain Tumors. *CNS Drugs* **2020**, *34*, 1121–1131. [CrossRef] [PubMed]
66. Kimura, A.; Jo, J.-I.; Yoshida, F.; Hong, Z.; Tabata, Y.; Sumiyoshi, A.; Taguchi, M.; Aoki, I. Ultra-small size gelatin nanogel as a blood brain barrier impermeable contrast agent for magnetic resonance imaging. *Acta Biomater.* **2021**, *125*, 290–299. [CrossRef]
67. She, D.; Huang, H.; Li, J.; Peng, S.; Wang, H.; Yu, X. Hypoxia-degradable zwitterionic phosphorylcholine drug nanogel for enhanced drug delivery to glioblastoma. *Chem. Eng. J.* **2021**, *408*, 127359. [CrossRef]
68. Abbott, N.J.; Patabendige, A.A.K.; Dolman, D.E.M.; Yusof, S.R.; Begley, D.J. Structure and function of the blood–brain barrier. *Neurobiol. Dis.* **2010**, *37*, 13–25. [CrossRef]
69. Furtado, D.; Bjornmalm, M.; Ayton, S.; Bush, A.I.; Kempe, K.; Caruso, F. Overcoming the Blood-Brain Barrier: The Role of Nanomaterials in Treating Neurological Diseases. *Adv. Mater.* **2018**, *30*, e1801362. [CrossRef] [PubMed]
70. Teleanu, D.M.; Chircov, C.; Grumezescu, A.M.; Volceanov, A.; Teleanu, R.I. Blood-Brain Delivery Methods Using Nanotechnology. *Pharmaceutics* **2018**, *10*, 269. [CrossRef]
71. Lin, F.-W.; Chen, P.-Y.; Wei, K.-C.; Huang, C.-Y.; Wang, C.-K.; Yang, H.-W. Rapid In Situ MRI Traceable Gel-forming Dual-drug Delivery for Synergistic Therapy of Brain Tumor. *Theranostics* **2017**, *7*, 2524–2536. [CrossRef] [PubMed]
72. McCrorie, P.; Mistry, J.; Taresco, V.; Lovato, T.; Fay, M.; Ward, I.; Ritchie, A.A.; Clarke, P.A.; Smith, S.J.; Marlow, M.; et al. Etoposide and olaparib polymer-coated nanoparticles within a bioadhesive sprayable hydrogel for post-surgical localised delivery to brain tumours. *Eur. J. Pharm. Biopharm.* **2020**, *157*, 108–120. [CrossRef]

73. Picone, P.; Sabatino, M.A.; Ditta, L.A.; Amato, A.; San Biagio, P.L.; Mulè, F.; Giacomazza, D.; Dispenza, C.; Di Carlo, M. Nose-to-brain delivery of insulin enhanced by a nanogel carrier. *J. Control. Release* **2018**, *270*, 23–36. [CrossRef]
74. Shatsberg, Z.; Zhang, X.; Ofek, P.; Malhotra, S.; Krivitsky, A.; Scomparin, A.; Tiram, G.; Calderon, M.; Haag, R.; Satchi-Fainaro, R. Functionalized nanogels carrying an anticancer microRNA for glioblastoma therapy. *J. Control. Release Off. J. Control. Release Soc.* **2016**, *239*, 159–168. [CrossRef] [PubMed]
75. Azadi, A.; Hamidi, M.; Rouini, M.R. Methotrexate-loaded chitosan nanogels as 'Trojan Horses' for drug delivery to brain: Preparation and in vitro/in vivo characterization. *Int. J. Biol. Macromol.* **2013**, *62*, 523–530. [CrossRef] [PubMed]
76. Jiang, L.; Zhou, Q.; Mu, K.; Xie, H.; Zhu, Y.; Zhu, W.; Zhao, Y.; Xu, H.; Yang, X. pH/temperature sensitive magnetic nanogels conjugated with Cy5.5-labled lactoferrin for MR and fluorescence imaging of glioma in rats. *Biomaterials* **2013**, *34*, 7418–7428. [CrossRef] [PubMed]
77. Danaei, M.; Dehghankhold, M.; Ataei, S.; Hasanzadeh Davarani, F.; Javanmard, R.; Dokhani, A.; Khorasani, S.; Mozafari, M.R. Impact of Particle Size and Polydispersity Index on the Clinical Applications of Lipidic Nanocarrier Systems. *Pharmaceutics* **2018**, *10*, 57. [CrossRef] [PubMed]

Article

Functionalized Poly(*N*-isopropylacrylamide)-Based Microgels in Tumor Targeting and Drug Delivery

Simona Campora [1,2,†], Reham Mohsen [3,4,†], Daniel Passaro [1], Howida Samir [3], Hesham Ashraf [3], Saif El-Din Al-Mofty [5], Ayman A. Diab [3], Ibrahim M. El-Sherbiny [5], Martin J. Snowden [4] and Giulio Ghersi [1,2,*]

1. Department of Biological, Chemical and Pharmaceutical Sciences and Technologies (STEBICEF), University of Palermo, Viale delle Scienze, Ed. 16, 90128 Palermo, Italy; simona.campora@unipa.it (S.C.); daniel.passaro@abbvie.com (D.P.)
2. Abiel s.r.l, c/o Arca Incubatore di Imprese, University of Palermo, Viale delle Scienze, Ed. 16 (Floor-2), 90128 Palermo, Italy
3. Faculty of Biotechnology, October University for Modern Sciences and Arts, Cairo 12451, Egypt; rmohsen@msa.edu.eg (R.M.); ho.samir@nu.edu.eg (H.S.); hesham.ashraf@msa.edu.eg (H.A.); adiab@msa.eun.eg (A.A.D.)
4. School of Science, University of Greenwich, Gillingham, Chatham, Kent, Canterbury ME4 4TB, UK; m.j.snowden@greenwich.ac
5. Center of Materials Science, Zewail City of Science and Technology, 6th October City, Giza 12588, Egypt; s-saifel-din.el-mofty@zewailcity.edu.eg (S.E.-D.A.-M.); ielsherbiny@zewailcity.edu.eg (I.M.E.-S.)

* Correspondence: giulio.ghersi@unipa.it
† These authors contributed equally to this work.

Abstract: Over the past several decades, the development of engineered small particles as targeted and drug delivery systems (TDDS) has received great attention thanks to the possibility to overcome the limitations of classical cancer chemotherapy, including targeting incapability, nonspecific action and, consequently, systemic toxicity. Thus, this research aims at using a novel design of Poly(*N*-isopropylacrylamide) p(NIPAM)-based microgels to specifically target cancer cells and avoid the healthy ones, which is expected to decrease or eliminate the side effects of chemotherapeutic drugs. Smart NIPAM-based microgels were functionalized with acrylic acid and coupled to folic acid (FA), targeting the folate receptors overexpressed by cancer cells and to the chemotherapeutic drug doxorubicin (Dox). The successful conjugation of FA and Dox was demonstrated by dynamic light scattering (DLS), Fourier-transform infrared (FTIR) spectroscopy, thermogravimetric analysis (TGA), UV-VIS analysis, and differential scanning calorimetry (DSC). Furthermore, viability assay performed on cancer and healthy breast cells, suggested the microgels' biocompatibility and the cytotoxic effect of the conjugated drug. On the other hand, the specific tumor targeting of synthetized microgels was demonstrated by a co-cultured (healthy and cancer cells) assay monitored using confocal microscopy and flow cytometry. Results suggest successful targeting of cancer cells and drug release. These data support the use of pNIPAM-based microgels as good candidates as TDDS.

Keywords: p(NIPAM)-co-5%AA microgels; folic acid; doxorubicin; cancer

1. Introduction

Cancer is one of the leading causes of death in the world. In 2020, the world health organization stated that the number of deaths caused by cancer reached ten million deaths worldwide [data from WHO] [1]. One of the most commonly used therapies is chemotherapy, which is delivered systematically in a non-targeted manner [2]. Over the past several decades, the development of engineered nano- and micro-systems for targeted drug delivery have received great attention thanks to their possibility to overcome the limitations of classical cancer chemotherapy, including poor solubility, targeting incapability, nonspecific action and, consequently, systemic toxicity [3,4]. For instance, the anticancer drug doxorubicin (Dox) showed several adverse effects, such as myelosuppression, which is the decrease in the ability of the bone marrow to produce new blood cells, vomiting, and in extreme cases, it can lead to liver dysfunction and heart diseases. All these adverse effects are due to the apoptosis of healthy cells along with cancer cells as a result of untargeted drug administration [5]. Recently, scientists have developed targeted drug delivery systems using smart particles against cancer cells to reduce the side effects of chemotherapy [6].

Ligand-mediated targeting is based on the conjugation of engineering particles to specific targeting molecules including small molecules, carbohydrates, antibodies, or peptides in order to bind to specific receptors present on the surface of cancer cells [7,8]. For instance, the low molecular weight, low production cost, and ease of nano- and micro-particles systems conjugation make small molecules optimal candidates as potential targeting ligands. Folate receptors are known to be overexpressed almost 100–300 times more in cancer cells than normal ones; this is to increase the cancer cells' uptake of folic acid (FA) used in different cellular metabolic pathways [9]. Accordingly, small particles can be conjugated with FA that binds specifically to folate receptors, to achieve targeted therapy [10].

Therefore, nanoparticles and microparticles have to be synthesized, engineered and optimized to raise the circulating half-life and to obtain a site-specific release of drugs at therapeutically optimal levels and dose regimes [11]. The composition, size, shape, surface properties, biocompatibility, and degradation profile should be carefully considered for the optimal design of the NPs for therapeutic purposes [12,13]. Depending on the aim and the particles nature, the drug can be encapsulated [14], conjugated by stacking interactions [15], or by chemical reactions [16,17] and the drug release can be induced in a stimuli-responsive way [18,19].

Among the different particle types, the nano- and micro-gels present many advantages, including high mechanical properties, stability, high water content, large flexible surface for the conjugation with a big amount of cargo protected in an aqueous environment, as well as biocompatibility [20]. They are constituted of polymer chains that form a matrix able to absorb and retain high quantity of aqueous solution (swelling capacity) [21,22].

In this contest, Poly(*N-iso*-propyl acrylamide)-*co*-Acrylic Acid (p(NIPAM)-co-5%AA) are smart polymeric microgels that change their physiochemical behavior in response to external stimuli such as temperature and pH change. These changes are instantaneous and reversible, as they return to their original status once the stimulus is removed [21,22]. As for other smart materials, p(NIPAM)-co-AA has been studied for different applications, such as tissue engineering scaffolds, cell culture supports, and bioseparation devices [23,24]. Smart particles can be used to reduce the adverse effects of the drug, increasing its efficiency, reducing the dosage, and consequently its cost [25]. The expected advantages of using p(NIPAM)-based particles have led researchers such as Guo et al. to suggest p(NIPAM)-co-AA as a choice for targeted cancer therapy since the pH of the microenvironment surrounding tumor cells is known to be more acidic than that surrounding healthy ones [26]. Moreover, p(NIPAM)-co-AA respond to acidic environment by contracting its size allowing ease of absorption, while in alkaline environment the p(NIPAM)-co-AA swells in size, allowing difficulty of absorption towards the cells [27,28].

In this research, a targeted drug delivery system for cancer cells was designed and developed through covalent bonding of p(NIPAM)-co-5%AA to FA, as the targeting agent, and to Doxorubicin as the anti-cancer drug, through 1-Ethyl-3-(3-dimethylaminopropyl)carbodiimide (EDC) and *N*-hydroxysuccinimide (NHS) coupling chemistry. The study was performed on HB2 (healthy breast cells) and MDA-MB 231 (breast cancer cells). In vitro characterization was used to evaluate the physicochemical behavior of the microgel particles through ultraviolet–visible (UV-Vis) spectroscopy, differential scanning calorimetry (DSC), and dynamic light scattering (DLS) to calculate the size distribution against temperature change. This is in addition to thermogravimetric analysis (TGA) and Fourier-transform infrared spectroscopy (FTIR) as confirmation of successful coupling reaction of EDC/NHS with each stage of folic acid conjugation and Dox conjugation. The cell biocompatibility of different concentrations of p (NIPAM)-co-5%AA, as well as p (NIPAM)-co-5%AA-co-FA and the cytotoxic effect of p (NIPAM)-co-5%AA-co-FA-co-Dox were tested. Finally, the specific tumor targeting experiments that test the suggested targeting behavior of the particles qualitatively and quantitatively were carried out. These are confocal microscopy and flow cytometry.

2. Results and Discussion

2.1. Synthesis of p(NIPAM)-co-5%AA Microgels and Conjugation with Folic Acid and Doxorubicin

A sequential synthesis and conjugation processes were performed to generate microgel particles decorated with the targeting molecule folic acid and the anticancer drug doxorubicin. p(NIPAM)-co-5%AA were synthesized by Surfactant Free Emulsion Polymerisation (SFEP) technique as described in materials and methods to avoid toxic surfactant contamination [28,29]. Successively, EDC-NHS protocol was adopted to first bind folic acid to some of the acrylic acids of p(NIPAM)-co-5%AA microgels and then doxorubicin to the remaining acrylic acid residues. The success of the protocol was demonstrated by the UV-VIS analysis in which it was evident the characteristic peak of folic acid (340 mm) on p(NIPAM)-co-5%AA-co-FA and both folic acid and doxorubicin (485 nm) peaks on p(NIPAM)-co-5%AA-co-FA-co-Dox (Figure 1). The amount of folic acid and doxorubicin conjugated was calculated by spectrophotometric analysis using the standard calibration curves (Figures S1 and S2).

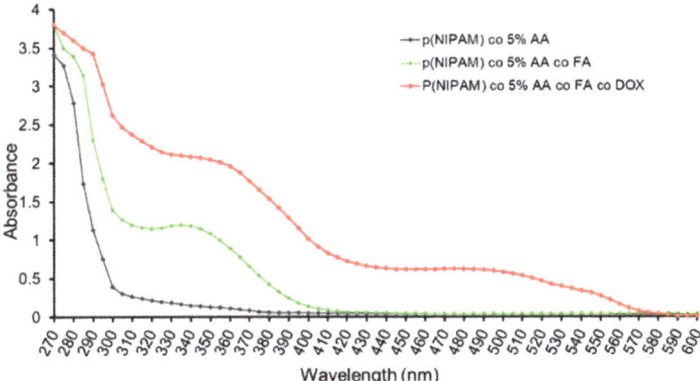

Figure 1. UV-VIS spectra of p(NIPAM)-co-5%AA, p(NIPAM)-co-5%AA-co-FA, and p(NIPAM)-co-5%AA-co-FA-co-Dox.

2.2. Size of Microgels

The effect of temperature change on the size of p(NIPAM)-co-5%AA, p(NIPAM)-co-5%AA-co-FA, and p(NIPAM)-co-5%AA-co-FA-co-DOX was studied by dynamic light scattering analysis (DLS) (Figure 2 and Figure S3). The size of the three microgel particles showed typical microgel behavior [30]. Below the VPTT (volume phase transition temperature) (34 °C), the particles were swollen and configure a large size. At 34 °C (VPTT), the three microgels underwent a sharp decrease in size as the hydrogen bonds between the polymer particles and water molecules break due to energy gained under higher temperature [30,31], causing the polymer–polymer interactions to dominate. Hence, water molecules were expelled from microgel particles, causing the microgel to collapse and deswell [30,32].

At 15 °C, p(NIPAM)-co-5%AA had an average diameter of 701 nm while that of p(NIPAM)-co-5%AA-co-FA had an average diameter of 451 nm particle size. This was because FA, being a large molecule with several hydrophobic aromatic moieties, tended to decrease the hydrophilicity of the particle and decrease the hydrogen bonding with water molecules, hence it contained less water than that of the AA one. Further conjugating the particles with Dox molecules had increased the length of the hairy layers, hence causing the particle to increase in size at an average diameter of 1500 nm. Doxorubicin, being another bulky molecule with several hydrophilic groups, had helped the microgels to swell and reach the micro-scale.

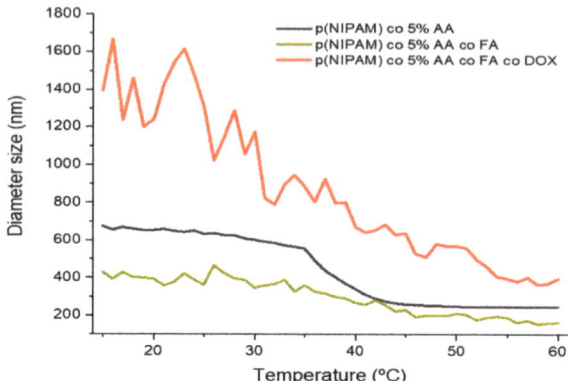

Figure 2. Size change of p(NIPAM)-based microgels against heating cycle temperature. The PDI for p(NIPAM)-co-5%AA, p(NIPAM)-co-5%AA-co-FA, and p(NIPAM)-co-5%AA-co-FA-co-Dox is 0.107, 0.482, and 0.531, respectively.

Attaching FA to the microgel particles, the microgel's VPTT was unaffected but the size of the microgel was reduced even further. In the case of p(NIPAM)-co-5%AA-co-FA-co-DOX, a rapid and sharp decrease in size was observed. At 50 °C, particles of the three microgels p(NIPAM)-co-5%AA, p(NIPAM)-co-5%AA-co-FA, and p(NIPAM)-co-5%AA-co-FA-co-Dox were deswollen to an average size of 247, 177, and 433 nm, respectively. Moreover, calculating the deswelling degrees between the minimum and maximum temperature is rather challenging. This is because the size of p(NIPAM)-co-5%AA-co-FA, tends to fluctuate greatly from 300 to 504 nm, then dropping back again to 400. The decrease in size of p(NIPAM)-co-5%AA-co-FA in comparison to p(NIPAM)-co-5% AA is due to the decreased hydrophobicity of the particles because of the hydrophobic rings in the molecular structure of folic acid. The hydrophobic structure of the molecule decreases the hydrogen bonding between the particle and water and hence decreases the amount of water entrapped within the particles. After adding Dox with a complex structure and large molecules, the particle size tends to increase due to elongated hairy structures [30].

It is worth mentioning that the overall PDI (polydispersity index) of p(NIPAM)-co-5%AA was 0.057, which indicated the highly satisfactory consistency between particle size and distribution. Attaching FA molecules to the above-mentioned microgels decreased this consistency and increased the overall PDI to reach 0.503, which was fairly satisfactory. However, the conjugation of the bulky Dox molecules had increased the overall PDI to 0.833. The reason for this increase in PDI was the fact that Dox is a bulky molecule. When Dox is chemically conjugated to p(NIPAM)-co-5%AA-co-FA, it can either attach to FA moiety or to the unreacted AA, which gives the microgel versatility to have free end FA moiety on the surface of the microgel to target the folate receptor.

2.3. Electrophoretic Mobility

Electrophoretic mobility (Em) of microgel particles is mainly affected by three factors: the size of microgels, solvent viscosity, and dielectric constant [33]. The latter two factors are needed to be kept at a minimum to measure the Em of microgel particles accurately across the temperature range, hence the usage of DI water as the dispersant [34]. The three microgels, p(NIPAM)-co-5%AA, p(NIPAM)-co-5%AA-co-FA, and p(NIPAM)-co-5%AA-co-FA-co-Dox, showed an increase in their magnitude of Em (|Em|) as the temperature increased from 15–60 °C (Figures 3 and S4).

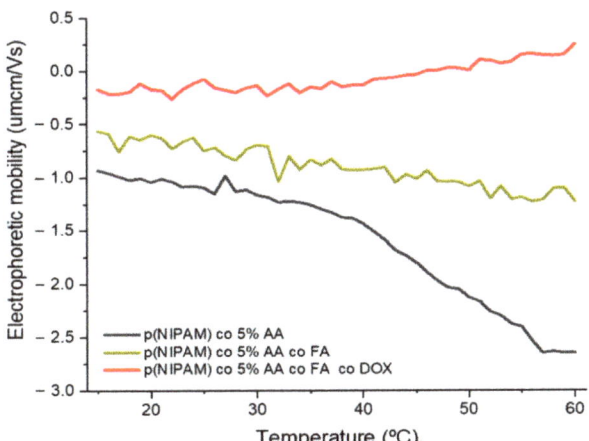

Figure 3. Electrophoretic mobility change of for p(NIPAM)-co-5%AA, p(NIPAM)-co-5%AA-co-FA, and p(NIPAM)-co-5%AA-co-FA-co-Dox versus temperature change (heating cycle).

At 15 °C, p(NIPAM)-co-5%AA had a negative electrophoretic mobility of average -0.946 μmcm/Vs. While that of p(NIPAM)-co-5%AA-co-FA average E_m is −0.401 μmcm/Vs, which showed that conjugating p(NIPAM)-co-5%AA to FA resulted in a decrease in its E_m. p(NIPAM)-co-5%AA-co-FA-co-Dox had an average E_m of -0.0364 μmcm/Vs, this was due to the positive charge density of Dox, as well as, the bulky structure of the particle that causes the negative charges from the sulphate ions to be masked [33,35].

At 37 °C, the particle size dramatically decreased, which causes an increase in the surface charge density, hence an increase in electrophoretic mobility. In the case of p(NIPAM)-co-5%AA, the increase in electrophoretic mobility around VPTT was sharp. This was because the negative charges were exposed, while in case of p(NIPAM)-co-5%AA-co-FA and p(NIPAM)-co-5%AA-co-FA-co-Dox, it was suggested that the complex structure of the particle had masked some of the charges causing the increase in E_m to be steep.

2.4. Thermogravimetric Analysis (TGA)

TGA (Thermogravimetric Analysis) analysis in Figure 4 shows the thermostability of microgel particles, in terms of mass percentage retained against temperature under ambient atmosphere. p(NIPAM) was thermally stable up till 250 °C where afterwards it started to decrease in mass. This was because the microgel gets burnt in the presence of oxygen until it reached a plateau at 400 °C and p(NIPAM) was turned to ashes (which is the remaining mass). p(NIPAM)-co-5%AA experienced a similar sigmoid curve as plain p(NIPAM), but showed higher thermal stability as it decreased in mass at 290 °C and reached a plateau at 440 °C.

p(NIPAM)-co-5%AA-co-FA, and p(NIPAM)-co-5%AA-co-FA-co-Dox exhibited similar behavior in thermal stability to one another. The steady decrease in mass over a wide range of temperatures indicates that FA led to an increase in thermal stability and slow decomposition for the p(NIPAM) microgels, this was due to the chemical conjugation of FA to p(NIPAM)-co-5%AA. FA is a thermal stable moiety and degrades slowly at high temperatures, and as such, FA sustained p(NIPAM)-co-5%AA-co-FA and p(NIPAM)-co-5%AA-co-FA-co-Dox microgels up to 40% of their masses at 600 °C [36]. It can then be concluded that FA had been chemically conjugated to p(NIPAM)-co-5%AA microgels due to the high thermal stability.

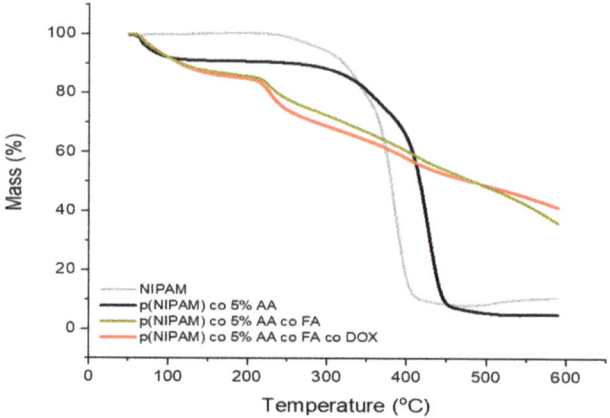

Figure 4. TGA curves of p(NIPAM), p(NIPAM)-co-5%AA, p(NIPAM)-co-5%AA-co-FA, and p(NIPAM)-co-5%AA-co-FA-co-Dox from ambient room temperature to 600 °C.

2.5. Differential Scanning Calorimetry (DSC)

The thermal behavior of p(NIPAM) and p(NIPAM)-co-5%AA undergoes two stages, these are melting of crystallization (micro-melting) and the melting point of the sample. Differential Scanning Calorimetry (DSC) showed that the first stage melting of crystallization occurs at 116 °C for p(NIPAM), while it occurred further in p(NIPAM)-co-5%AA at 153 °C. Furthermore, a series of endothermic peaks at 411 °C for p(NIPAM) and 404 °C for p(NIPAM)-co-5%AA indicating their melting points was registered. Further heating exhibited two-step exothermic peaks for p(NIPAM), but one for p(NIPAM)-co-5%AA.

p(NIPAM)-co-5%AA-co-FA and p(NIPAM)-co-5%AA-co-FA-co-Dox exhibited a lower crystallization melting point at an endothermic peak of 116 °C for p(NIPAM)-co-5%AA-co-FA and 131 °C for p(NIPAM)-co-5%AA-co-FA-co-Dox. Melting points of p(NIPAM)-co-5%AA-co-FA and p(NIPAM)-co-5%AA-co-FA-co-Dox were 154 and 145 °C, respectively and did not exhibit any exothermic peaks like the other two microgels (Figure 5). This indicates that the change in thermal behavior in p(NIPAM)-co-5%AA-co-FA and p(NIPAM)-co-5%AA-co-FA-co-Dox was due to the moieties that were chemically conjugated to p(NIPAM)-co-5%AA. Moreover, the existence of only one melting point in each p(NIPAM)-co-5%AA-co-FA and p(NIPAM)-co-5%AA-co-FA-co-Dox indicated the purity of the sample and that nothing else was co-existing with these microgels.

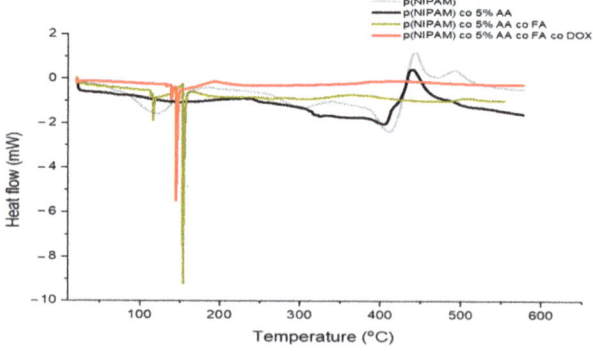

Figure 5. The DSC graph of p(NIPAM), p(NIPAM)-co-5%AA, p(NIPAM)-co-5%AA-co-FA, and p(NIPAM)-co-5%AA-co-FA-co-Dox under nitrogen atmosphere at a temperature range of rtp-600 °C.

2.6. Fourier Transform Infra-Red Spectroscopy (FTIR)

The FTIR (Fourier Transform Infra-Red Spectroscopy) spectra of the three microgels are shown in Figure 6, while the peaks and their assignments are mentioned in Table 1. The FTIR of p(NIPAM)-co-5%AA showed a peak at 3417 cm^{-1} of the hydroxyl group of the carboxylic acid and the C=O in the carboxylic acid group. The sulphate ions were expressed at 1130 cm^{-1}. It was also worth noting that some peaks that were available in p(NIPAM) were shifted in p(NIPAM)-co-5%AA, these include 3283, 2972, 2933, 2876, 1632, 1538, 1457, and 1386 cm^{-1}.

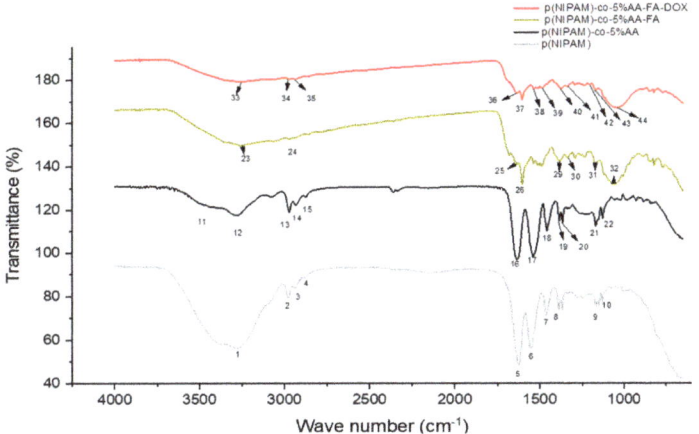

Figure 6. FTIR of p(NIPAM), p(NIPAM)-co-5%AA, p(NIPAM)-co-5%AA-co-FA, and p(NIPAM)-co-5%AA-co-FA-co-Dox showing the peaks that signify the chemical conjugates of each moiety.

p(NIPAM)-co-5%AA-co-FA had additional functional groups due to the presence of FA, such as the aromatic ring in FA, the aryl stretch 1603 cm^{-1} and aryl C=C at 1487 cm^{-1} and a heterocyclic ring containing secondary amine 1339 cm^{-1}.

Finally, p(NIPAM)-co-5%AA-co-FA-co-Dox had few additional functional groups that are expressed exclusively for Dox in its spectrum including the 13C-H and COH stretch of Dox occurring at 1377 and 1209 cm^{-1}, which are very unique to Dox [37].

The FTIR results of p(NIPAM)-co-5%AA-co-FA and p(NIPAM)-co-5%AA-co-FA-co-Dox, and the shift in wavenumbers that were observed in the spectra (Figure 5 and Table 1) were confirmatory results that the moieties were chemically conjugated and that FA and Dox were not ionically interacting with the p(NIPAM)-co-5%AA microgels, as it would have diffused out through the dialysis step.

2.7. Biocompatibility of p(NIPAM)-co-5%AA and p(NIPAM)-co-5%AA-co-FA Microsystems

Viability assay was initially performed on cells treated with microgels without any anticancer drug conjugated, used as a control, in order to verify their biocompatibility. Therefore, CCK-8 (Cell counting kit-8) assay was performed on normal (HB2) and tumor (MDA-MB 231) cells treated for 24h with different concentrations (15, 31, 46, 62, 77, and 93 µg/mL) of p(NIPAM)-co-5%AA or p(NIPAM)-co-5%AA-co-FA microgels. Cells treated with doxorubicin (5, 10. 15, 20, 25, and 30 µM) were used as positive control. As reported in Figure 7, microgel particles alone or conjugated with folic acid do not alter the cell viability of both normal and tumor cells (viability of around 100%), also if used at high concentration (92.88 µg/mL). Furthermore, cell viability was also maintained at a higher concentration (100 µg/mL) of p(NIPA)-co-5%AA until 48 h of treatment and the cell morphology was not altered as suggested by acridine orange assay (Figures S5 and S6).

2.8. Qualitative Uptake of p(NIPAM)-co-5%AA-co-FA-co-Dox

Once established the biocompatibility of microgels, cell internalization uptake was initially investigated by fluorescence microscopy by incubating MDA-MB 231 cells with a fluorescence variant of microgel particles over time, as reported in supporting information (Figures S7 and S8). The green fluorescence relative to microgels appeared localized in specific areas, probably corresponding to the Golgi apparatus or the endoplasmic reticulum after 1 h of incubation (Figures S7c–c" and S8).

Table 1. FTIR peaks of p(NIPAM), p(NIPAM)-co-5%AA, p(NIPAM)-co-5%AA-co-FA, and p(NIPAM)-co-5%AA-co-FA-co-Dox and their assignments with references.

Polymer	Peak No.	Peak (cm^{-1})	Bond Type	Reference
p(NIPAM)	1	3279	secondary amine	[38]
	2	2978	CH_3 asymmetric stretch	[38]
	3	2938	CH_2 asymmetric stretch	[38]
	4	2880	C-H stretch	[38]
	5	1625	amide I secondary	[39]
	6	1551	amide II	[39]
	7	1462	CH_2 bend	[38,39]
	8	1389	CH_3 bend	[38,39]
	9	1171	C-N stretch secondary amine	[38]
	10	1131	sulfate ion	[38]
p(NIPAM)-co-5%AA	11	3417	O-H group	[38,39]
	12	3283	secondary NH	[38,39]
	13	2972	CH_3 asymmetric stretch	[38]
	14	2933	CH_2 assymetric stretch	[38]
	15	2876	C-H stretch	[38]
	16	1632	amide I secondary	[39]
	17	1538	amide II	[39]
	18	1457	C-H2 bend	[38,39]
	19	1386	C-H3 bend	[38,39]
	20	1367	Carboxylate	[38]
	21	1171	C-N stretch secondary amine	[38]
	22	1130	sulfate ion	[38]
p(NIPAM)-co-5%AA-co-FA	23	3250	secondary amine	[38]
	24	2972	CH_3 asymmetric stretch	[38]
	25	1635	amide I secondary	[39]
	26	1603	aromatic ring stretch	[38]
	27	1534	amide II	[39]
	28	1487	aryl C=C	[39]
	29	1387	Carboxylate group	[38]
	30	1339	aromatic secondary amine C-N	[38]
	31	1173	C-N secondary amine	[38]
	32	1058	sulfate ion	[38]
p(NIPAM)-co-5%AA-co-FA-co-Dox	33	3251	secondary amine	[38]
	34	2969	CH_3 asymmetric stretch	[38]
	35	2938	CH_2 asymmetric stretch	[38]
	36	1635	amide I secondary	[39]
	37	1604	aromatic ring stretch	[38]
	38	1531	amide II	[39]
	39	1487	aryl C=C	[39]
	40	1377	C13-H	[37]
	41	1342	aromatic secondary amine C-N	[38]
	42	1209	COH of Dox	[37]
	43	1171	C-N secondary amine	[38]
	44	1047	sulphate ion	[38]

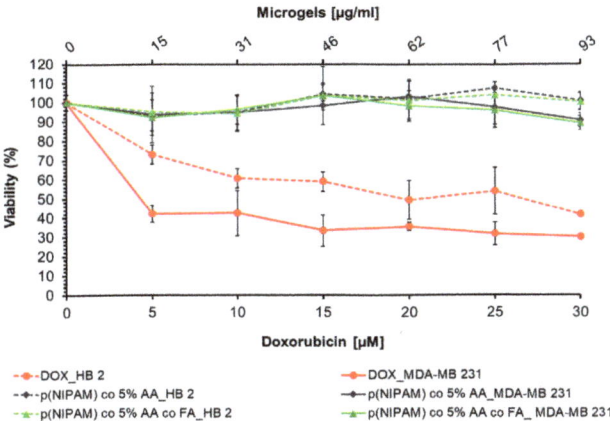

Figure 7. Viability assay on HB2 and MDA-MB 231 cells incubated for 24 h with p(NIPAM)-co-5%AA, p(NIPAM)-co-5%AA-co-FA (5; 10; 15; 20; 25; 30 µg/mL). Cells treated with doxorubicin were used as positive control, while untreated cells were used as negative control.

Fluorescence microscopy was also adopted to investigate the specific tumor targeting of microgels functionalized with folic acid. The folate receptor (FR) is overexpressed in the majority of human tumors, like breast, and, in particular, MDA-MB231 cells produce high FR concentration [40]. Therefore, a co-culture experiment was performed by seeding HB2 and green-labelled MDA-MB 231 cells together and incubating them with p(NIPAM)-co-5%AA-co-FA-co-Dox microgels or doxorubicin alone as control (identified by the doxorubicin red auto-fluorescence, Figures 8c–c''' and S9c–c'''). Nuclei of both cells were stained with DAPI (blue fluorescence, Figures 8a–a''' and S9a–a''') so that HB2 healthy cells were identified by blue fluorescence alone, while MDA-MB 231 tumor cells were individuated by both blue and green fluorescence. Following the microgel particles cellular uptake over time, it was evident the presence of the red fluorescence (corresponding to doxorubicin conjugated to the particles) exclusively in tumor cells already at the shortest incubation time (30 min, Figure 8a–d) and more and more at the following incubation times (1, 2, and 4 h, Figure 8a'–d',a''–d'',a'''–d'''). On the other hand, red fluorescence was totally absent in correspondence of HB2 cells (white arrows in Figure 8d–d''), suggesting a specific tumor targeting of p(NIPAM)-co-5%AA-co-FA-co-Dox microparticles. The red fluorescence, relative to doxorubicin, began to appear in HB-2 cytoplasm in 4 h, as expected by static in vitro system. On the contrary, the soluble form of the doxorubicin was inside both normal and tumor cells already after 30 min of treatment, suggesting that microparticles, conjugated with folic acid, were responsible for the selectively for cancer cells (Figure S9). The co-localization of the blue (nuclei) and the red (doxorubicin) fluorescence in tumor cells (Figure 8) suggested that the drug was released from the microgels and entered into the nuclei, which can intercalate into the DNA causing cell death. On the other hand, microgels fluorescence signal was always localized in the cytoplasm (Figures S7 and S8).

2.9. Quantitative Uptake Study

Differential microgel particles cellular uptake between normal and tumor cells was furthermore investigated by the quantitative flow cytometric analysis, following the red autofluorescence of conjugated doxorubicin (Figures 9 and S10). Initially (30 min), there were no significant differences in p(NIPAM)-co-5%AA-co-FA-co-Dox internalization between HB2 (breast healthy cells) and MDA-MB-231 (breast cancer cells). After 1 h of incubation, the uptake gap started to increase, suggesting a specific tumor targeting due to the conjugated folic acid, reaching the maximum value after 4 h of treatment: the microgels internalization in tumor cells was 60% against the 14% of internalization into normal cells.

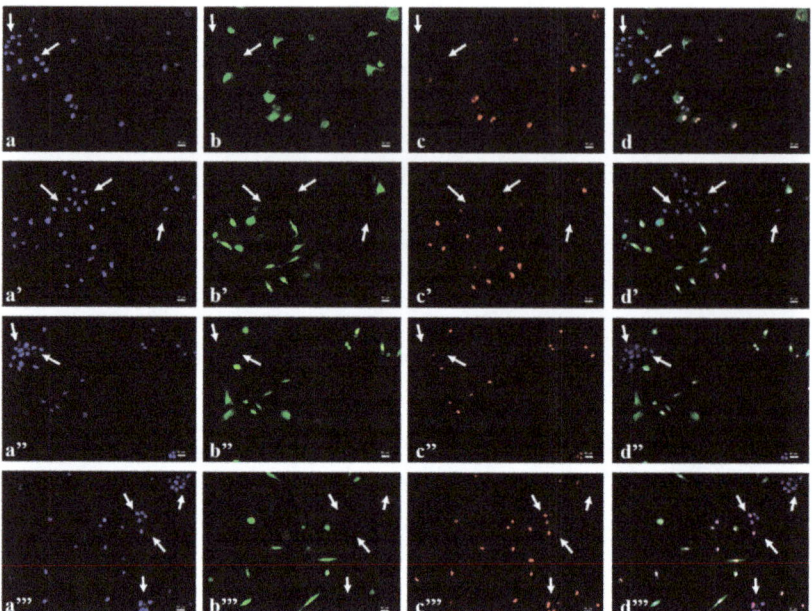

Figure 8. Fluorescence images of co-culture of HB2 (blue) and MDA-MB231 (blue and green) cells incubated with p(NIPAM)-co-5%AA-co-FA-co-Dox (10 µM) (red) for 30 min (**a–d**); 1 h (**a′–d′**); 2 h (**a″–d″**), and 4 h (**a‴–d‴**). Blue: nuclei (DAPI); Green: MDA-MB 231 cells (CellTrace CFSE); Red: doxorubicin of p(NIPAM)-co-5%AA-co-FA-co-Dox microgels. Magnification 20×. Scale bar: 50 µm.

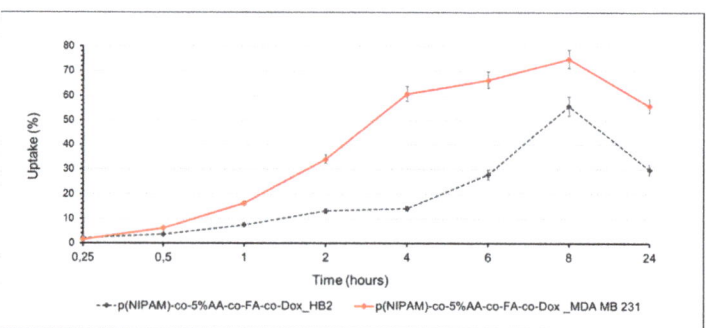

Figure 9. Uptake percentage of p(NIPAM)-co-5%AA-co-FA-co-Dox (doxorubicin conjugated concentration of 20 µM) by HB2 and MDA-MB 321 cells during different incubation times.

After 6 and 8 h, the amount of p(NIPAM)-co-5%AA-co-FA-co-Dox inside MDA-MB 231 cells increased slowly (66 and 75%, respectively), suggesting the reaching of the maximum cell internalization. By contrast, it increased inside normal cells, as expected for longer incubation time in a static in vitro system. In summary, the particle uptake ratio at 0.5, 1, 2, 4, 6, 8, and 24 h was 1.7, 2.2, 2.6, 4.3, 2.3, 1.3, and 1.8, respectively. This showed that the maximum difference in particle uptake was a ratio of 4.3 after 4 h of incubation, suggesting that p(NIPAM)-co-5%AA-co-FA-co-Dox targeted MDA-MB 321 cancer cells due to the recognition between folate and its receptor. On the contrary, in HB2 healthy cells, which present lower FR expression, the microparticles uptake was time-delayed, suggesting again a specific particles tumor targeting. The decrease registered at 24 h of incubation for both normal and tumor cells (30% and 56%, respectively) was correlated to the death of cells that initially had internalized particles.

2.10. Cytotoxicity Assay

The cytotoxic effect of doxorubicin conjugated to microparticles was evaluated on normal HB 2 and MDA-MB 231 tumor cells by a viability assay.

The selected doxorubicin concentrations corresponded exactly to the amount of drug conjugated to microgels analyzed in biocompatibility assay (Figure 7): 5, 10, 15, 20, 25, and 30 µM of the drug to 15, 31, 46, 62, 77, and 93 µg/mL of microgels, respectively.

As shown in Figure 10, 5 µM of the drug conjugated to p(NIPAM)-co-5%AA-co-FA-co-Dox induces cell mortality on MDA-MB 231 cells (48% of mortality) and the viability decreases in a concentration-dependent way, reaching the maximum efficiency at 15 µM, so that, at higher concentration, the plateau state was registered (around 37% at 20, 25, and 30 µM of Dox). These data suggest that conjugation protocol does not alter the structure and functionality of conjugated drug and, furthermore, that microsystems, can release the drug inside cells. On the contrary, the viability of healthy cells after incubation was around 66% for all the drug concentrations used, confirming again the specific targeting of p(NIPAM)-co-5%AA-co-FA-co-Dox to tumor cells. The small mortality of 33% registered in this case was due to the long treatment time in a static system (24 h). Doxorubicin alone was used as a positive control.

The differences in toxicity among different cell lines and microgels was probably due to the specific targeting of microgels to tumor cells, recognizing the folate receptor overexpressed by MDA-MB 231 cells. This brought a diverse cell internalization between tumor and normal cells as suggested by flow cytometry analysis, and therefore, to a distinct cytotoxic effect. It is worth mentioning that the biocompatibility of p(NIPAM) was previously tested by Mohsen et al. [41] when it showed cell viability over 90% at concentrations up to 3 mg/mL.

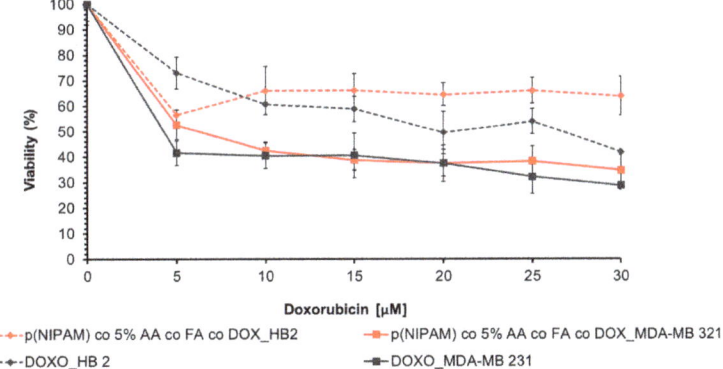

Figure 10. Cell viability assay on HB2 (normal) and MDA-MB 231 (cancer) cells incubated for 24 h with different concentrations of doxorubicin conjugated with p(NIPAM)-co-5%AA-co-FA-co-Dox. Cells incubated with the equivalent concentrations of doxorubicin were used as positive control.

3. Conclusions

Although in the last years, cancer research has seen significant progress in the understanding, diagnosis, treatment, and prevention, low selectivity of the chemotherapeutic agents and consequently high side effects often occur. In this context, a novel drug delivery system that aims to specifically target cancer cells was designed and synthesized. Based on the tumor characteristic, p(NIPAM)-co-5%AA microgel particles were covalently conjugated to folic acid that is overexpressed in the majority of tumor cells (targeting agent) and to the anti-cancer drug Doxorubicin through EDC/NHS coupling reaction. The advantage of covalently tethering DOX, rather than loading it by self-assembly, is that the amount of DOX conjugated to the microgel is taken up almost completely. While the other self-assembly systems have either low entrapment efficiency (in case of synthetic polymers), or are not feasible to scale up (such as micelles) [42,43]. Moreover, tethering the DOX and conjugating it with a targeting moiety, ensures that DOX targets only cancer cells and shall be intracellularly released upon degradation of the microgel particles by relevant enzymes. Unlike other self-assembly systems, the DOX can be released in the bloodstream. Accordingly, it is suggested that calculating the needed doses of covalently tethered Dox can be easier and more accurate than a physically entrapped one.

The new delivery system was then characterized and tested for targeting ability and capability to release the conjugated drug inside cells.

Several characterization studies were carried out, including UV-Vis analysis, DLS, TGA, DSC, and FTIR to demonstrate the successful conjugation of FA and Dox to p(NIPAM)-co-5%AA microgel and that the new microgels retain microgel behavior [44].

The appearance of the typical FA and Doxo peaks in UV-VIS analysis (Figure 1) and the variation in size (DLS analysis, Figure 2) demonstrated a variety of microgel composition due to FA and Doxo conjugation. These data were confirmed by not only the variation of TGA curves (Figure 4), but also by the alteration of the DSC profiles of the microgels (Figure 5), shifting both the melting point and the crystallization melting point; furthermore, any exothermic peaks (that are present in p(NIPA) and p(NIPAM)-co-5%AA) were not registered. At the same time, also the FTIR profiles changed probably due to the different functional groups of the folic acid and doxorubicin. Taken together, these data confirmed the success of the conjugation, as demonstrated also by cytotoxic assay performed on normal and tumor cells (Figure 10) and the targeting studies (Figures 8 and 9).

The uptake and localization studies of p(NIPAM)-co-5%AA-co-FA-co-Dox were performed using flow cytometry and fluorescence microscopy, while viability assay was carried out to investigate the cytotoxicity of the drug conjugated to developed microgels. Co-culture experiment demonstrated the drug release and the specific targeting of the microcomplex exclusively to the tumor cells by an active targeting that probably could be increased in vivo by a passive targeting based on the enhanced permeability and retention effect (EPR effect). Besides, viability assay results show higher cell viability for healthy cells incubated with p(NIPAM)-co-5%AA-co-FA-co-Dox than the cancer ones. Also, it is shown that at higher concentrations (25 µm and above), healthy cells were more viable when incubated with p(NIPAM)-co-5%AA-co-FA-co-Dox than when incubated with soluble form Dox. Therefore, these data suggest that p(NIPAM)-co-5%AA-co-FA-co-Dox are good candidates as delivery systems to increase the specific tumor targeting probably reducing general side effects, even if more in vivo studies need to be carried out.

4. Materials and Methods

4.1. Synthesis of p(NIPAM)-co-5%AA

A Surfactant Free Emulsion Polymerisation (SFEP) technique was used for the preparation of p(NIPAM)-co-5%AA as described previously and in accordance with literature [27–29,41]. Briefly, a three-neck lid was then fitted to the reaction vessel, which was placed onto a hot plate stirrer and heated to 70 °C with continuous stirring under N2 atmosphere. Potassium persulphate initiator (0.5 g) was dissolved in 800 mL of distilled water. The crosslinker N,N'-methylenebisacrylamide 99% (0.5 g) (BS, Sigma Aldrich, Gillingham, UK), N-isopropylacrylamide (NIPAM, Sigma Aldrich, Gillingham, UK) 97% monomer (4.75 g) and acrylic acid (AA, Sigma Aldrich, Gillingham, UK) co-monomer (0.25 g) were dissolved in 200 mL of distilled water while stirring gently with a magnetic stirrer. After all the reagents were dissolved, they were transferred into the reaction vessel containing the initiator. The reaction was run for 6 h with constant stirring and under nitrogen. After 6 h, the microgel dispersion was allowed to cool down to room temperature, then dialyzed (MW cut-off was 12–14,000 kDa) in fresh distilled water for 7 days.

4.2. Conjugation of p(NIPAM)-co-5%AA with Folic Acid

Folic acid (FA, Sigma Aldrich, Milano, Italy) was conjugated with p(NIPAM)-co-5%AA microgel particles by EDC/NHS protocol [45]. Briefly, p(NIPAM)-co-5%AA microgels were suspended in 2-(N-morpholino) ethanesulfonic acid (MES, Sigma Aldrich, Milano, Milano, Italy) buffer solution (0.1 M, pH 5 with NaOH) at the final concentration of 5 mg/mL and sonicated for 20 min on ice bath in order to homogenize the solution. 1-ethyl-3-(3-dimethylaminopropyl) carbodiimide hydrochloride (EDC, Sigma Aldrich, Milano, Italy) was added 10 times more than NPs (w/w), mixed by vortex, and then N-hydroxysulfosuccinimide (Sulfo-NHS, Sigma Aldrich, Milano, Italy) powder was put (NPs/SulfoNHS = 4.5 w/w) [46,47]. The solution was then left for 30 min in agitation at room temperature and FA was added 10 times more than NPs (w/w) and mixed by a vortex. The solution of p(NIPAM)-co-5%AA and FA was then diluted with complete Phosphate Buffered Saline (PBS, Sigma-Aldrich, Milano, Italy) to reach a final NPs concentration of 1 mg/mL, the pH was adjusted to 7 using sodium bicarbonate and the solution was left for 2 h in agitation at room temperature.

The microparticles suspension was sonicated for 20 min at 37 °C and dialyzed to get rid of the unconjugated folic acid using a nitrocellulose tube (100 kDa cut-off). The dialysis buffer (distilled H_2O) was changed twice a day for one week. Samples were sterilized by filtering with 0.22 µm filter and analyzed by spectrophotometric analysis [microplate reader DU-730 Life Science spectrophotometer (Beckman Coulter, Milano, Italy)] at 340 nm in order to determine the amount of folic acid conjugated to the microgel particles using a calibration curve (0.05; 0.10; 0.15; 0.20; 0.25; 0.30; 0.35; 0.40; 0.45; 0.50 µg/mL).

4.3. Conjugation of p(NIPAM)-co-5%AA-co-FA with Doxorubicin

After the freeze-drying process, p(NIPAM)-co-5%AA-co-FA were solubilized (1 mg/mL) on MES Buffer (0.1 M, pH 5 with NaOH) and sonicated on an ice bath for 20 min. EDC (10 times more than NPs w/w) and Sulfo-NHS (NPs/SulfoNHS = 4.5 w/w) were then added to the microparticles solution and mixed well by vortex and left in agitation at room temperature for 30 min. Doxorubicin (Dox, Sigma-Aldrich, Milano, Italy) powder was added to the solution (NPs/Dox = 1.2 w/w) and the final pH was adjusted to 7 using sodium bicarbonate. After 2.5 h of agitation at room temperature, the solution was sonicated for 20 min at 37 °C and put in a nylon membrane dialysis tube (14 KDa cut-off) in order to get rid of the unconjugated Dox. The dialysis buffer (distilled H_2O) was changed twice a day for one week. Spectrophotometric analysis [microplate reader DU-730 Life Science spectrophotometer (Beckman Coulter, Milano, Italy)] was then performed for the p(NIPAM)-co-5%AA-co-FA-co-Dox solution at 485 nm to determine the amount of Dox conjugated to the microgel particles using a standard curve (5; 10; 20; 40; 60; 80; 100 µM).

4.4. Dynamic light Scattering (DLS) and Electrophoretic Mobility

p(NIPAM)-co-5%AA, p(NIPAM)-co-5%AA-co-FA, and p(NIPAM)-co-5%AA-co-FA-co-Dox were suspended in distilled water by 0.5% (w/v) using distilled water in a ratio of 1:2. The DLS software was programmed to measure the size [Zetasizer NS series (Malvern, Gillingham, UK)] and electrophoretic mobility in triplicates from 15 to 60 °C with a heating and cooling cycle.

4.5. Thermogravimetric Analysis (TGA)

Freeze-dried p(NIPAM), p(NIPAM)-co-5%AA, p(NIPAM)-co-5%AA-co-FA, and p(NIPAM)-co-5%AA-co-FA-co-Dox were weighed on platinum pans by the instrument [TGA Q50 (TA instruments, New Castle, DE, USA]. The system was heated under ambient air from room temperature to 600 °C at 10 °C/min.

4.6. Differential Scanning Calorimetry (DSC)

Known masses of freeze-dried p(NIPAM), p(NIPAM)-co-5%AA, p(NIPAM)-co-5%AA-co-FA, and p(NIPAM)-co-5%AA-co-FA-co-Dox were placed in Tzero aluminum pans and placed on the heater unit. The empty pan is placed in the reference heating unit and the system is heated from room temperature to 600 °C at 10 °C/min under nitrogen purge of 50 mL/min. [DSC Q20 (TA instruments, USA)].

4.7. Fourier-Transform Infrared Spectroscopy (FTIR)

The suspensions of p(NIPAM), p(NIPAM)-co-5%AA, p(NIPAM)-co-5%AA-co-FA and p(NIPAM)-co-5%AA-co-FA-co-Dox were freeze dried. The powders obtained were placed directly on diamond iTR of FTIR spectroscopy from 600 to 4000 cm^{-1} [FTIR Nicolet iS20 (thermoscientfic, Tewksbury, MA, USA)].

4.8. UV–Visible Spectra

UV–Visible spectra of p(NIPAM)-co-5%AA, p(NIPAM)-co-5%AA-co-FA, and p(NIPAM)-co-5%AA-co-FA-co-Dox were obtained using the range 270–600 nm at 5 nm increments, using 200 µL of each sample solution in 96 well plate (Synergy™ HT Multidetection microplate reader spectrophotometer (BioTek, Milano, Italy).

4.9. Cell Culture of HB2 and MDA-MB 231

MDA-MB 231 human breast cancer cells were grown in Dulbecco's Modified Eagle's Medium (DMEM, Sigma-Aldrich, Milano, Italy) high glucose (HG-DMEM) with 10% (*v/v*) Fetal bovine serum (FBS, Euroclone, Celbar, Pero (MI) Italy), 2 mM L-Glutamine (Euroclone, Celbar, Pero (MI) Italy), 100 units per mL penicillin G (Euroclone, Celbar, Pero (MI) Italy), 100 mg mL^{-1} streptomycin, while HB2 human mammary epithelial cells were grown in DMEM low glucose (LG-DMEM) with 10% (*v/v*) FBS, 4 mM L-Glutamine, 100 units per mL penicillin G, 100 mg mL^{-1} streptomycin, 5 mg mL^{-1} hydrocortisone (Sigma-Aldrich, Milano, Italy), and 10 µg mL^{-1} bovine insulin (Sigma-Aldrich, Milano, Italy). All cells were cultivated at 37 °C, in a humidified atmosphere of 5% CO_2 and maintained in sterile conditions.

4.10. Viability of Cells Treated with Microgels

Viability assay was performed on MDA-MB 231 or HB2 cells incubated with p(NIPAM)-co-5%AA or p(NIPAM)-co-5%AA-co-FA microgel particles (Biocompatible assay) or with p(NIPAM)-co-5%AA-co-FA-co-Dox (Cytotoxic assay). Cells were seeded on 96-well plates at the density of 1×10^4 cells/well and grown in the opportune medium at 37 °C for 24 h. Therefore, cells were treated with p(NIPAM)-co-5%AA or p(NIPAM)-co-5%AA-co-FA (15, 31, 46, 62, 77 and 93 µg/mL) or p(NIPAM)-co-5%AA-co-FA-co-Dox (5; 10; 15; 20; 25; 30 µM of conjugated drug) for 24 h and cell viability was detected by using Cell Counting Kit-8 (CCK-8, Sigma-Aldrich). In particular, water-soluble tetrazolium salt (WST-8) was added to each sample (1:10 dilution in complete medium) and incubated at 37 °C for 2 h to allow for its reduction by mitochondrial dehydrogenases of the living cells into soluble formazan dye that is directly proportional to the number of living cells. Spectrophotometric analysis [microplate reader DU-730 Life Science spectrophotometer (Beckman Coulter, Milano, Italy)] at 450 nm was then performed to determine the percentage of viable cells relative to the negative control (untreated cells). Cells treated with Doxorubicin were considered as a positive control.

4.11. Specific Targeting Cell Uptake

MDA-MB 231 cells (10^5 cells per mL) were harvested by centrifugation and the cell pellet was incubated with 25 mM Molecular Probe CellTrace CFSE fluorescent stain (CellTrace CFSE Cell Proliferation Kit, Life Technologies, Italy) for 30 min at 37 °C.

For co-culture preparation, pre-labelled MDA-MB 231 and unlabeled HB2 cells were mixed (ratio 1:1) and seeded with a density of 8×10^4 cells per well into 12-well plates containing sterile coverslips in complete LG-DMEM for grown 24 h at 37 °C.

In sterile conditions, cells were incubated with 10 µM of p(NIPAM)-co-5%AA-co-FA-co-Dox microgels for 15 min, 30 min, 1 h, 2 h, and 4 h. At the end of each incubation time, the cells were washed with PBS and then fixed with 3.7% formaldehyde (in PBS) for 5 min at room temperature, followed by three washes with PBS. Nuclei were stained in the dark with DAPI solution (dilution of 1:10,000 in water) for 15 min at room temperature. Samples were analyzed by fluorescence microscopy (Leica, Buccinasco (MI), Italy) and confocal microscope (FLUOVIEW FV10i-LIV, Olympus, Italy).

4.12. Quantitative Uptake by Flow Cytometry

MDA-MB 321 and HB2 cells were grown in 6 well plates until confluent state at 37 °C in a humidified atmosphere of 5% and then incubated with p(NIPAM)-co-5%AA-co-FA-co-Dox (final Doxorubicin concentration of 20 µM) for 15 min, 30 min, 1 h, 2 h, 4 h, 6 h, 8 h, and 24 h. Untreated cells were used as the negative control for background fluorescence. Subsequently, the samples were washed with PBS without Ca^{2+} and Mg^{2+}, detached by Trypsin-EDTA 1× (Sigma-Aldrich) treatment and collected by centrifugation at 1000 rpm for 5'. The pellets were re-suspended in 500µL of PBS and analyzed by FACS-Canto cytometer (Germany) detecting the red (Dox) fluorescence emission (585 nm). For each sample were collected 1×10^5 events investigated by BD FACS Diva software.

Supplementary Materials: The following are available online at https://www.mdpi.com/article/10.3390/gels7040203/s1, Figure S1: Folic acid calibration standard curve and calculation of folate conjugated to p(NIPAM)-co-5%AA [p(NIPAM)-co-5%AA-co-FA]; Figure S2: Doxorubicin calibration standard curve and calculation of drug conjugated to p(NIPAM)-co-5%AA-co-FA [p(NIPAM)-co-5%AA-co-FA-co-Dox]; Figure S3: Cooling cycles of p(NIPAM)-co-5%AA, p(NIPAM)-co-5%AA-co-FA and p(NIPAM)-co-5%AA-co-FA-co-Dox in contrast with cooling cycles shown in Figure 2; Figure S4: Cooling cycles of p(NIPAM)-co-5%AA, p(NIPAM)-co-5%AA-co-FA and p(NIPAM)-co-5%AA-co-FA-co-Dox in respect to their electrophoretic mobility in contrast with their heating cycles shown in Figure 3; Figure S5: Cell viability of HB 2 and MDA-MB 231 cells treated with 0; 12.5; 25; 50 and 100 µg/mL of p(NIPAM)-co-5%AA for 24 h (a) and 48 h (b). Cells incubated with doxorubicin were used as positive control; Figure S6: Acridine orange assay on MDA-MB 231 cells treated with 12.5 (c) or 100 µg/mL (d) of p(NIPAM)-co-5%AA for 24 h. Untreated cells and cells incubated with doxorubicin were used as negative (a) and positive (b) control, respectively; Figure S7: Confocal microscopy of MDA-MB 231 cells treated with 100 µg/mL of p(NIPAM)-co-5%AA-co-LY for 15' (a–a''); 30' (b–b''); 1 h (c–c''); 2 h (d–d''); 4 h (e–e''); 6 h (f–f''); 8 h (g–g'') and 24 h (h–h''). Red: bromide ethidium (DNA and RNA). Green: p(NIPAM)-co-5%AA-co-LY. Magnification 60×; Figure S8: Confocal microscopy of MDA-MB 231 cells treated for 1 h with 100 µg/mL of p(NIPAM)-co-5%AA-co-LY. Red: bromide ethidium (DNA and RNA). Green: p(NIPAM)-co-5%AA-co-LY. Magnification 160×; Figure S9: Fluorescence images of co-culture of HB2 (blue) and MDA-MB231 (blue and green) cells incubated with Doxorubicin (10 µM) (red) for 30 min (a–d); 1 h (a'–d'); 2 h (a''–d'') and 4 h (a'''–d'''). Blue: nuclei (DAPI); Green: MDA-MB 231 cells (CellTrace CFSE); Red: doxorubicin. Magnification 40×; Figure S10: Cytograms of flow cytometric analysis of HB2 and MDA-MB 231 cells incubated with 25 µM of doxorubicin conjugated to microgel (p(NIPAM)-co-5%AA-co-FA-co-Dox).

Author Contributions: Conceptualization, S.C., R.M. and G.G.; Methodology, S.C. and R.M.; Validation, D.P., H.S. and H.A.; Formal analysis, S.C. and R.M.; Investigation, S.C., R.M., D.P., H.S., H.A., S.E.-D.A.-M., A.A.D., I.M.E.-S. and M.J.S.; Resources, G.G. and R.M.; Data curation, S.C., R.M. and G.G.; Writing—original draft preparation, S.C. and R.M.; Writing—review and editing, G.G.; Visualization, S.C., R.M. and G.G.; Supervision, G.G.; Project administration, S.C. and G.G.; Funding acquisition, G.G. All authors have read and agreed to the published version of the manuscript.

Funding: This research was funded by Horizon 2020 project "Future Formulations: Developing Future Pharmaceuticals Through Advanced Analysis and Intersectoral Exchange" project n 691128 and the APC was funded by University of Palermo on GG research funding (R4D15-P8MSRI07_MARGINE).

Conflicts of Interest: The authors declare no conflict of interest.

References

1. Sung, H.; Ferlay, J.; Siegel, R.L.; Laversanne, M.; Soerjomataram, I.; Jemal, A.; Bray, F. Global Cancer Statistics 2020: GLOBOCAN Estimates of Incidence and Mortality Worldwide for 36 Cancers in 185 Countries. CA. *Cancer J. Clin.* **2021**, *71*, 209–249. [CrossRef]

2. Wei, G.; Wang, Y.; Yang, G.; Wang, Y.; Ju, R. Recent progress in nanomedicine for enhanced cancer chemotherapy. *Theranostics* **2021**, *11*, 6370–6392. [CrossRef] [PubMed]
3. Lengyel, M.; Kállai-Szabó, N.; Antal, V.; Laki, A.J.; Antal, I. Microparticles, Microspheres, and Microcapsules for Advanced Drug Delivery. *Sci. Pharm.* **2019**, *87*, 20. [CrossRef]
4. Suhail, M.; Fang, C.-W.; Khan, A.; Minhas, M.U.; Wu, P.-C. Fabrication and In Vitro Evaluation of pH-Sensitive Polymeric Hydrogels as Controlled Release Carriers. *Gels* **2021**, *7*, 110. [CrossRef] [PubMed]
5. Tacar, O.; Sriamornsak, P.; Dass, C.R. Doxorubicin: An update on anticancer molecular action, toxicity and novel drug delivery systems. *J. Pharm. Pharmacol.* **2013**, *65*, 157–170. [CrossRef] [PubMed]
6. Shrestha, B.; Wang, L.; Brey, E.M.; Uribe, G.R.; Tang, L. Smart Nanoparticles for Chemo-Based Combinational Therapy. *Pharmaceutics* **2021**, *13*, 853. [CrossRef] [PubMed]
7. Erel-Akbaba, G.; Carvalho, L.A.; Tian, T.; Zinter, M.; Akbaba, H.; Obeid, P.J.; Chiocca, E.A.; Weissleder, R.; Kantarci, A.G.; Tannous, B.A. Radiation-induced targeted nanoparticle-based gene delivery for brain tumor therapy. *ACS Nano* **2019**, *13*, 4028–4040. [CrossRef]
8. Adamo, G.; Campora, S.; Ghersi, G. Functionalization of Nanoparticles in Specific Targeting and Mechanism Release. In *Nanostructures for Novel Therapy*; Synthesis, Characterization and Applications Micro and Nano Technologies; Elsevier: Amsterdam, The Netherlands, 2017; pp. 57–80. [CrossRef]
9. Fernández, M.; Javaid, F.; Chudasama, V. Advances in targeting the folate receptor in the treatment/imaging of cancers. *Chem. Sci.* **2018**, *9*, 790. [CrossRef]
10. Vinothini, K.; Rajendran, N.K.; Ramu, A.; Elumalai, N.; Rajan, M. Folate receptor targeted delivery of paclitaxel to breast cancer cells via folic acid conjugated graphene oxide grafted methyl acrylate nanocarrier. *Biomed. Pharmacother.* **2019**, *110*, 906–917. [CrossRef]
11. Mandracchia, D.; Tripodo, G. CHAPTER 1 Micro and Nano-drug Delivery Systems. In *Silk-Based Drug Delivery Systems*; Royal Society of Chemistry: Cambridge, UK, 2020; pp. 1–24. [CrossRef]
12. Sun, T.; Zhang, Y.S.; Pang, B.; Hyun, D.C.; Yang, M.; Xia, Y. Engineered nanoparticles for drug delivery in cancer therapy. *Angew. Chem. Int. Ed.* **2014**, *53*, 12320–12364. [CrossRef]
13. Hale, S.J.M.; Perrins, R.D.; García, C.E.; Pace, A.; Peral, U.; Patel, K.R.; Robinson, A.; Williams, P.; Ding, Y.; Saito, G.; et al. DM1 Loaded Ultrasmall Gold Nanoparticles Display Significant Efficacy and Improved Tolerability in Murine Models of Hepatocellular Carcinoma. *Bioconjug. Chem.* **2019**, *30*, 703–713. [CrossRef]
14. Chen, M.; Quan, G.; Sun, Y.; Yang, D.; Pan, X.; Wu, C. Nanoparticles-encapsulated polymeric microneedles for transdermal drug delivery. *J. Control. Release* **2020**, *325*, 163–175. [CrossRef] [PubMed]
15. Campora, S.; Mauro, N.; Griffiths, P.; Giammona, G.; Ghersi, G. Graphene nanosystems as supports in siRNA Delivery. *Chem. Eng. Trans.* **2018**, *64*, 415–420. [CrossRef]
16. Li, Y.; Geng, J.; Titmarsh, H.; Megia-Fernandez, A.; Dhaliwal, K.; Frame, M.; Bradley, M. Rapid Polymer Conjugation Strategies for the Generation of pH-Responsive, Cancer Targeting, Polymeric Nanoparticles. *Biomacromolecules* **2018**, *19*, 2721–2730. [CrossRef] [PubMed]
17. Mauro, N.; Campora, S.; Scialabba, C.; Adamo, G.; Licciardi, M.; Ghersi, G.; Giammona, G. Self-organized environment-sensitive inulin-doxorubicin conjugate with a selective cytotoxic effect towards cancer cells. *RSC Adv.* **2015**, *5*, 32421–32430. [CrossRef]
18. Adamo, G.; Grimaldi, N.; Campora, S.; Sabatino, M.A.; Dispenza, C.; Ghersi, G. Glutathione-sensitive nanogels for drug release. *Chem. Eng. Trans.* **2014**, *38*, 457–462. [CrossRef]
19. Mauro, N.; Campora, S.; Ada Mo, G.; Scialabba, C.; Ghersi, G.; Giammona, G. Polyaminoacid-doxorubicin prodrug micelles as highly selective therapeutics for targeted cancer therapy. *RSC Adv.* **2016**, *6*, 77256–77266. [CrossRef]
20. De Lima, C.S.A.; Balogh, T.S.; Varca, J.P.R.O.; Varca, G.H.C.; Lugão, A.B.; Camacho-Cruz, L.A.; Bucio, E.; Kadlubowski, S.S. An Updated Review of Macro, Micro, and Nanostructured Hydrogels for Biomedical and Pharmaceutical Applications. *Pharmaceutics* **2020**, *12*, 970. [CrossRef] [PubMed]
21. Almeida, H.; Amaral, M.H.; Lobão, P. Temperature and pH stimuli-responsive polymers and their applications in controlled and selfregulated drug delivery. *J. Appl. Pharm. Sci.* **2012**, *2*, 1–10. [CrossRef]
22. Kocak, G.; Tuncer, C.; Bütün, V. PH-Responsive polymers. *Polym. Chem.* **2017**, *8*, 144–176. [CrossRef]
23. Lanzalaco, S.; Armelin, E. Poly(N-isopropylacrylamide) and Copolymers: A Review on Recent Progresses in Biomedical Applications. *Gels* **2017**, *3*, 36. [CrossRef]
24. Xu, X.; Liu, Y.; Fu, W.; Yao, M.; Ding, Z.; Xuan, J.; Li, D.; Wang, S.; Xia, Y.; Cao, M. Poly(N-isopropylacrylamide)-based thermoresponsive composite hydrogels for biomedical applications. *Polymers* **2020**, *12*, 580. [CrossRef]
25. Zahin, N.; Anwar, R.; Tewari, D.; Kabir, M.T.; Sajid, A.; Mathew, B.; Uddin, M.S.; Aleya, L.; Abdel-Daim, M.M. Nanoparticles and its biomedical applications in health and diseases: Special focus on drug delivery. *Environ. Sci. Pollut. Res.* **2020**, *27*, 19151–19168. [CrossRef]
26. Guo, M.; Yan, Y.; Liu, X.; Yan, H.; Liu, K.; Zhang, H.; Cao, Y. Multilayer nanoparticles with a magnetite core and a polycation inner shell as pH-responsive carriers for drug delivery. *Nanoscale* **2010**, *2*, 434–441. [CrossRef] [PubMed]
27. Kwok, M.; Li, Z.; Ngai, T. Controlling the Synthesis and Characterization of Micrometer-Sized PNIPAM Microgels with Tailored Morphologies. *Langmuir* **2013**, *29*, 9581–9591. [CrossRef]

28. Zhou, S.; Chu, B. Synthesis and Volume Phase Transition of Poly(methacrylic acid-co-N-isopropylacrylamide) Microgel Particles in Water. *J. Phys. Chem. B* **1998**, *102*, 1364–1371. [CrossRef]
29. Ruscito, A.; Chiessi, E.; Toumia, Y.; Oddo, L.; Domenici, F.; Paradossi, G. Microgel Particles with Distinct Morphologies and Common Chemical Compositions: A Unified Description of the Responsivity to Temperature and Osmotic Stress. *Gels* **2020**, *6*, 34. [CrossRef]
30. Dowding, P.J.; Vincent, B.; Williams, E. Preparation and swelling properties of poly(NIPAM) "minigel" particles prepared by inverse suspension polymerization. *J. Colloid Interface Sci.* **2000**, *221*, 268–272. [CrossRef]
31. Maeda, Y.; Higuchi, T.; Ikeda, I. Change in hydration state during the coil-globule transition of aqueous solutions of poly(af-isopropylacrylamide) as evidenced by FTIR spectroscopyi. *Langmuir* **2000**, *16*, 7503–7509. [CrossRef]
32. Farooqi, Z.H.; Khan, H.U.; Shah, S.M.; Siddiq, M. Stability of poly(N-isopropylacrylamide-co-acrylic acid) polymer microgels under various conditions of temperature, pH and salt concentration. *Arab. J. Chem.* **2017**, *10*, 329–335. [CrossRef]
33. Daly, E.; Saunders, B.R. Temperature-dependent electrophoretic mobility and hydrodynamic radius measurements of poly(N-isopropylacrylamide) microgel particles: Structural insights. *Phys. Chem. Chem. Phys.* **2000**, *2*, 3187–3193. [CrossRef]
34. Korson, L.; Drost-Hansen, W.; Millero, F.J. Viscosity of water at various temperatures. *J. Phys. Chem.* **1969**, *73*, 34–39. [CrossRef]
35. Poudel, L.; Wen, A.M.; French, R.H.; Parsegian, V.A.; Podgornik, R.; Steinmetz, N.F.; Ching, W.Y. Electronic structure and partial charge distribution of doxorubicin in different molecular environments. *ChemPhysChem* **2015**, *16*, 1451–1460. [CrossRef]
36. Vora, A.; Riga, A.; Dollimore, D.; Alexander, K.S. Thermal stability of folic acid. *Thermochim. Acta* **2002**, *392–393*, 209–220. [CrossRef]
37. Das, G.; Nicastri, A.; Coluccio, M.L.; Gentile, F.; Candeloro, P.; Cojoc, G.; Liberale, C.; De Angelis, F.; Di Fabrizio, E. FT-IR, Raman, RRS measurements and DFT calculation for doxorubicin. *Microsc. Res. Tech.* **2010**, *73*, 991–995. [CrossRef] [PubMed]
38. Coates, J. Interpretation of Infrared Spectra, A Practical Approach. In *Encyclopedia of Analytical Chemistry*; John Wiley & Sons, Ltd.: Hoboken, NJ, USA, 2006.
39. Abraham, R.J. Organic Structure Analysis. Phillip Crews, Jaime Rodriguez, Marcel Jaspars. Oxford University Press, New York, 1998, pp. 552. Price $85.00. ISBN 0 19 510102 2 (cloth). *Magn. Reson. Chem.* **2001**, *39*, 367. [CrossRef]
40. Sambi, M.; Decarlo, A.; Malardier-Jugroot, C.; Szewczuk, M.R. Next-generation multimodality of nanomedicine therapy: Size and structure dependence of folic acid conjugated copolymers actively target cancer cells in disabling cell division and inducing apoptosis. *Cancers* **2019**, *11*, 1698. [CrossRef]
41. Mohsen, R.; Alexander, B.D.; Richardson SC, W.; Mitchell, J.C.; Diab, A.A.; Snowden, M.J. Design, Synthesis, Characterization and Toxicity Studies of Poly (N-Iso- Propylacrylamide-co-Lucifer Yellow) Particles for Drug Delivery Applications. *J. Nanomed. Nanotechnol.* **2016**, *7*, 363–372. [CrossRef]
42. Bresseleers, J.; Bagheri, M.; Storm, G.; Metselaar, J.M.; Hennink, W.E.; Meeuwissen, S.A.; Hest, J.C.M. van Scale-Up of the Manufacturing Process To Produce Docetaxel-Loaded mPEG-b-p(HPMA-Bz) Block Copolymer Micelles for Pharmaceutical Applications. *Org. Process Res. Dev.* **2019**, *23*, 2707–2715. [CrossRef] [PubMed]
43. Yadav, S.; Sharma, A.K.; Kumar, P. Nanoscale Self-Assembly for Therapeutic Delivery. *Front. Bioeng. Biotechnol.* **2020**, *8*, 127. [CrossRef]
44. Yang, L.; Fan, X.; Zhang, J.; Ju, J. Preparation and Characterization of Thermoresponsive Poly(N-Isopropylacrylamide) for Cell Culture Applications. *Polymers* **2020**, *12*, 389. [CrossRef] [PubMed]
45. Adamo, G.; Grimaldi, N.; Campora, S.; Bulone, D.; Bondì, M.L.; Al-Sheikhly, M.; Sabatino, M.A.; Dispenza, C.; Ghersi, G. Multi-functional nanogels for tumor targeting and redox-sensitive drug and siRNA delivery. *Molecules* **2016**, *21*, 1594. [CrossRef]
46. Nguyen, D.H.; Bae, J.W.; Choi, J.H.; Lee, J.S.; Park, K.D. Bioreducible cross-linked Pluronic micelles: pH-triggered release of doxorubicin and folate-mediated cellular uptake. *J. Bioact. Compat. Polym.* **2013**, *28*, 341–354. [CrossRef]
47. Hashemkhani, M.; Muti, A.; Sennaroğlu, A.; Yagci Acar, H. Multimodal image-guided folic acid targeted Ag-based quantum dots for the combination of selective methotrexate delivery and photothermal therapy. *J. Photochem. Photobiol. B Biol.* **2020**, *213*, 112082. [CrossRef] [PubMed]

Review

Phytochemical-Based Nano-Pharmacotherapeutics for Management of Burn Wound Healing

Abdul Qadir [1], Samreen Jahan [1], Mohd Aqil [1], Musarrat Husain Warsi [2,*], Nabil A. Alhakamy [3], Mohamed A. Alfaleh [3,4], Nausheen Khan [5] and Athar Ali [6]

1. Department of Pharmaceutics, School of Pharmaceutical Education & Research, Jamia Hamdard, New Delhi 110062, India; aqkhan90@gmail.com (A.Q.); samreenj1996@gmail.com (S.J.); aqilmalik@yahoo.com (M.A.)
2. Department of Pharmaceutics and Industrial Pharmacy, College of Pharmacy, Taif University, Al-Haweiah, Taif 21974, Saudi Arabia
3. Department of Pharmaceutics, Faculty of Pharmacy, King Abdulaziz University, Jeddah 21589, Saudi Arabia; nalhakamy@kau.edu.sa (N.A.A.); maalfaleh@kau.edu.sa (M.A.A.)
4. Vaccines and Immunotherapy Unit, King Fahd Medical Research Center, King Abdulaziz University, Jeddah 21589, Saudi Arabia
5. Department of Pharmacognosy and Phytochemistry, School of Pharmaceutical Education & Research, Jamia Hamdard, New Delhi 110062, India; nausheenkhan070@gmail.com
6. Centre for Transgenic Plant Development, Department of Biotechnology, Jamia Hamdard, New Delhi 110062, India; atharbiotech@gmail.com
* Correspondence: mvarsi@tu.edu.sa or mhwarsi@gmail.com

Abstract: Medicinal plants have been used since ancient times for their various therapeutic activities and are safer compared to modern medicines, especially when properly identifying and preparing them and choosing an adequate dose administration. The phytochemical compounds present in plants are progressively yielding evidence in modern drug delivery systems by treating various diseases like cancers, coronary heart disease, diabetes, high blood pressure, inflammation, microbial, viral and parasitic infections, psychotic diseases, spasmodic conditions, ulcers, etc. The phytochemical requires a rational approach to deliver the compounds to enhance the efficacy and to improve patients' compatibility. Nanotechnology is emerging as one of the most promising strategies in disease control. Nano-formulations could target certain parts of the body and control drug release. Different studies report that phytochemical-loaded nano-formulations have been tested successfully both in vitro and in vivo for healing of skin wounds. The use of nano systems as drug carriers may reduce the toxicity and enhance the bioavailability of the incorporated drug. In this review, we focus on various nano-phytomedicines that have been used in treating skin burn wounds, and how both nanotechnology and phytochemicals are effective for treating skin burns.

Keywords: burn; injury; phytochemical; nanotechnology; wound healing

1. Introduction

Skin is the largest visible and vulnerable organ of the human body. It protects our body from environmental changes and dehydration [1,2]. There are certain skin conditions, such as burns and other substantial loss of the outer layer of the skin (epidermis), which acts as the barricade that prevents the skin from degeneration and microbial incursion and balances the fluid levels of the body. In such conditions, both nutritional and electrolytes constituents get demolished. Hence, skin wounds can drastically impact human health [3]. Various diseases, such as eczema, herpes zoster, rosacea, and psoriasis, can cause harm to skin; however, burns are the major cause of skin damage [4]. According to the World Health Organization (WHO), an estimated 180,000 deaths are caused by burns annually [5]. A burn injury may result from hot and cold materials and vulnerability to chemicals and radiations. Burn wounds are of three types and classified by the profundity: (1) superficial (first degree), (2) partial thickness (second degree), and (3) full thickness (third degree) [6,7]. Healing of burn wounds is a complicated process, and it proceeds through various phases, including inflammation, proliferation, and remodeling. These phases should occur in the proper order and time sequence for better would healing as changes in any of the phases may cause a delay in the healing process [8,9]. After epidermal injury, platelet activation leads to control of blood loss and results in clot formation, which is the first step in wound healing mechanism [10]. Nanotechnology offers an excellent outlook to fast-track persistent wound healing by altering the different phases of healing with high payloads of phytoconstituents [11]. Over the last two centuries, the use of plants extracts in wound healing has increased due to the presence of active compounds in these extracts [12]. Herbal medicines are widely accepted because of their efficacy and low level of adverse effects [13]. For skin-related diseases and other disorders, utilization of herbal constituents is accepted by 80% of population [14]. Using different plant extracts, various studies have been performed to observe the pharmacological action of constituents on various disease. From 2011 to present day, the use of herbal medicines has increased from $18 million to $26 billion; it is also estimated that 50% of approved herbal drugs are provided worldwide [15,16].

Nanocarriers with herbal drugs have gathered significant recognition for their potential and distinctive attributes in numerous domains of human activity [17].

The combination of nanotechnology with natural drugs would be a novel development for enhancing the medicinal effect of these natural drugs [18]. To increase the acceptability of these compounds by patients and to prevent the need for repeated administration, the phytochemical needs an approach that can encourage the delivery of active components in a sustained release format. Novel drug delivery systems help achieve the required therapeutic effects with reduced adverse events and enhance the bioavailability of herbal constituents [19,20].

Comprehensive searches were done on Google Scholar and PubMed databases pertaining to herbal-based pharmaceuticals for burn wound, nano-drug delivery applications in burn healings, and past to present evolution of nano-phytomedicines for the management of burn wound healing. We focused mainly on the last 15 years of works in this area, although some older references were also included to provide validity to the review.

2. Wound Healing Process

Would healing is a complicated process, and it progresses through various phases, including inflammation, proliferation, and remodeling. Involvement of fibroblasts, leukocytes, and monocytes in the healing process aid in reconstituting the destructed skin (Figure 1). Vitamins E and C play crucial roles in wound healing as these are key factors in this process. Vitamin K prevents severe bleeding, carotenoids restore the skin epithelial layer and tissues, and phytosterols have antimicrobial and anti-inflammatory effects [21]. Would healing also involves the use of biochemical genetic reprogramming to reinstate the skin health. Recent research has shown that the use of phytochemicals has active constituents that have the capability to induce wound healing with less side effects [22].

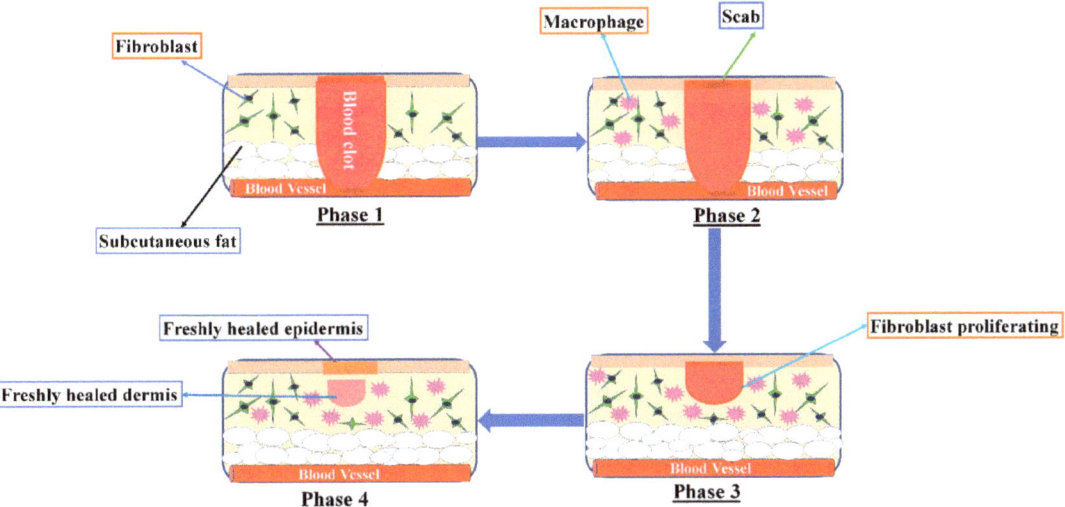

Figure 1. Different phases of the wound-healing process.

2.1. Inflammatory Phase

The inflammatory phase starts 4–5 days after the skin damage has already occurred. Once injury occurs, intravascular platelets mediate homeostasis, which is responsible for clot formation over the wound to stop further bleeding [23]. Post-homeostasis, thrombin activates platelets, which releases multiple growth factors, such as epidermal growth factor (EGF), insulin growth factor-1 (IGF-1), platelet derived growth factor (PDGF), fibroblast growth factor (FGF), and transforming growth factor (EGF, IGF-1, PDGF, FGF, and TGF, respectively) [24,25]. These growth factors activate neutrophils, monocytes, leukocytes, and macrophages, which act as a shield for skin to prevent further damage and initiate the wound healing process [26,27].

2.2. The Proliferative Phase

The proliferative phase involves cell proliferation and migration and takes 3–15 days to activate [28]. Once PDGF is released by platelets during the inflammatory phase, formation of new blood vessels and capillaries occurs [29]. Following this step, angiogenesis and migration of fibroblasts takes place to form granulation tissues [30]. Once the fibroblasts process is complete, a new extracellular matrix (ECM) consisting of collagen and proteoglycans is produced. Some fibroblasts undergo fission into myofibroblasts, which helps with the contraction of the wounded area [31]. Following this step, activation of keratinocytes, which migrate to the injured area and completes the last step, that of re-epithelialization, occurs [32].

2.3. Re-Modeling Stage

The re-modeling stage continues to change over the first several weeks to several years after the wound occurs. Collagen I replace collagen III, which consists of newly synthesized ECM, and the fresh collagen fibers develop into an assembled lattice composition that enhances the intensity of healed tissues [33,34].

3. The Impact of Antibiotics and Antioxidant Properties of Plants

Since ancient times, various plants have been used for treatment of different diseases and are currently in use worldwide. Due to their antimicrobial activities, natural constituents, such as aminoglycosides [35], beta-lactams [36], glycopeptides [37], quinolones [38], sulfonamides [39], and tetracyclines [40] have been utilized for wound healing treatment. For wound therapy, the benefits of plant extracts or phytochemical have been recognized as has the existence of antioxidants in numerous plants extracts. The presence of oxygen free radicals causes disruption in the wound healing process. Stress caused by oxidation slows down the healing process and causes additional damage to tissues. An antioxidant confers protection from oxygen free radicals by reducing their effects; this process assists in the wound healing process. Active compounds possessing antimicrobial effects play an important role as they neutralize free oxygen radicals and enhance the wound healing process [41,42]. Antioxidants have been used to hasten healing activity by extending antioxidant effects throughout the healing process. It seems the existence of antioxidants is essential to facilitate recuperation from persistent skin injury; Table 1 shows plants with both antimicrobial and antioxidant properties [43]. Natural metabolites influence the wound healing process by introducing various growth factors, such as EGF and FGF, which affect cellular movement [44]. Animal studies have indicated that herbal compounds encourage anti-inflammatory and antimicrobial activities for wound treatment by promoting regeneration of skin cells and displacing connective tissues [45]. Table 1 shows the names of natural compounds having antioxidant and antibiotic activities toward wound healing.

The efficacy of *Centella asiatica* has been studied broadly in animal models. This herb helps heal the incision-induced injuries. In one study, it was concluded that the level of antioxidants increases extensively in the presence of this herb, leading to improved healing activity [46]. Asiaticoside is extracted from *Centella asiatica* and produces a better capacity for injury healing process in both chronic and immediate healing as it has fibroblast proliferating activity [47].

Leaf extracts from *Chromolaena odorata* contain flavonoids and this herb has been used widely in the wound healing activity due to the free radical approach, which has shown a conclusive promotion in healing activity [48]. This extract was shown to cause improvement in the fibroblast proliferation and keratinocyte and endothelial cell activity, and to stimulate keratinocyte migration [48].

Quercus infectoria Olivier possesses anti-inflammatory, anti-bacterial, and antioxidant properties. The ethanolic extract of gallic acid, ellagic acid, and syringic acid form the active constituents of tannins, which might be the reason for antioxidant effects resulting in enhanced healing activity. In this study, the incision wound animal model showed better healing activity after stimulation with antioxidants, which caused enhancement of the superoxide dismutase and catalase levels, both of which are influential antioxidant enzymes [49].

Wounds, burns, and internal and external ulcers can be treated by *Buddleja globasa* (common name: orange bell Buddleja). This herb is also used traditionally in Chile for the treatment of ulcers and burns. This herb was tested for its capability to stimulate fibroblast growth and antioxidant activity in vitro. Testing was specific as the effect of the aqueous solution of *B. globasa* on these two processes is considered as the first stage in the tissue repair cycle [50]. It was proven that the damage caused by oxygen free radical causes a delay in the healing process; thus, an antioxidant was needed to reverse this delay [51]. This process was achieved because of the presence of flavonoids and caffeic acid in the extract. Buddleja leaves have other applications in the skin layer formation that is part of wound healing [52].

Curcumin is a derivative of *Curucuma longa*, which is also known as turmeric and Haldi. *C. longa* has numerous biological properties, which consist of antioxidant, initiation of enzyme detoxification, and prevention of degenerative diseases [53]. Dermal application of these substances to patients resulted in enhanced wound healing activity and provided tissue defense from oxidation-induced damage [54].

Table 1. Phytochemicals having antibiotic, antimicrobial, and antioxidants activity on wound healing.

Medicinal Plants	Family	Active Ingredients	Activities	Reference
Acacia Senegal L.	Leguminosae	Saponins, alkaloids, and malic acid	Antimicrobial, anti-inflammatory, and antioxidant activity	[55]
Acalypha indica L.	Euphorbiaceae	Flavanoids, alkaloids, saponins	Antioxidant and antimicrobial activity	[55]
Allophylus rubifolius L.	Sapindaceae	-	Antioxidant, antibacterial, and anti-inflammatory activity	[55]
Anagallis arvensis L.	Primulaceae	flavonoids, saponins, glycosides, alkaloids, and anthraquinones	Antioxidant, anti-inflammatory, and antimicrobial activity	[56]
Anogeissus dhofarica L.	Combretaceae	-	Antioxidant and antimicrobial activity	[41]
Aloevera L.	Liliacae	Saponins, acemannan, and anthraquinone	Antimicrobial activity	[57]
Anethum graveolens L.	Umbelliferae	-	Anti-inflammatory and antibacterial activity	[58]
Aristolochia bracteolate L.	Aristolochiaceae	-	Antimicrobial and antioxidant activity	[59]
Alternanthera brasiliana L.	Amaranthaceae	-	Antimicrobial activity	[60]
Achillea millefolium L.	Asteraceae	Isovaleric acid, salicylic acid, sterols, flavonoids, tannins, and coumarins	Antimicrobial activity	[15]
Acanthus polystachyus L.	Acanthaceae	Tannins, flavonoids, saponins, polyphenols, terpenoids, glycosides, and anthraquinones	Anti-inflammatory and antioxidant activity	[61]
Becium dhofarense L.	Lamiaceae	-	Antioxidant activity	[55]
Bridelia ferruginea L.	Phyllanthaceae	Flavonoids, tannins, saponins, and terpenoids	Antioxidant and antibacterial activity	[62]
Buddleja globosa L.	Scrophulariaceae	-	Antioxidant activity	[50]
Centella asiatica L.	Araliaceae	Flavonoids	Antioxidant activity	[46,63,64]
Chromolaena odorata L.	Asteraceae	Alkaloids, flavonoids, flavanone, essential oils, phenolics, saponins, tannins, and terpenoids	Antimicrobial, antioxidant, and anti-inflammatory activity	[48]
Clerodendrum infortunatum L.	Lamiaceae	Flavonoids	Antioxidant and antimicrobial activity	[65]
Combretum smeathmanii L.	Combretaceae	Alkaloids, coumarins, flavonoids, saponins, terpenes, and sterols	Antioxidant and antimicrobial activity	[66]
Cordia perrottettii L.	Boraginaceae	-	Antioxidant activity	[55]
Curcuma longa L.	Zingiberaceae	Glycosides, tannins, and flavonoids	Anti-inflammatory, antimicrobial, and antioxidant activity	[53,54,67]
Crassocephalum crepidioides L.	Asteraceae	Phenolic, flavonoid, and essential oil	Anti-inflammatory and antioxidant activity	[68]
Cinnamomum verum L.	Lauraceae	Tannins	Anti-inflammatory, antimicrobial, and antioxidant activity	[69]
Dendrophthoe falcata L.	Loranthaceae	-	Antioxidant and antimicrobial activity	[70]
Eucalyptus globulus L.	Myrtaceae	Alkaloids, flavonoids, saponin, tannin, carbohydrates, and glycosides etc.	Anti-inflammatory activity	[71]
Ficus asperifolia L.	Moraceae	Flavonoids, phenolics, alkaloids, and tannins	Antioxidant and antimicrobial activity	[72]

Table 1. Cont.

Medicinal Plants	Family	Active Ingredients	Activities	Reference
Gossypium arboreum L.	Malvaceae	Alkaloids, phenolic compounds, terpenoids, tannins, saponins flavonoids, cardiac glycosides, and protein	Antioxidant and antimicrobial activity	[72]
Gunnera perpensa L.	Gunneraceae	Alkaloids, ellagic acids, flavonoids, phenols, benzoquinones, proanthocyanidins, tannins, and minerals	Anti-inflammatory, antioxidant, and antibacterial activity	[73]
Hippophae rhamnoides L.	Elaeagnaceae	Flavonoids, tannins, triterpenes, glycerides of palmitic, stearic, oleic acids, vitamins (C, E, K), and amino acids	Anti-inflammatory, antimicrobial, and antioxidant activity	[74]
Holoptelea integrifolia L.	Ulmaceae	Terpenoids, saponins, tannins, phenols, alkaloids, flavonoids, glycosides, and quinines	Anti-inflammatory, antibacterial, and antioxidant activity	[75]
Memecylon edule L.	Melastomataceae	-	Anti-inflammatory activity	[76]
Moringa peregrina L.	Moringaceae	Proteins, vitamins, beta-carotene, amino acid, and phenolics	Anti-inflammatory, antimicrobial and antioxidant activity	[53,77]
Olea europaea L.	Oleaceae	Flavonoids, iridoids, secoiridoids, flavanones, biophenols, triterpenes, benzoic acid	Antioxidant activity	[53,78]
Phyllanthus muellerianus L.	Phyllanthaceae	Isoquercitrin, rutin, astragalin, phaselic acid, gallic acid, caffeic acid, methylgallate	Anti-inflammatory and antioxidant activity	[71]
Plagiochasma appendiculatum L.	Aytoniaceae	-	Antioxidant and antimicrobial activity	[79]
Pluchea Arabica L.	Asteraceae	Anthocyanins, phenolic acids, flavonoids, and carotenoids	Antioxidant activity	[53]
Quercus infectoria L.	Fagaceae	Tannin	Anti-inflammatory and antibacterial	[49]
Rhizophora mangle L.	Rhizophoraceae	Triterpenes, tannins and their glycosides	Antimicrobial and antioxidant activity	[80]
Secamone afzelii L.	Apocynaceae	alkaloids, tannins, cardiac glycosides and saponins	Antioxidant activity	[50]
Trigonella foenum L.	Papilionaceae	Carbohydrates, proteins, lipids, fibers, flavonoids, alkaloids, and saponins	Anti-inflammatory and antioxidant activity	[81]

4. Nanotechnology Involvement in Wound Healing Enhancement

Nano-drug delivery systems enormously influence the potential of drugs' medicinal effects and also protect the drugs from deterioration. Wound healing and skin re-formation involves various nano-delivery systems, such as those contained in organic nanoparticles, lipid nanoparticles, liposomes, polymeric nanoparticles, nanohydrogels, and nanofibers. These nano-systems show better efficacy compared to conventional systems (Figure 2).

In the past decades, a steady increase in the filing of patents based on herbal nano-formulations has been recorded. The key factor behind this increase is the capability of nano-formulations to overcome solubility drawbacks and bioavailability problems faced by conventional systems. One of most frequently filed patents is for curcumin, the multifunctional phytoceutical, extensively used in the treatment of tumors, cancers, and skin disorders. Other herbal based nano-formulations patents include carotenoids (nano-particles), silymarin (nano-particles), Panax ginseng (liquid mixture), *Syzygium cumini*, *Tinospora cordifolia*, *Trigonella foenum*-graecum, *Withania somnifera* (nano-emulsion, nanoen-capsulation, nano-dispersion, or synergistic liquid mixture), and Arbutin (emulsified nanoparticles) [82].

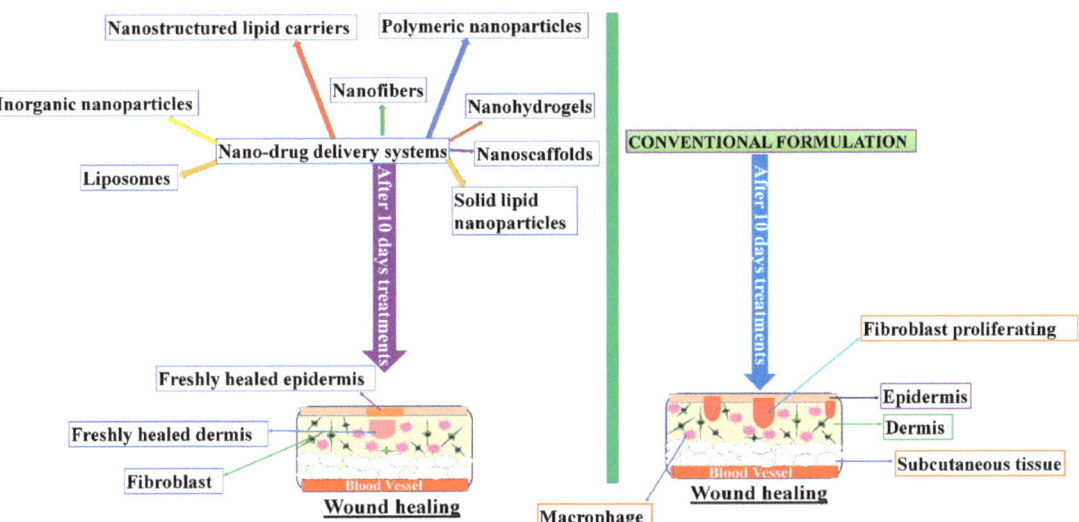

Figure 2. A comparison of nano-drug delivery systems involved in the skin regeneration and wound treatment with conventional drug delivery systems.

4.1. Nanofibers

Nanofibers are formed by unbreakable polymer chains of natural and synthetic compounds, which act as a sheet of nanofibers when placed on the skin to improve the tissues [83]. Nanofibers imitate collagen fibrils in the ECM, which can be formed from the synthetic or natural compounds and have numerous qualities that provide benefits to the wound healing process [83]. Nanofibers are beneficial for wound healing because they have a permeable construction and great orifice connection. Nanofibers have the capability to keep moisture at a suitable level. The synthesis of nanofibers with phytochemicals in nanofibrous materials has yielded tremendous results in the area of wound healing as these fibers have the capability to reduce the incision mark because of their porosity, which allows movement of oxygen [84]. Emodin (1,3,8-trihydroxy-6-methyl-anthraquinone) is an anthraquinone derivative that is found in the roots of *Rheum officinale* L. and is used extensively for wound healing as it has antimicrobial and anti-inflammatory activities. It produced a positive result when used for acute skin injuries [85]. The nanofibers of emodin in polyvinylpyrrolidone were harmless, anti-allergenic, bioactive, and dissolved at a rapid rate when compared to the pure compound. Re-epitheliazation was shown to have occurred at the wounded area, which hastened the healing process [86]. To increase the composition of collagen in human cells to 100%, emodin was incorporated in cellulose acetate nanostructure fibers [87]. The development of herbal constituents in cellulose acetate nanofibers promotes wound healing by using biomaterials as an interactive dressing material. Asiaticoside is extracted from *C. asiatica*, and the incorporation of trisachharide triterpene into cellulose acetate nanofibers produces an antioxidative effect during the early stages of the injury healing [88]. Increases in types I and III pro-collagen mRNAs were shown to enhance skin fibroblasts by elevating the protein levels [89]. Curcumin incorporated into cellulose acetate caused an improvement in fibroblast proliferation, enhanced collagen synthesis, and protected the dermal fibroblast cells from oxidative stress caused by hydrogen peroxide (H_2O_2) [90].

The active constituent of turmeric curcumin (1,7-bis (4-hydroxy-3-methoxyphenyl)-1,6-heptadiene-3,5-dione) is a polyphenolic compound, which is obtained from *C. longa* L. Curcumin is an active ingredient and its use is widely accepted for wound healing because it possesses various properties, such as anti-inflammatory, antibacterial, and antioxidant ones [91]. For epidermal injury healing, curcumin has been used in various in-vivo animal models [92]. Re-epithelization occurs during the early stages and enhances coagulation synthesis because it releases TGFβ1, which leads to increases in the number of blood vessels and cell granulation [93]. Liakos et al. [94] suggested that essential oils, such as cinnamon, lemongrass, and peppermint, can be used as antimicrobial agents. These electro-spun cellulose-based nanofibrous dressings were shown to prevent the *Escherichia coli* growth and required lesser quantities of oils. These dressings did not show any kind of cytotoxic effects and appears to be safe to use.

It has been reported that the curcumin loaded poly(ε-caprolactone)/gum tragacanth (PCL/GT) led to an improvement in the mechanical properties and tensile strengths of nanofibers and had a positive impact on collagen content for the treatment of diabetic wounds. By 15 days after injury, this moiety led to rapid wound healing by causing regeneration of the epithelial layer [95]. Bromelain-loaded chitosan nanofibers produced favorable wound healing results. It was observed for the second degree burn and has positive impact. The chitosan 2% w/v bromelain showed better physiochemical results compared to chitosan 4% w/v bromelain and was effective in reducing burn-induced injuries [96]. Bixin-loaded polycaprolactone (PCL) nanofibers maintained and accelerated wound healing activity in excisional wounds and effectively reduced the scar tissue area on the diabetic mice [97]. Alfalfa nanofibers yielded better results with respect to skin regeneration as these nanofibers possess antibacterial activity and bioactive phytoestrogens that work as a building block for the dressings for regenerative wounds [98].

4.2. Polymeric Nanoparticles (PNPs)

Polymeric nanoparticles are biocompatible colloidal systems that have risen in importance for both biomedical and bioengineering applications [99]. They are generally integrated by charged polymers and connected by interactivity of cationic and anionic chains of groups [100]. When drugs are incorporated into polymeric systems, this process prevents the deterioration caused by proteases found in the injury and delivered in stages to lower the frequency of administration [101]. Polymeric nanomaterials are widely utilized because of their antibacterial and wound healing activities [102]. For regeneration of skin injury, keratinocyte growth factor (KGF) is an impressive and potent growth factor [103]. It was observed that the KGF consists of self-assembled nanovesicles that enhances healing of the injured tissue cells of the skin by enhancing epithelization and skin re-modeling [104,105]. Recently, PNPs have been formulated by poly-lactic-co-glycolic acid (PLGA) and some of the other combinations in polymeric systems, including alginate, gelatin, and chitosan [106]. PLGA is approved by the Food and Drug Administration (FDA) for use in PNPs. The size of the PGLA NPS is 1–200 nm, which provides the benefits of biodegradability, biocompatibility, and being innocuous [107]. PLGA particles are generally formulated by emulsification of lipophilic compounds utilizing numerous surfactants and organic solvents [108]. The development of EGF-loaded nanoparticles for injury healing using PGLA yielded a positive response with respect to fibroblast proliferation and enhancement of the healing activity in the full thickness wounded skin. EGF plays an importance role in mediating the de-differentiation of keratinocytes into an epithelial linage and to reestablishing the epithelial barrier [109]. One of the studies also suggested that PGLA might produce a biocompatible system for growth factor delivery. To reduce lactate levels and enhance wound healing activity, the peptide defense host, known as LL37, was incorporated into PGLA nanoparticles [110]. Natural polymers, such as chitosan, have been chiefly considered for wound healing activity because of antibacterial and biocompatibility activities [111]. Chitosan is cationic in nature and has been utilized for the inhibition of microbial-induced infections [112]. Chitosan nanovesicles (150–300 nm) are generally formulated utilizing the method of ionic gelation [113]. Nowadays, chitosan is widely accepted in wound treatment, and it can also be utilized as a prophylactic agent to inhibit the infection development and enhance healing activity [114,115]. Studies have shown that polylactic acid-loaded chitosan magnetic eugenol nanospheres had improved prevention and development of biofilm compared to pure chitosan, whilst performing endothelial proliferation [116]. Most of the studies concerning this topic have reported that nanovesicles containing chitosan and analogs might enhance healing activity by improving inflammatory cell function and restoring fibroblasts and osteoblast functions [117]. In two different studies, it was observed that chitosan-loaded nanovesicles improved the coagulation by binding to red blood cells (RBC) and ameliorating the function of inflammatory cells. In another study, the chitosan nanovesicles were used as the compounds in bandages outlined for the skin wound and, hence, enhanced healing activity in both humans and animals [115,118].

4.3. Dendrimers

Dendrimers are nanoscale (1–10 nm) systems with homogeneous structures that are monodispersed in polymer macromolecule that can be used for both therapeutic and diagnostic purposes. Subunits of phenyl acetylene were used to develop dendrimers [119,120]. In addition, functional groups present on the surface of dendrimers can operate as antibacterial agents. Dendrimers cause detachment of contaminated tissues and may extend the phase of inflammation and slow injury diminution in addition to promoting re-epithelization and better wound healing activity [121]. The interaction between positively and negatively charged groups present on dendrimers and on the bacterial cell wall would lead to the bacterial structure disturbance [121]. In another study, silver-loaded dendrimer NSs were observed to show anti-inflammatory and anti-microbial activities in a synergistic manner. These properties were also shown to prevent inflammation and enhance healing activity [122].

4.4. Metallic Nanoparticles

Metal-based nanoparticles are widely utilized as they produce antibacterial, antimicrobial, and anti-inflammatory effects. The chemical and physical structures of nanoparticles are important for determining the propensity of a nanoparticle to enter and/or bind to target cells with the capacity to interact with their biological machinery and elicit a response. The metal-based nanoparticles are widely accepted in medicine, and the most acceptable metallic nanoparticles are silver- and gold-based nanostructures. Herbal plants are widely accepted in the development of metallic nanoparticles because of their low levels of side effects and more therapeutic effects as compared to the conventional dosage form [123]. Most of the herbal extracts, such as *Cladophora fascicularis* [124], *Aerva lanata* [125], *Hippophae rhamnoides* [126], *Eucommia ulmoides* [127], Black tea leaf [128], *Averrhoa bilimbi* [129], *Salicornia brachiate* [130], *Abelmoschus esculentus* [131], olive leaf [132], *Ipomoea carnea* [133], *geranium* [134], and *Cissus arnotiana* [135] have been incorporated into metallic nanoparticles

Silver nanoparticles are widely used as they possess antimicrobial, antibacterial, and anti-inflammatory properties [136]. The solubility and bioactivity of the silver particles at the wounded area depend on the size of silver particles; the smaller the size is, the stronger the contact with the will skin be. Silver nanoparticle vesicle sizes range from 1 to 100 nm. In one study, the silver–silver chloride nanoparticles combined with lower grapheme oxide nanovesicles induced an escalation of the healing process because it generated a higher number of oxygen free radicals rather than free the silver ions. A positive impact on the antibacterial activity on both Gram-negative and -positive bacteria has been shown, and, hence, these particles can enhance wound healing activity as shown in in-vivo studies in mice [137]. ACTICOAT is an alternate form of silver antimicrobial barrier wound dressing, which prevents the complication of prior agents. It slows down the bacterial activity, which leads to a reduction in inflammation and causes an improvement in the healing process [138]. The plant-based bio-prepared nanoparticles reveal potential for wound remedy and bacterial infection prevention [139]. Different methods for the preparation of silver nanoparticles are used. Photochemical and chemical reduction are the two most widely used methods [140]. Different plant extracts have been incorporated into silver nanoparticles for wound healing containing alkaloids, glycoside, corticosteroids and essential oils [141]. *Cassia roxburghii* prepared silver nanoparticles show the potential for wound healing enhancement as these particles have significant antibacterial and antifungal activities [142].

The active constituent of *Drosera binata* is naphthoquinones, primarily plumbagin. *D. binata* silver nanoparticles show better antibacterial activity against *Staphylococcus aureus* without affecting human keratinocytes. It was also inconclusive as to whether it is *D. binata* extract or its pure form (3-chloroplumbagin) that would have effective results for antibiotics and, hence, enhance wound healing [143]. Extracts of grape pomace were also combined with silver nitrate, and grape-silver nanoparticle-stabilized liposomes were developed by Castangia et al. The resulting nano-formulation showed potential to offer a significant shield of keratinocytes and fibroblasts to combat oxidative stress, thus, avoiding cell damage and death [144]. The other highly acceptable nanoparticles in different applications, such as wound treatment, re-epithelization, and particularly drug delivery, include gold nanoparticles [145].

Their chemical stability and capability to absorb near-infrared (NIR) light combined with their positive impact and antibacterial activity will strengthen the wound healing process [146]. Gold nanoparticles have the potential to penetrate bacterial tissues and cause alterations in the cell membrane, which causes inhibition of bacterial activity [147], and also prevents bacteria from developing reactive oxygen species [148].

Gold nanoparticles are synthesized with collagen, gelatin, and chitosan to yield effective injury recovery activity and also helps to achieve the biocompatibility [149]. Chitosan-loaded gold nanoparticles showed enhanced results in the healing process as these particles increase free radical scavenging and improve biocompatibility; in the model, these particles enhance the formation of cells and lead to an improvement in hemostasis by increasing the healing activity in comparison to pure chitosan [150]. The resulting metabolites from *Indigofera aspalathoides* Vahl. (Papilionaceae), which is also known as Shivanarvembu, are extracted from plants and used for wound healing. The histopathology results demonstrate that the *I. aspalathoides* silver nanoparticles have a better effect on wound healing in mice. When treated with plant extract, the granulation tissue which possesses fibroblasts, collagen fibers, minimal edema, and newly developed blood vessels were noted [151]. The other forms of metallic nanoparticles are gold and copper oxide nanovesicles that improve wound healing, which leads to fast injury healing and slows down the infection development. Both silver and gold nanoparticles are formed by incorporating *Coleous forskohlii* root extracts. These particles exhibit antimicrobial activity and antioxidant activities and have a positive effect on re-epithelization at the site of wound, which enhances connective tissue formation and causes an increase in proliferation and remodeling rates of dermal cells [152]. The development of both titanium dioxide and copper oxide nanoparticles of *Moringa oleifera* and *Ficus religiosa* leaf extracts, respectively, were shown to enhance wound healing and decrease the removal wound site in rats [153].

4.5. Nanohydrogels

For wound treatment, nanohydrogels are considered to be effective carriers as they possess three-dimensional polymeric networks. Due to their permeable network, they have the capability to absorb the liquid, which helps the wound to keep hydrated and enhance the wound healing process by keeping the proper oxygen level. Due to their effectiveness, compatibility, and showing beneficial results on skin revitalization, nanohydrogels have become widely accepted [154].

To improve wound healing activity, the gellan cholesterol nanohydrogel is immersed in baicalin. The baicalin-loaded nanohydrogels manifest ideal efficacy for skin repair and also act as inflammation inhibitors when applied to an epidermal inflammation mice model in in-vivo studies [155]. The freshly developed nanocrystal bacterial cellulose hydrogels instantly stick to fibroblasts, support human dermal fibroblast morphology, restrict the relocation of cells, enhance the proliferation of cells, and influence the nine expressions of genes connected to healing of injury. These genes include interleukins 6 and 10, granulocyte-macrophage colony-stimulating factor, matrix metalloproteinase 2 (IL-6 and -10, GM-CSF, MMP-2, respectively), and TGF-β; hence, nanohydrogels play an important role in skin regeneration [156].

4.6. Liposomes

Liposomes appear to be an important vehicle for topical delivery; they are harmless and environmentally safe and possess high drug loading efficiency, long-term stability, biological acceptability with skin in addition to having the capability to incorporate both hydrophobic and hydrophilic drugs in water and bilayer cavities [157]. Liposomes successfully shield the injury site and build a humid habitat at the site of injury, which is beneficial for the healing of the wounded skin. Taking all these characteristics into consideration, liposomes have become widely accepted in skin regeneration and injury treatment [158]. A study on propylene glycol liposome nanocarriers demonstrated numerous merits in comparison to other nano-systems. This system showed the tendency to enhance the stability, retention, and permeation in the tissues of skin [159]. It surmised that propylene glycol ameliorate the elasticity of vesicle containing bilayer of phospholipids. Hence, it improved the permeation into the skin. Moreover, the particles size of liposomes should be 150 nm for better drug perforation into the skin layers [160]. Liposomes with silk fibroin hydrogels were prepared to stabilize the basic fibroblast growth factor (bFGF) that maintained the activity of proliferation of cells on wound fluids; it also enhances the healing process by inspiring angiogenesis [161]. Rabelo et al. assessed the gelatin-membrane consisting of usnic acid-loaded liposomes and obtained encouraging results for wound healing. These results showed that the membrane of liposomes prominently manages the second-grade infection on porcine model [162]. Furthermore, with improved collagen, accumulation on cellularized granulation tissue was discovered in the treated group of liposomal membrane, which when compared to one of the commercial products improved the granulation tissue maturation and repaired the scars [163]. Argan-liposomes and argan-hyalurosomes have been successfully developed by incorporating neem oil into them. These formulations were extremely biocompatible and could protect skin cells from oxidative stress effectively with improved efficacy of oil. Moreover, formulations stimulate wound closure substantially more effectively than oil dispersion [164]. The efficacy of mangiferin (employed in cure of skin lesions) was enhanced by modifying transferosomes with propylene glycol and glycerol. Improved deposition of mangiferin was observed in epidermal and dermal layer and fibroblasts were protected from oxidative stress and intensified their propagation [165].

4.7. Inorganic Nanoparticles

Inorganic nanoparticles are those derived from the inorganic materials and include carbon-, metal-, and ceramic-based nanovesicles that accelerate tissue repair and remodeling. These particles deliver assistance in the region of medicines, counting cancer, imaging, and drug delivery; however, their utilization in tissue regulation and skin remodeling is new, it also provides adhesion in tissue and enhanced antimicrobial activity in injury healing [166].

4.8. Lipid Nanoparticles

Lipid nanoparticles were designed to overcome the stability limitation of liposomes due to the lipid bilayer. Lipid nanovesicles consist of two types: (1) solid lipid nanoparticles (SLNs) and (2) nanostructured lipid carriers (NLCs). The preparation of lipid nanovesicles amid lipids molecules does not include the use of any potentially harmful biotic solvents [167]. In a study, both SLN- and NLC-loaded rh-EGF (epidermal growth factor) for chronic injury treatment were formulated by the emulsification followed by an ultra-sonication method; however, the NLC process included no organic solvent and showed better entrapment efficiency. The results of both formulations show capabilities to enhance cell proliferation when compared with free rh-EGF and considerably enhance the healing activity for wound closure, re-establish the process of inflammation, and facilitate re-epithelization [168].

In another study, development of SLNs with the elastase inhibitor serpin A1 and antimicrobial peptide LL37 had a synergistic impact on injury healing. SLNs promoted the closure of injury in cells of fibroblasts and keratinocytes. Moreover, it also led to improvement in the activity of antibacterial against *S. aureus* and *E. coli* when compared with the LL37- and A1-treated groups [169].

5. Future Perspective and Conclusions

The main aim of this review article was to describe the advantages of using nano-systems for use in the wound healing process. The distinctive physiochemical properties of nano-systems make them a perfect candidate for the application of wound healing process. The wound therapy process by nanotechnological systems demonstrates better therapeutic effect compared to the conventional therapy for wound healing. Nanotechnological systems can change one or more than one phase of wound healing during the process, as it possesses antibacterial, anti-inflammatory, and anti-proliferation activities. Worldwide, the research has been conducted on natural and herbal compounds due to their more therapeutic effects and lesser side effects. There is a need for the development of improved systems for the delivery of drugs at the target site with a dose that does not alter the existing treatment of disease. The herbal compounds have great potential and, hence, a better future, especially when incorporated into the nanocarriers for chronic wound treatment as they have shown promising results. Herbal medicine-based novel drug delivery systems have acknowledged the approaches in the field of pharmaceuticals, which will improve the health of the people. It is also concluded that the incorporation of herbal compound in the nano-vehicle will aggrandize the magnitude of the existing delivery system. Anyhow, various approaches have been employed for the privileged application of nanocarriers in wound healing therapy. The main concerns for the nano-vehicles are toxicity because they may cause possible side effects in the human body. Hence, this requires to be rectified at the starting point for further progression of wound healing therapies in clinical trials. In in vivo models, there is slighter comprehension regarding non-material mediated wound healing processes and this is one of the problems observed. The studies of non-material-wound healing processes are based on in-vitro studies or mainly depend on single aim bacteria. The in-vivo wound healing application is required for the in-depth studies utilizing both Gram-positive and -negative bacterial strains. Subsequently, the main focus should be on improving and enhancing target efficiency for more efficacious wound healing. Therefore, the investigators should target producing a nanomaterial that is biocompatible and biodegradable and has the capability to correct all the phases of the wound healing process.

Author Contributions: Conceptualization, A.Q., M.A. and M.H.W.; methodology, S.J., A.Q.; software, A.A.; validation, A.A., N.A.A. and M.A.A.; formal analysis, A.Q., M.H.W. and N.K.; investigation, A.A., N.K.; resources, N.A.A., A.Q.; data curation, N.K., S.J.; writing—original draft preparation, A.Q.; writing—review and editing, M.H.W., M.A.A. and N.A.A.; visualization, M.H.W.; supervision, M.A.; project administration, M.A.A., N.A.A.; funding acquisition, M.A.A., N.A.A. All authors have read and agreed to the published version of the manuscript.

Funding: The Deanship of Scientific Research (DSR) at King Abdulaziz University, Jeddah, Saudi Arabia has funded this project, under grant no. (FP-032-43).

Institutional Review Board Statement: Not applicable.

Informed Consent Statement: Not applicable.

Data Availability Statement: Not applicable.

Conflicts of Interest: The authors declare no conflict of interest.

References

1. Wang, W.; Lu, K.J.; Yu, C.H.; Huang, Q.L.; Du, Y.Z. Nano-drug delivery systems in wound treatment and skin regeneration. *J. Nanobiotechnology* **2019**, *17*, 82. [CrossRef]
2. Pereira, R.F.; Carvalho, A.; Gil, M.H.; Mendes, A.; Bartolo, P. Influence of *Aloe vera* on water absorption and enzymatic in vitro degradation of alginate hydrogel films. *Carbohydr. Polym.* **2013**, *98*, 311–320. [CrossRef]
3. World Health Organization Burns—Key Facts. Available online: https://www.who.int/news-room/fact-sheets/detail/burns (accessed on 4 January 2020).
4. Souto, E.B.; Ribeiro, A.F.; Ferreira, M.I.; Teixeira, M.C.; Shimojo, A.A.; Soriano, J.L.; Naveros, B.C.; Durazzo, A.; Lucarini, M.; Souto, S.B.; et al. New nanotechnologies for the treatment and repair of skin burns infections. *Int. J. Mol. Sci.* **2020**, *21*, 393. [CrossRef] [PubMed]
5. Deitch, E.A. The management of burns. *N. Engl. J. Med.* **1990**, *323*, 1249–1253. [PubMed]
6. Wild, T.; Rahbarnia, A.; Kellner, M.; Sobotka, L.; Eberlein, T. Basics in nutrition and wound healing. *Nutrition* **2010**, *26*, 862–866. [CrossRef] [PubMed]
7. Hermans, M.H. Results of an internet survey on the treatment of partial thickness burns, full thickness burns, and donor sites. *J. Burn Care Res.* **2007**, *28*, 835–847. [CrossRef] [PubMed]
8. Roshangar, L.; Kheirjou, R.; Ranjkesh, R. Skin Burns: Review of Molecular Mechanisms and Therapeutic Approaches. *Wounds Compend. Clin. Res. Pract.* **2019**, *31*, 308–315.
9. Guo, S.A.; DiPietro, L.A. Factors affecting wound healing. *J. Dent. Res.* **2010**, *89*, 219–229. [CrossRef]
10. Ghosh, P.K.; Gaba, A. Phyto-extracts in wound healing. *J. Pharm. Pharm. Sci.* **2013**, *16*, 760–820. [CrossRef]
11. Sandhiya, V.; Ubaidulla, U. A review on herbal drug loaded into pharmaceutical carrier techniques and its evaluation process. *Future J. Pharm. Sci.* **2020**, *6*, 1–16. [CrossRef]
12. Bhatt, D.; Jethva, K.; Patel, S.; Zaveri, M. Novel drug delivery systems in herbals for cancer. *World J. Pharm. Res.* **2016**, *5*, 368–378.
13. Ferreira, V.F.; Pinto, A.C. A fitoterapia no mundo atual. *Química Nova* **2010**, *33*, 1829. [CrossRef]
14. Vickers, A.; Zollman, C. Herbal medicine. *Br. Med. J.* **1999**, *319*, 1050–1053. [CrossRef] [PubMed]
15. Ali, S.I.; Gopalakrishnan, B.; Venkatesalu, V. Pharmacognosy, phytochemistry and pharmacological properties of *Achillea millefoliumh* L.: A review. *Phytother. Res.* **2017**, *31*, 1140–1161. [CrossRef] [PubMed]
16. Watkins, R.; Wu, L.; Zhang, C.; Davis, R.; Xu, B. Natural product-based nanomedicine: Recent advances and issues. *Int. J. Nanomed.* **2015**, *10*, 6055–6074.
17. Xu, R.; Luo, G.; Xia, H.; He, W.; Zhao, J.; Liu, B.; Tan, J.; Zhou, J.; Liu, D.; Wang, Y.; et al. Novel bilayer wound dressing composed of silicone rubber with particular micropores enhanced wound re-epithelialization and contraction. *Biomaterials* **2014**, *40*, 1–11. [CrossRef] [PubMed]
18. Singh, R.P.; Singh, S.G.; Naik, H.; Jain, D.; Bisla, S. Herbal excipients in novel drug delivery system. *Int. J. Compr. Pharm.* **2011**, *2*, 1–7.
19. Sungthongjeen, S.; Pitaksuteepong, T.; Somsiri, A.; Sriamornsak, P. Studies on pectins as potential hydrogel matrices for controlled-release drug delivery. *Drug Dev. Ind. Pharm.* **1999**, *25*, 1271–1276. [CrossRef]
20. Portou, M.; Baker, D.; Abraham, D.; Tsui, J. The innate immune system, toll-like receptors and dermal wound healing: A review. *Vasc. Pharmacol.* **2015**, *71*, 31–36. [CrossRef]
21. Andritoiu, C.V.; Andriescu, C.E.; Ibanescu, C.; Lungu, C.; Ivanescu, B.; Vlase, L.; Havarneanu, C.; Popa, M. Effects and Characterization of Some Topical Ointments Based on Vegetal Extracts on Incision, Excision, and Thermal Wound Models. *Molecules* **2020**, *25*, 5356. [CrossRef]
22. Hajialyani, M.; Tewari, D.; Sobarzo-Sánchez, E.; Nabavi, S.M.; Farzaei, M.H.; Abdollahi, M. Natural product-based nanomedicines for wound healing purposes: Therapeutic targets and drug delivery systems. *Int. J. Nanomed.* **2018**, *13*, 5023–5043. [CrossRef]
23. Martin, P. Wound Healing–Aiming for Perfect Skin Regeneration. *Science* **1997**, *276*, 75–81. [CrossRef]
24. Braund, R.; Hook, S.; Medlicott, N.J. The role of topical growth factors in chronic wounds. *Curr. Drug Deliv.* **2007**, *4*, 195–204. [CrossRef] [PubMed]
25. Gainza, G.; Villullas, S.; Pedraz, J.L.; Hernandez, R.M.; Igartua, M. Advances in drug delivery systems (DDSs) to release growth factors for wound healing and skin regeneration. *Nanomed. Nanotechnol. Biol. Med.* **2015**, *11*, 1551–1573. [CrossRef] [PubMed]
26. Kiritsy, C.P.; Lynch, S.E. Role of growth factors in cutaneous wound healing: A review. *Crit. Rev. Oral Biol. Med.* **1993**, *4*, 729–760. [CrossRef] [PubMed]
27. Eming, S.A.; Krieg, T.; Davidson, J.M. Inflammation in wound repair: Molecular and cellular mechanisms. *J. Investig. Dermatol.* **2007**, *127*, 514–525. [CrossRef]

28. Singer, A.J.; Clark, R.A. Cutaneous wound healing. *N. Engl. J. Med.* **1999**, *341*, 738–746. [CrossRef]
29. Velnar, T.; Bailey, T.; Smrkolj, V. The wound healing process: An overview of the cellular and molecular mechanisms. *J. Int. Med. Res.* **2009**, *37*, 1528–1542. [CrossRef]
30. Malinda, K.M.; Sidhu, G.S.; Banaudha, K.K.; Gaddipati, J.P.; Maheshwari, R.K.; Goldstein, A.L.; Kleinman, H.K. Thymosin α1 stimulates endothelial cell migration, angiogenesis, and wound healing. *J. Immunol.* **1998**, *160*, 1001–1006.
31. Li, B.; Wang, J.H.C. Fibroblasts and myofibroblasts in wound healing: Force generation and measurement. *J. Tissue Viability* **2011**, *20*, 108–120. [CrossRef]
32. Montesinos, M.C.; Gadangi, P.; Longaker, M.; Sung, J.; Levine, J.; Nilsen, D.; Reibman, J.; Li, M.; Jiang, C.-K.; Hirschhorn, R.; et al. Wound Healing Is Accelerated by Agonists of Adenosine A2 (Gαs-linked) Receptors. *J. Exp. Med.* **1997**, *186*, 1615–1620. [CrossRef]
33. Ehrlich, H.P.; Keefer, K.A.; Myers, R.L.; Passaniti, A. Vanadate and the absence of myofibroblasts in wound contraction. *Arch. Surg.* **1999**, *134*, 494–501. [CrossRef] [PubMed]
34. Stadelmann, W.K.; Digenis, A.G.; Tobin, G.R. Physiology and healing dynamics of chronic cutaneous wounds. *Am. J. Surg.* **1998**, *176*, 26S–38S. [CrossRef]
35. Pawar, H.V.; Tetteh, J.; Boateng, J.S. Preparation, optimisation and characterisation of novel wound healing film dressings loaded with streptomycin and diclofenac. *Colloids Surf. B Biointerfaces* **2013**, *102*, 102–110. [CrossRef]
36. Sabitha, M.; Rajiv, S. Preparation and characterization of ampicillin-incorporated electrospun polyurethane scaffolds for wound healing and infection control. *Polym. Eng. Sci.* **2014**, *55*, 541–548. [CrossRef]
37. Lan, Y.; Li, W.; Guo, R.; Zhang, Y.; Xue, W.; Zhang, Y. Preparation and characterisation of vancomycin-impregnated gelatin microspheres/silk fibroin scaffold. *J. Biomater. Sci. Polym. Ed.* **2013**, *25*, 75–87. [CrossRef] [PubMed]
38. Pásztor, N.; Rédai, E.; Szabó, Z.-I.; Sipos, E. Preparation and characterization of levofloxacin-loaded nanofibers as potential wound dressings. *Acta Med. Marisiensis* **2017**, *63*, 66–69. [CrossRef]
39. Mohseni, M.; Shamloo, A.; Aghababaei, Z.; Vossoughi, M.; Moravvej, H. Antimicrobial wound dressing containing silver sulfadiazine with high biocompatibility: In vitro study. *Artif. Organs* **2016**, *40*, 765–773. [CrossRef]
40. Adhirajan, N.; Shanmugasundaram, N.; Shanmuganathan, S.; Babu, M. Collagen-based wound dressing for doxycycline delivery: In-vivo evaluation in an infected excisional wound model in rats. *J. Pharm. Pharmacol.* **2009**, *61*, 1617–1623. [CrossRef]
41. Parihar, A.; Parihar, M.S.; Milner, S.; Bhat, S. Oxidative stress and anti-oxidative mobilization in burn injury. *Burns* **2008**, *34*, 6–17. [CrossRef]
42. Suntar, I.; Akkol, E.K.; Nahar, L.; Sarker, S.D. Wound healing and antioxidant properties: Do they coexist in plants? *Free Radic. Antioxid.* **2012**, *2*, 1–7. [CrossRef]
43. Blass, S.C.; Goost, H.; Tolba, R.H.; Stoffel-Wagner, B.; Kabir, K.; Burger, C.; Stehle, P.; Ellinger, S. Time to wound closure in trauma patients with disorders in wound healing is shortened by supplements containing antioxidant micronutrients and glutamine: A PRCT. *Clin. Nutr.* **2012**, *31*, 469–475. [CrossRef] [PubMed]
44. Schultz, G.S.; Sibbald, R.G.; Falanga, V.; Ayello, E.A.; Dowsett, C.; Harding, K.; Romanelli, M.; Stacey, M.C.S.; Teot, L.; Vanscheidt, W. Wound bed preparation: A systematic approach to wound management. *Wound Repair Regen.* **2003**, *11*, S1–S28. [CrossRef]
45. Tsala, D.E.; Amadou, D.; Habtemariam, S. Natural wound healing and bioactive natural products. *Phytopharmacology* **2013**, *4*, 532–560.
46. Ruszymah, B.H.I.; Chowdhury, S.R.; Manan, N.A.B.A.; Fong, O.S.; Adenan, M.I.; Bin Saim, A. Aqueous extract of *Centella asiatica* promotes corneal epithelium wound healing in vitro. *J. Ethnopharmacol.* **2012**, *140*, 333–338. [CrossRef] [PubMed]
47. Maquart, F.X.; Bellon, G.; Gillery, P.; Wegrowski, Y.; Borel, J.P. Stimulation of collagen synthesis in fibroblast cultures by a triterpene extracted from *Centella asiatica*. *Connect. Tissue Res.* **1990**, *24*, 107–120. [CrossRef] [PubMed]
48. Thang, P.T.; Teik, L.S.; Yung, C.S. Anti-oxidant effects of the extracts from the leaves of *Chromolaena odorata* on human dermal fibroblasts and epidermal keratinocytes against hydrogen peroxide and hypoxanthine–xanthine oxidase induced damage. *Burns* **2001**, *27*, 319–327. [CrossRef]
49. Umachigi, S.P.; Jayaveera, K.N.; Kumar, C.K.A.; Kumar, G.S.; Swamy, B.M.V.; Kumar, D.V.K. Studies on wound healing properties of *Quercus infectoria*. *Trop. J. Pharm. Res.* **2008**, *7*, 913–919. [CrossRef]
50. Mensah, A.Y.; Sampson, J.; Houghton, P.; Hylands, P.; Westbrook, J.; Dunn, M.; Hughes, M.; Cherry, G. Effects of *Buddleja globosa* leaf and its constituents relevant to wound healing. *J. Ethnopharmacol.* **2001**, *77*, 219–226. [CrossRef]
51. Shukla, A.; Rasik, A.M.; Dhawan, B.N. Asiaticoside-induced elevation of antioxidant levels in healing wounds. *Phytother. Res. Int. J. Devoted Pharmacol. Toxicol. Eval. Nat. Prod. Deriv.* **1999**, *13*, 50–54. [CrossRef]
52. Yamasaki, T.; Li, L.; Lau, B.H.S. Garlic compounds protect vascular endothelial cells from hydrogen peroxide-induced oxidant injury. *Phytother. Res.* **1994**, *8*, 408–412. [CrossRef]
53. Ravindran, P.N.; Babu, K.N.; Sivaraman, K. (Eds.) *Turmeric: The Genus Curcuma*; CRC Press: Boca Raton, FL, USA, 2007.
54. Gopinath, D.; Ahmed, M.; Gomathi, K.; Chitra, K.; Sehgal, P.; Jayakumar, R. Dermal wound healing processes with curcumin incorporated collagen films. *Biomaterials* **2004**, *25*, 1911–1917. [CrossRef]
55. Marwah, R.G.; Fatope, M.O.; Al Mahrooqi, R.; Varma, G.B.; Al Abadi, H.; Al-Burtamani, S.K.S. Antioxidant capacity of some edible and wound healing plants in Oman. *Food Chem.* **2007**, *101*, 465–470. [CrossRef]
56. Qadir, M.I. Medicinal and cosmetological importance of Aloe vera. *Int. J. Nat. Ther.* **2009**, *2*, 21–26.

57. Pattanayak, S.; Sunita, P. Wound healing, anti-microbial and antioxidant potential of *Dendrophthoe falcata* (L.f) Ettingsh. *J. Ethnopharmacol.* **2008**, *120*, 241–247. [CrossRef]
58. Altameme, H.J.; Hameed, I.H.; Hamza, L.F. *Anethum graveolens*: Physicochemical properties, medicinal uses, antimicrobial effects, antioxidant effect, anti-inflammatory and analgesic effects: A review. *Int. J. Pharm. Qual. Assur.* **2017**, *8*, 88–91.
59. Shirwaikar, A.; Somashekar, A.; Udupa, A.; Udupa, S.; Somashekar, S. Wound healing studies of *Aristolochia bracteolata* Lam. with supportive action of antioxidant enzymes. *Phytomedicine* **2003**, *10*, 558–562. [CrossRef] [PubMed]
60. Barua, C.C.; Talukdar, A.; Begum, S.A.; Sarma, D.K.; Fathak, D.C.; Barua, A.G.; Bora, R.S. Wound healing activity of methanolic extract of leaves of *Alternanthera brasiliana* Kuntz using in vivo and in vitro model. *Indian J. Exp. Boil.* **2009**, *47*, 1001–1005.
61. Demilew, W.; Adinew, G.M.; Asrade, S. Evaluation of the wound healing activity of the crude extract of leaves of *Acanthus polystachyus* Delile (Acanthaceae). *Evid.-Based Complement. Altern. Med.* **2018**, *2018*, 1–9. [CrossRef]
62. Adetutu, A.; Morgan, W.A.; Corcoran, O. Antibacterial, antioxidant and fibroblast growth stimulation activity of crude extracts of *Bridelia ferruginea* leaf, a wound-healing plant of Nigeria. *J. Ethnopharmacol.* **2011**, *133*, 116–119. [CrossRef]
63. Shukla, A.; Rasik, A.; Jain, G.; Shankar, R.; Kulshrestha, D.; Dhawan, B. In vitro and in vivo wound healing activity of asiaticoside isolated from *Centella asiatica*. *J. Ethnopharmacol.* **1999**, *65*, 1–11. [CrossRef]
64. Chen, Y.J.; Dai, Y.S.; Chen, B.F.; Chang, A.; Chen, H.C.; Lin, Y.C.; Chang, K.H.; Lai, Y.L.; Chung, C.H.; Lai, Y.J. The effect of tetrandrine and extracts of *Centella asiatica* on acute radiation dermatitis in rats. *Biol. Pharm. Bull.* **1999**, *22*, 703–706. [CrossRef]
65. Gouthamchandra, K.; Mahmood, R.; Manjunatha, H. Free radical scavenging, antioxidant enzymes and wound healing activities of leaves extracts from *Clerodendrum infortunatum* L. *Environ. Toxicol. Pharmacol.* **2010**, *30*, 11–18. [CrossRef]
66. Agyare, C.; Asase, A.; Lechtenberg, M.; Niehues, M.; Deters, A.; Hensel, A. An ethnopharmacological survey and in vitro confirmation of ethnopharmacological use of medicinal plants used for wound healing in Bosomtwi-Atwima-Kwanwoma area, Ghana. *J. Ethnopharmacol.* **2009**, *125*, 393–403. [CrossRef]
67. Koca, U.; Süntar, I.; Akkol, E.K.; Yılmazer, D.; Alper, M. Wound repair potential of *Olea europaea* L. leaf extracts revealed by in vivo experimental models and comparative evaluation of the extracts' antioxidant activity. *J. Med. Food* **2011**, *14*, 140–146. [CrossRef] [PubMed]
68. Arawande, J.O.; Komolafe, E.A.; Imokhuede, B. Nutritional and phytochemical compositions of fireweed (*Crassocephalum crepidioides*). *J. Agric. Technol.* **2013**, *9*, 439–449.
69. Mathew, S.; Abraham, T.E. In vitro antioxidant activity and scavenging effects of *Cinnamomum verum* leaf extract assayed by different methodologies. *Food Chem. Toxicol.* **2006**, *44*, 198–206. [CrossRef]
70. Atun, S.; Handayani, S.; Rakhmawati, A.; Purnamaningsih, N.A.; Naila, B.I.; Lestari, A. Study of potential phenolic compounds from stems of *Dendrophthoe falcata* (Loranthaceae) plant as antioxidant and antimicrobial agents. *Orient. J. Chem.* **2018**, *34*, 2342–2349. [CrossRef]
71. Velmurugan, C.; Geetha, C.; Shajahan, S.; Vijayakumar, S.; Kumar, P.L. Wound healing potential of leaves of *Eucalyptus citriodoral* in rats. *World J. Pharm. Sci.* **2014**, *2*, 62–71.
72. Annan, K.; Houghton, P.J. Antibacterial, antioxidant and fibroblast growth stimulation of aqueous extracts of *Ficus asperifolia* Miq. and *Gossypium arboreum* L., wound-healing plants of Ghana. *J. Ethnopharmacol.* **2008**, *119*, 141–144. [CrossRef]
73. Steenkamp, V.; Mathivha, E.; Gouws, M.; van Rensburg, C.J. Studies on antibacterial, antioxidant and fibroblast growth stimulation of wound healing remedies from South Africa. *J. Ethnopharmacol.* **2004**, *95*, 353–357. [CrossRef] [PubMed]
74. Upadhyay, N.; Kumar, R.; Mandotra, S.; Meena, R.; Siddiqui, M.; Sawhney, R.; Gupta, A. Safety and healing efficacy of Sea buckthorn (*Hippophae rhamnoides* L.) seed oil on burn wounds in rats. *Food Chem. Toxicol.* **2009**, *47*, 1146–1153. [CrossRef] [PubMed]
75. Reddy, B.S.; Reddy, R.K.K.; Naidu, V.; Madhusudhana, K.; Agwane, S.B.; Ramakrishna, S.; Diwan, P.V. Evaluation of antimicrobial, antioxidant and wound-healing potentials of *Holoptelea integrifolia*. *J. Ethnopharmacol.* **2008**, *115*, 249–256. [CrossRef] [PubMed]
76. Nualkaew, S.; Rattanamanee, K.; Thongpraditchote, S.; Wongkrajang, Y.; Nahrstedt, A. Anti-inflammatory, analgesic and wound healing activities of the leaves of *Memecylon edule* Roxb. *J. Ethnopharmacol.* **2009**, *121*, 278–281. [CrossRef] [PubMed]
77. Muhammad, A.A.; Pauzi, N.A.S.; Arulselvan, P.; Abas, F.; Fakurazi, S. In vitro wound healing potential and identification of bioactive compounds from *Moringa oleifera* Lam. *BioMed Res. Int.* **2013**, *2013*, 974580. [CrossRef]
78. Jain, S.; Shrivastava, S.; Nayak, S.; Sumbhate, S. Recent trends in *Curcuma longa* Linn. *Pharmacogn. Rev.* **2007**, *1*, 119–128.
79. Singh, M.; Govindarajan, R.; Nath, V.; Rawat, A.K.S.; Mehrotra, S. Antimicrobial, wound healing and antioxidant activity of *Plagiochasma appendiculatum* Lehm. et Lind. *J. Ethnopharmacol.* **2006**, *107*, 67–72. [CrossRef]
80. Berenguer, B.; Sánchez, L.; Quílez, A.; López-Barreiro, M.; de Haro, O.; Gálvez, J.; Martín, M. Protective and antioxidant effects of *Rhizophora mangle* L. against NSAID-induced gastric ulcers. *J. Ethnopharmacol.* **2006**, *103*, 194–200. [CrossRef]
81. Taranalli, A.D.; Kuppast, I.J. Study of wound healing activity of seeds of *Trigonella foenum* graecum in rats. *Indian J. Pharm. Sci.* **1996**, *58*, 117.
82. Jadhav, N.R.; Powar, T.; Shinde, S.; Nadaf, S. Herbal nanoparticles: A patent review. *Asian J. Pharm.* **2014**, *8*, 1–12. [CrossRef]
83. Hromadka, M.; Collins, J.B.; Reed, C.; Han, L.; Kolappa, K.K.; Cairns, B.A.; Andrady, T.; van Aalst, J.A. Nanofiber applications for burn care. *J. Burn Care Res.* **2008**, *29*, 695–703. [CrossRef]

84. Cerchiara, T.; Abruzzo, A.; Palomino, R.A.Ñ.; Vitali, B.; De Rose, R.; Chidichimo, G.; Ceseracciu, L.; Athanassiou, A.; Saladini, B.; Dalena, F.; et al. Spanish Broom (*Spartium junceum* L.) fibers impregnated with vancomycin-loaded chitosan nanoparticles as new antibacterial wound dressing: Preparation, characterization and antibacterial activity. *Eur. J. Pharm. Sci.* **2017**, *99*, 105–112. [CrossRef] [PubMed]
85. Tang, T.; Yin, L.; Yang, J.; Shan, G. Emodin, an anthraquinone derivative from *Rheum officinale* Baill, enhances cutaneous wound healing in rats. *Eur. J. Pharmacol.* **2007**, *567*, 177–185. [CrossRef] [PubMed]
86. Dai, X.Y.; Nie, W.; Wang, Y.C.; Shen, Y.; Li, Y.; Gan, S.J. Electrospun emodin polyvinylpyrrolidone blended nanofibrous membrane: A novel medicated biomaterial for drug delivery and accelerated wound healing. *J. Mater. Sci. Mater. Med.* **2012**, *23*, 2709–2716. [CrossRef]
87. Panichpakdee, J.; Pavasant, P.; Supaphol, P. Electrospun cellulose acetate fiber mats containing emodin with potential for use as wound dressing. *Chiang Mai J. Sci.* **2016**, *43*, 1249–1259.
88. Suwantong, O.; Ruktanonchai, U.; Supaphol, P. In vitro biological evaluation of electrospun cellulose acetate fiber mats containing asiaticoside or curcumin. *J. Biomed. Mater. Res. Part A* **2010**, *94*, 1216–1225.
89. Panichpakdee, J.; Pavasant, P.; Supaphol, P. Electrospinning of asiaticoside/2-hydroxypropyl-β-cyclodextrin inclusion complex-loaded cellulose acetate fiber mats: Release characteristics and potential for use as wound dressing. *Polym. Korea* **2014**, *38*, 338–350. [CrossRef]
90. Suwantong, O.; Ruktanonchai, U.; Supaphol, P. Electrospun cellulose acetate fiber mats containing asiaticoside or *Centella asiatica* crude extract and the release characteristics of asiaticoside. *Polymer* **2008**, *49*, 4239–4247. [CrossRef]
91. Momtazi, A.A.; Haftcheshmeh, S.M.; Esmaeili, S.-A.; Johnston, T.P.; Abdollahi, E.; Sahebkar, A. Curcumin: A natural modulator of immune cells in systemic lupus erythematosus. *Autoimmun. Rev.* **2018**, *17*, 125–135. [CrossRef]
92. Prasad, R.; Kumar, D.; Kant, V.; Tandan, S.K.; Kumar, D. Curcumin enhanced cutaneous wound healing by modulating cytokines and transforming growth factor in excision wound model in rats. *Int. J. Curr. Microbiol. Appl. Sci.* **2017**, *6*, 2263–2273. [CrossRef]
93. Sidhu, G.S.; Singh, A.K.; Thaloor, D.; Banaudha, K.K.; Patnaik, G.K.; Srimal, R.C.; Maheshwari, R.K. Enhancement of wound healing by curcumin in animals. *Wound Repair Regen.* **1998**, *6*, 167–177. [CrossRef] [PubMed]
94. Liakos, I.; Rizzello, L.; Hajiali, H.; Brunetti, V.; Carzino, R.; Pompa, P.P.; Athanassiou, A.; Mele, E. Fibrous wound dressings encapsulating essential oils as natural antimicrobial agents. *J. Mater. Chem. B* **2015**, *3*, 1583–1589. [CrossRef]
95. Huang, S.; Fu, X. Naturally derived materials-based cell and drug delivery systems in skin regeneration. *J. Control. Release* **2010**, *142*, 149–159. [CrossRef]
96. Tokuda, M.; Yamane, M.; Thickett, S.C.; Minami, H.; Zetterlund, P.B. Synthesis of polymeric nanoparticles containing reduced graphene oxide nanosheets stabilized by poly(ionic liquid) using miniemulsion polymerization. *Soft Matter* **2016**, *12*, 3955–3962. [CrossRef]
97. Yun, Y.H.; Goetz, D.J.; Yellen, P.; Chen, W. Hyaluronan microspheres for sustained gene delivery and site-specific targeting. *Biomaterials* **2003**, *25*, 147–157. [CrossRef]
98. Korrapati, P.S.; Karthikeyan, K.; Satish, A.; Krishnaswamy, V.R.; Venugopal, J.R.; Ramakrishna, S. Recent advancements in nanotechnological strategies in selection, design and delivery of biomolecules for skin regeneration. *Mater. Sci. Eng. C* **2016**, *67*, 747–765. [CrossRef]
99. Gardner, J.C.; Wu, H.; Noel, J.G.; Ramser, B.J.; Pitstick, L.; Saito, A.; Nikolaidis, N.M.; McCormack, F.X. Keratinocyte growth factor supports pulmonary innate immune defense through maintenance of alveolar antimicrobial protein levels and macrophage function. *Am. J. Physiol. Cell. Mol. Physiol.* **2016**, *310*, L868–L879. [CrossRef] [PubMed]
100. Feng, Z.G.; Pang, S.F.; Guo, D.J.; Yang, Y.T.; Liu, B.; Wang, J.W.; Zheng, K.Q.; Lin, Y. Recombinant keratinocyte growth factor 1 in tobacco potentially promotes wound healing in diabetic rats. *BioMed Res. Int.* **2014**, *2014*, 1–9. [CrossRef]
101. Koria, P.; Yagi, H.; Kitagawa, Y.; Megeed, Z.; Nahmias, Y.; Sheridan, R.; Yarmush, M.L. Self-assembling elastin-like peptides growth factor chimeric nanoparticles for the treatment of chronic wounds. *Proc. Natl. Acad. Sci. USA* **2010**, *108*, 1034–1039. [CrossRef]
102. Ye, M.; Kim, S.; Park, K. Issues in long-term protein delivery using biodegradable microparticles. *J. Control. Release* **2010**, *146*, 241–260. [CrossRef] [PubMed]
103. Chereddy, K.K.; Vandermeulen, G.; Préat, V. PLGA based drug delivery systems: Promising carriers for wound healing activity. *Wound Repair Regen.* **2016**, *24*, 223–236. [CrossRef]
104. Zhang, Y.; Wischke, C.; Mittal, S.; Mitra, A.; Schwendeman, S.P. Design of controlled release PLGA microspheres for hydrophobic fenretinide. *Mol. Pharm.* **2016**, *13*, 2622–2630.
105. Yüksel, E.; Karakeçili, A.; Demirtaş, T.T.; Gümüşderelioğlu, M. Preparation of bioactive and antimicrobial PLGA membranes by magainin II/EGF functionalization. *Int. J. Biol. Macromol.* **2016**, *86*, 162–168.
106. Chereddy, K.K.; Her, C.-H.; Comune, M.; Moia, C.; Lopes, A.; Porporato, P.E.; Vanacker, J.; Lam, M.C.; Steinstraesser, L.; Sonveaux, P.; et al. PLGA nanoparticles loaded with host defense peptide LL37 promote wound healing. *J. Control. Release* **2014**, *194*, 138–147. [PubMed]
107. Matica, M.A.; Aachmann, F.L.; Tøndervik, A.; Sletta, H.; Ostafe, V. Chitosan as a wound dressing starting material: Antimicrobial properties and mode of action. *Int. J. Mol. Sci.* **2019**, *20*, 5889. [CrossRef]
108. Dai, T.; Tegos, G.P.; Burkatovskaya, M.; Castano, A.P.; Hamblin, M.R. Chitosan acetate bandage as a topical antimicrobial dressing for infected burns. *Antimicrob. Agents Chemother.* **2009**, *53*, 393–400. [CrossRef] [PubMed]

109. Karimi, M.; Zangabad, P.S.; Ghasemi, A.; Amiri, M.; Bahrami, M.; Malekzad, H.; Asl, H.G.; Mahdieh, Z.; Bozorgomid, M.; Ghasemi, A.; et al. Temperature-responsive smart nanocarriers for delivery of therapeutic agents: Applications and recent advances. *ACS Appl. Mater. Interfaces* **2016**, *8*, 21107–21133. [CrossRef] [PubMed]
110. Shrestha, A.; Hamblin, M.R.; Kishen, A. Characterization of a conjugate between Rose Bengal and chitosan for targeted antibiofilm and tissue stabilization effects as a potential treatment of infected dentin. *Antimicrob. Agents Chemother.* **2012**, *56*, 4876–4884. [CrossRef]
111. Dai, T.; Tanaka, M.; Huang, Y.; Hamblin, M.R. Chitosan preparations for wounds and burns: Antimicrobial and wound-healing effects. *Expert Rev. Anti-Infective Ther.* **2011**, *9*, 857–879. [CrossRef]
112. Holban, A.M.; Grumezescu, V.; Grumezescu, A.M.; Vasile, B.; Truşcă, R.; Cristescu, R.; Socol, G.; Iordache, F. Antimicrobial nanospheres thin coatings prepared by advanced pulsed laser technique. *Beilstein J. Nanotechnol.* **2014**, *5*, 872–880. [CrossRef]
113. Baxter, R.M.; Dai, T.; Kimball, J.; Wang, E.; Hamblin, M.R.; Wiesmann, W.P.; McCarthy, S.J.; Baker, S.M. Chitosan dressing promotes healing in third degree burns in mice: Gene expression analysis shows biphasic effects for rapid tissue regeneration and decreased fibrotic signaling. *J. Biomed. Mater. Res. Part A* **2013**, *101*, 340–348. [CrossRef] [PubMed]
114. Karimi, M.; Avci, P.; Ahi, M.; Gazori, T.; Hamblin, M.R.; Naderi-Manesh, H. Evaluation of chitosan-tripolyphosphate nanoparticles as a p-shRNA delivery vector: Formulation, optimization and cellular uptake study. *J. Nanopharm. Drug Deliv.* **2013**, *1*, 266–278. [CrossRef] [PubMed]
115. Abbasi, E.; Aval, S.F.; Akbarzadeh, A.; Milani, M.; Nasrabadi, H.T.; Joo, S.W.; Hanifehpour, Y.; Nejati-Koshki, K.; Pashaei-Asl, R. Dendrimers: Synthesis, applications, and properties. *Nanoscale Res. Lett.* **2014**, *9*, 247. [CrossRef]
116. Kalomiraki, M.; Thermos, K.; Chaniotakis, N.A. Dendrimers as tunable vectors of drug delivery systems and biomedical and ocular applications. *Int. J. Nanomed.* **2016**, *11*, 1.
117. Nusbaum, A.G.; Gil, J.; Rippy, M.K.; Warne, B.; Valdes, J.; Claro, A.; Davis, S.C. Effective method to remove wound bacteria: Comparison of various debridement modalities in an in vivo porcine model. *J. Surg. Res.* **2012**, *176*, 701–707. [CrossRef] [PubMed]
118. Kumar, P.S.; Raj, N.M.; Praveen, G.; Chennazhi, K.P.; Nair, S.V.; Jayakumar, R. In vitro and in vivo evaluation of microporous chitosan hydrogel/nanofibrin composite bandage for skin tissue regeneration. *Tissue Eng. Part A* **2013**, *19*, 380–392. [CrossRef] [PubMed]
119. El-Rafie, H.; El-Rafie, M.; Zahran, M. Green synthesis of silver nanoparticles using polysaccharides extracted from marine macro algae. *Carbohydr. Polym.* **2013**, *96*, 403–410. [CrossRef]
120. Rajasekar, P.; Palanisamy, S.; Anjali, R.; Vinosha, M.; Thillaieswari, M.; Malaikozhundan, B.; Boomi, P.; Saravanan, M.; You, S.; Prabhu, N.M. *Cladophora fascicularis* Mediated Silver Nanoparticles: Assessment of Their Antibacterial Activity Against *Aeromonas hydrophila*. *J. Clust. Sci.* **2019**, *31*, 673–683. [CrossRef]
121. Joseph, S.; Mathew, B. Microwave assisted facile green synthesis of silver and gold nanocatalysts using the leaf extract of *Aerva lanata*. *Spectrochim. Acta Part A Mol. Biomol. Spectrosc.* **2015**, *136*, 1371–1379. [CrossRef]
122. Nasrollahzadeh, M.; Sajadi, S.M.; Maham, M. Green synthesis of palladium nanoparticles using *Hippophae rhamnoides* Linn leaf extract and their catalytic activity for the Suzuki–Miyaura coupling in water. *J. Mol. Catal. A Chem.* **2015**, *396*, 297–303. [CrossRef]
123. Guo, M.; Li, W.; Yang, F.; Liu, H. Controllable biosynthesis of gold nanoparticles from a *Eucommia ulmoides* bark aqueous extract. *Spectrochim. Acta Part A Mol. Biomol. Spectrosc.* **2015**, *142*, 73–79. [CrossRef]
124. Begum, N.; Mondal, S.; Basu, S.; Laskar, R.A.; Mandal, D. Biogenic synthesis of Au and Ag nanoparticles using aqueous solutions of Black Tea leaf extracts. *Colloids Surf. B Biointerfaces* **2009**, *71*, 113–118. [CrossRef]
125. Isaac, R.; Sakthivel, G.; Murthy, C. Green synthesis of gold and silver nanoparticles using *Averrhoa bilimbi* fruit extract. *J. Nanotechnol.* **2013**, *2013*, 906592. [CrossRef]
126. Ahmed, K.B.A.; Subramanian, S.; Sivasubramanian, A.; Veerappan, G.; Veerappan, A. Preparation of gold nanoparticles using *Salicornia brachiata* plant extract and evaluation of catalytic and antibacterial activity. *Spectrochim. Acta Part A Mol. Biomol. Spectrosc.* **2014**, *130*, 54–58. [CrossRef] [PubMed]
127. Mollick, M.R.; Bhowmick, B.; Mondal, D.; Maity, D.; Rana, D.; Dash, S.K.; Chattopadhyay, S.; Roy, S.; Sarkar, J.; Acharya, K.; et al. Anticancer (in vitro) and antimicrobial effect of gold nanoparticles synthesized using *Abelmoschus esculentus* (L.) pulp extract via a green route. *RSC Adv.* **2014**, *4*, 37838–37848. [CrossRef]
128. Khalil, M.M.H.; Ismail, E.H.; El-Magdoub, F. Biosynthesis of Au nanoparticles using olive leaf extract. *Arab. J. Chem.* **2012**, *5*, 431–437. [CrossRef]
129. Abbasi, T.; Anuradha, J.; Ganaie, S.; Abbasi, S. Gainful utilization of the highly intransigent weed ipomoea in the synthesis of gold nanoparticles. *J. King Saud Univ.-Sci.* **2015**, *27*, 15–22. [CrossRef]
130. Franco-Romano, M.; Gil, M.; Palacios-Santander, J.; Delgado-Jaén, J.; Naranjo-Rodríguez, I.; de Cisneros, J.H.-H.; Cubillana-Aguilera, L. Sonosynthesis of gold nanoparticles from a geranium leaf extract. *Ultrason. Sonochemistry* **2014**, *21*, 1570–1577. [CrossRef] [PubMed]
131. Rajeshkumar, S.; Menon, S.; Kumar, S.V.; Tambuwala, M.M.; Bakshi, H.A.; Mehta, M.; Satija, S.; Gupta, G.; Chellappan, D.K.; Thangavelu, L.; et al. Antibacterial and antioxidant potential of biosynthesized copper nanoparticles mediated through *Cissus arnotiana* plant extract. *J. Photochem. Photobiol. B Biol.* **2019**, *197*, 111531. [CrossRef] [PubMed]
132. Jain, P.; Pradeep, T. Potential of silver nanoparticle-coated polyurethane foam as an antibacterial water filter. *Biotechnol. Bioeng.* **2005**, *90*, 59–63. [CrossRef] [PubMed]

133. Zhou, Y.; Chen, R.; He, T.; Xu, K.; Du, D.; Zhao, N.; Cheng, X.; Yang, J.; Shi, H.; Lin, Y. Biomedical potential of ultrafine Ag/AgCl nanoparticles coated on graphene with special reference to antimicrobial performances and burn wound healing. *ACS Appl. Mater. Interfaces* **2016**, *8*, 15067–15075. [CrossRef]
134. Dunn, K.; Edwards-Jones, V. The role of Acticoat™ with nanocrystalline silver in the management of burns. *Burns* **2004**, *30*, S1–S9. [CrossRef]
135. Thomas, R.; Soumya, K.R.; Mathew, J.; Radhakrishnan, E.K. *Electrospun polycaprolactone* membrane incorporated with biosynthesized silver nanoparticles as effective wound dressing material. *Appl. Biochem. Biotechnol.* **2015**, *176*, 2213–2224. [CrossRef]
136. Rivas, L.; Sanchez-Cortes, S.; Garcia-Ramos, J.; Morcillo, G. Growth of silver colloidal particles obtained by citrate reduction to increase the Raman enhancement factor. *Langmuir* **2001**, *17*, 574–577. [CrossRef]
137. Srinivas Reddy, K.; Reddy, C.S.; Sanjeeva Kumar, A. Antimicrobial potential of Cassia roxburghii leaves. *J. Pharm. Res.* **2011**, *4*, 4278–4279.
138. Balakumaran, M.; Ramachandran, R.; Balashanmugam, P.; Mukeshkumar, D.; Kalaichelvan, P. Mycosynthesis of silver and gold nanoparticles: Optimization, characterization and antimicrobial activity against human pathogens. *Microbiol. Res.* **2015**, *182*, 8–20. [CrossRef]
139. Krychowiak, M.; Grinholc, M.; Banasiuk, R.; Krauze-Baranowska, M.; Głód, D.; Kawiak, A.; Królicka, A. Combination of silver nanoparticles and *Drosera binata* extract as a possible alternative for antibiotic treatment of burn wound infections caused by resistant *Staphylococcus aureus*. *PLoS ONE* **2014**, *9*, e115727.
140. Naskar, A.; Bera, S.; Bhattacharya, R.; Roy, S.S.; Jana, S. Effect of bovine serum albumin immobilized Au–ZnO–graphene nanocomposite on human ovarian cancer cell. *J. Alloy. Compd.* **2018**, *734*, 66–74. [CrossRef]
141. Niska, K.; Zielinska, E.; Radomski, M.W.; Inkielewicz-Stepniak, I. Metal nanoparticles in dermatology and cosmetology: Interactions with human skin cells. *Chem.-Biol. Interact.* **2018**, *295*, 38–51. [CrossRef]
142. Rajendran, N.K.; Kumar, S.S.D.; Houreld, N.N.; Abrahamse, H. A review on nanoparticle based treatment for wound healing. *J. Drug Deliv. Sci. Technol.* **2018**, *44*, 421–430. [CrossRef]
143. Vijayakumar, V.; Samal, S.K.; Mohanty, S.; Nayak, S.K. Recent advancements in biopolymer and metal nanoparticle-based materials in diabetic wound healing management. *Int. J. Biol. Macromol.* **2019**, *122*, 137–148. [CrossRef] [PubMed]
144. Castangia, I.; Marongiu, F.; Manca, M.L.; Pompei, R.; Angius, F.; Ardu, A.; Fadda, A.M.; Manconi, M.; Ennas, G. Combination of grape extract-silver nanoparticles and liposomes: A totally green approach. *Eur. J. Pharm. Sci.* **2017**, *97*, 62–69. [CrossRef]
145. Akturk, O.; Kismet, K.; Yasti, A.C.; Kuru, S.; Duymus, M.E.; Kaya, F.; Caydere, M.; Hucumenoglu, S.; Keskin, D. Collagen/gold nanoparticle nanocomposites: A potential skin wound healing biomaterial. *J. Biomater. Appl.* **2016**, *31*, 283–301. [CrossRef]
146. Volkova, N.; Yukhta, M.; Pavlovich, O.; Goltsev, A. Application of cryopreserved fibroblast culture with Au nanoparticles to treat burns. *Nanoscale Res. Lett.* **2016**, *11*, 22. [CrossRef] [PubMed]
147. Arunachalam, K.D.; Annamalai, S.K.; Arunachalam, A.M.; Kennedy, S. Green Synthesis of Crystalline Silver Nanoparticles Using *Indigofera aspalathoides*-Medicinal Plant Extract for Wound Healing Applications. *Asian J. Chem.* **2013**, *25*, S311–S314.
148. Dhayalan, M.; Denison, M.I.J.; Ayyar, M.; Gandhi, N.N.; Krishnan, K.; Abdulhadi, B. Biogenic synthesis, characterization of gold and silver nanoparticles from *Coleus forskohlii* and their clinical importance. *J. Photochem. Photobiol. B Biol.* **2018**, *183*, 251–257. [CrossRef]
149. Sivaranjani, V.; Philominathan, P. Synthesize of Titanium dioxide nanoparticles using *Moringa oleifera* leaves and evaluation of wound healing activity. *Wound Med.* **2016**, *12*, 1–5. [CrossRef]
150. Bhattacharya, M.; Malinen, M.M.; Laurén, P.; Lou, Y.-R.; Kuisma, S.W.; Kanninen, L.; Lille, M.; Corlu, A.; GuGuen-Guillouzo, C.; Ikkala, O.; et al. Nanofibrillar cellulose hydrogel promotes three-dimensional liver cell culture. *J. Control. Release* **2012**, *164*, 291–298. [CrossRef] [PubMed]
151. Manconi, M.; Manca, M.L.; Caddeo, C.; Cencetti, C.; di Meo, C.; Zoratto, N.; Nacher, A.; Fadda, A.M.; Matricardi, P. Preparation of gellan-cholesterol nanohydrogels embedding baicalin and evaluation of their wound healing activity. *Eur. J. Pharm. Biopharm.* **2018**, *127*, 244–249. [CrossRef]
152. Lokhande, G.; Carrow, J.K.; Thakur, T.; Xavier, J.R.; Parani, M.; Bayless, K.J.; Gaharwar, A.K. Nanoengineered injectable hydrogels for wound healing application. *Acta Biomater.* **2018**, *70*, 35–47. [CrossRef]
153. Chen, J.; Cheng, N.; Li, J.; Wang, Y.; Guo, J.-X.; Chen, Z.-P.; Cai, B.-C.; Yang, T. Influence of lipid composition on the phase transition temperature of liposomes composed of both DPPC and HSPC. *Drug Dev. Ind. Pharm.* **2012**, *39*, 197–204. [CrossRef]
154. Manca, M.L.; Matricardi, P.; Cencetti, C.; Peris, J.E.; Melis, V.; Carbone, C.; Escribano, E.; Zaru, M.; Fadda, A.M.; Manconi, M. Combination of argan oil and phospholipids for the development of an effective liposome-like formulation able to improve skin hydration and allantoin dermal delivery. *Int. J. Pharm.* **2016**, *505*, 204–211. [CrossRef]
155. Zhao, Y.-Z.; Lu, C.-T.; Zhang, Y.; Xiao, J.; Zhao, Y.-P.; Tian, J.-L.; Xu, Y.-Y.; Feng, Z.-G.; Xu, C.-Y. Selection of high efficient transdermal lipid vesicle for curcumin skin delivery. *Int. J. Pharm.* **2013**, *454*, 302–309. [CrossRef]
156. El Maghraby, G.M.; Barry, B.W.; Williams, A.C. Liposomes and skin: From drug delivery to model membranes. *Eur. J. Pharm. Sci.* **2008**, *34*, 203–222. [CrossRef]
157. Xu, H.-L.; Chen, P.-P.; Zhuge, D.-L.; Zhu, Q.-Y.; Jin, B.-H.; Shen, B.-X.; Xiao, J.; Zhao, Y.-Z. Liposomes with Silk Fibroin Hydrogel Core to Stabilize bFGF and Promote the Wound Healing of Mice with Deep Second-Degree Scald. *Adv. Health Mater.* **2017**, *6*, 1700344. [CrossRef]

158. Nunes, P.S.; Rabelo, A.S.; de Souza, J.C.C.; Santana, B.V.; da Silva, T.M.M.; Serafini, M.R.; Menezes, P.D.P.; Lima, B.D.S.; Cardoso, J.; Alves, J.C.S.; et al. Gelatin-based membrane containing usnic acid-loaded liposome improves dermal burn healing in a porcine model. *Int. J. Pharm.* **2016**, *513*, 473–482. [CrossRef] [PubMed]
159. Jahromi, M.A.M.; Zangabad, P.S.; Basri, S.M.M.; Zangabad, K.S.; Ghamarypour, A.; Aref, A.R.; Karimi, M.; Hamblin, M.R. Nanomedicine and advanced technologies for burns: Preventing infection and facilitating wound healing. *Adv. Drug Deliv. Rev.* **2017**, *123*, 33–64. [CrossRef] [PubMed]
160. Urie, R.; Ghosh, D.; Ridha, I.; Rege, K. Inorganic Nanomaterials for Soft Tissue Repair and Regeneration. *Annu. Rev. Biomed. Eng.* **2018**, *20*, 353–374. [CrossRef]
161. Almeida, A.; Souto, E. Solid lipid nanoparticles as a drug delivery system for peptides and proteins. *Adv. Drug Deliv. Rev.* **2007**, *59*, 478–490. [CrossRef] [PubMed]
162. Gainza, G.; Pastor, M.; Aguirre, J.J.; Villullas, S.; Pedraz, J.L.; Hernandez, R.M.; Igartua, M. A novel strategy for the treatment of chronic wounds based on the topical administration of rhEGF-loaded lipid nanoparticles: In vitro bioactivity and in vivo effectiveness in healing-impaired db/db mice. *J. Control. Release* **2014**, *185*, 51–61. [CrossRef]
163. Fumakia, M.; Ho, E.A. Nanoparticles encapsulated with LL37 and serpin A1 promotes wound healing and synergistically enhances antibacterial activity. *Mol. Pharm.* **2016**, *13*, 2318–2331. [CrossRef] [PubMed]
164. Manca, M.; Manconi, M.; Meloni, M.; Marongiu, F.; Allaw, M.; Usach, I.; Peris, J.; Escribano-Ferrer, E.; Tuberoso, C.; Gutierrez, G.; et al. Nanotechnology for Natural Medicine: Formulation of Neem Oil Loaded Phospholipid Vesicles Modified with Argan Oil as a Strategy to Protect the Skin from Oxidative Stress and Promote Wound Healing. *Antioxidants* **2021**, *10*, 670. [CrossRef] [PubMed]
165. Allaw, M.; Pleguezuelos-Villa, M.; Manca, M.L.; Caddeo, C.; Aroffu, M.; Nacher, A.; Diez-Sales, O.; Saurí, A.R.; Ferrer, E.E.; Fadda, A.M.; et al. Innovative strategies to treat skin wounds with mangiferin: Fabrication of transfersomes modified with glycols and mucin. *Nanomedicine* **2020**, *15*, 1671–1685. [CrossRef]
166. Ranjbar-Mohammadi, M.; Rabbani, S.; Bahrami, H.; Joghataei, M.T.; Moayer, F. Antibacterial performance and in vivo diabetic wound healing of curcumin loaded gum tragacanth/poly(ε-caprolactone) electrospun nanofibers. *Mater. Sci. Eng. C* **2016**, *69*, 1183–1191. [CrossRef] [PubMed]
167. Bayat, S.; Amiri, N.; Pishavar, E.; Kalalinia, F.; Movaffagh, J.; Hashemi, M. Bromelain-loaded chitosan nanofibers prepared by electrospinning method for burn wound healing in animal models. *Life Sci.* **2019**, *229*, 57–66. [CrossRef]
168. García, A.D.P.; Cassini-Vieira, P.; Ribeiro, C.C.; Jensen, C.E.D.M.; Barcelos, L.S.; Cortes, M.E.; Sinisterra, R.D. Efficient cutaneous wound healing using bixin-loaded PCL nanofibers in diabetic mice. *J. Biomed. Mater. Res. Part B Appl. Biomater.* **2016**, *105*, 1938–1949. [CrossRef] [PubMed]
169. Ahn, S.; Ardoña, H.A.M.; Campbell, P.H.; Gonzalez, G.M.; Parker, K.K. Alfalfa nanofibers for dermal wound healing. *ACS Appl. Mater. Interfaces* **2019**, *11*, 33535–33547.

Review

Multifunctional Polymeric Nanogels for Biomedical Applications

Tisana Kaewruethai [1,2], Chavee Laomeephol [1,3], Yue Pan [4] and Jittima Amie Luckanagul [1,3,*]

1. Department of Pharmaceutics and Industrial Pharmacy, Faculty of Pharmaceutical Sciences, Chulalongkorn University, Phayathai Road, Bangkok 10330, Thailand; 6076126733@student.chula.ac.th (T.K.); papomchavee@gmail.com (C.L.)
2. Department of Biochemistry and Microbiology, Faculty of Pharmaceutical Sciences, Chulalongkorn University, Phayathai Road, Bangkok 10330, Thailand
3. Biomaterial Engineering for Medical and Health Research Unit, Chulalongkorn University, Phayathai Road, Bangkok 10330, Thailand
4. Guangdong Provincial Key Laboratory of Malignant Tumor Epigenetics and Gene Regulation, Medical Research Center, Sun Yat-Sen Memorial Hospital, Sun Yat-Sen University, Guangzhou 510120, China; panyue@mail.sysu.edu.cn
* Correspondence: Jittima.L@pharm.chula.ac.th; Tel.: +662-218-8400; Fax: +662-218-8401

Abstract: Currently, research in nanoparticles as a drug delivery system has broadened to include their use as a delivery system for bioactive substances and a diagnostic or theranostic system. Nanogels, nanoparticles containing a high amount of water, have gained attention due to their advantages of colloidal stability, core-shell structure, and adjustable structural components. These advantages provide the potential to design and fabricate multifunctional nanosystems for various biomedical applications. Modified or functionalized polymers and some metals are components that markedly enhance the features of the nanogels, such as tunable amphiphilicity, biocompatibility, stimuli-responsiveness, or sensing moieties, leading to specificity, stability, and tracking abilities. Here, we review the diverse designs of core-shell structure nanogels along with studies on the fabrication and demonstration of the responsiveness of nanogels to different stimuli, temperature, pH, reductive environment, or radiation. Furthermore, additional biomedical applications are presented to illustrate the versatility of the nanogels.

Keywords: polymeric nanogels; stimuli-responsive; functionalized polymer; core-shell nanogels

Citation: Kaewruethai, T.; Laomeephol, C.; Pan, Y.; Luckanagul, J.A. Multifunctional Polymeric Nanogels for Biomedical Applications. *Gels* **2021**, *7*, 228. https://doi.org/10.3390/gels7040228

Academic Editors: Chien-Chi Lin, Emanuele Mauri and Filippo Rossi

Received: 30 September 2021
Accepted: 13 November 2021
Published: 23 November 2021

Publisher's Note: MDPI stays neutral with regard to jurisdictional claims in published maps and institutional affiliations.

Copyright: © 2021 by the authors. Licensee MDPI, Basel, Switzerland. This article is an open access article distributed under the terms and conditions of the Creative Commons Attribution (CC BY) license (https://creativecommons.org/licenses/by/4.0/).

1. Introduction

Drug delivery systems (DDSs are platforms that protect loaded active ingredients from degradation due to physiological conditions, as well as allowing the payloads to perform their desired activities at the target sites. Furthermore, some chemicals and biological drugs are cytotoxic to some extent, which is referred to as an undesired off-target effect. DDSs, such as inorganic nanotubes [1], liposomes [2], and inorganic [3] or polymeric nanoparticles [4], have been developed to overcome these disadvantages. A polymeric nanogel is classified as containing small nanoparticles (ca. 20–200 nm) that is fabricated using hydrophilic polymers as the main component. The crosslinked networks of the polymeric chains encapsulate and enhance the colloidal stability of the loaded molecules, and the hydrophilicity of the polymers entraps large amounts of water, which facilitate the diffusion and mass-exchange with physiological milieu, resulting in a controlled or sustained release of the payload.

Most drugs exhibit low solubility and stability; polymer modification, i.e., a smart polymer, has been used to overcome these limitations. Smart polymers refer to a modification or functionalization of the polymers with side chains or ligands to adjust the polymer hydrophobicity [5–7]. Additionally, smart polymers can be designed to display distinct stimuli-responsive behaviors, e.g., thermo-, pH-, and redox-responsiveness [8–14]. These environmentally sensitive features have been extensively investigated because site-specific interactions of the DDS can increase specificity, targeted drug release [15–18], and diminish side effects [19,20]. Moreover, the improved properties of smart nanogels can be used as diagnostic or theranostic devices [21–24]. Smart polymers can be designed to specifically respond to various stimuli, light, radiation, an electric field, or temperature, which could be advantageous for target-specific applications [25–27]. Therefore, smart polymeric nanogel systems are promising drug delivery platforms with specificity and safety that can be used in a wide range of biomedical applications, such as an effective therapy or an accurate diagnostic system.

2. Core-Shell Structure of Polymeric Nanogels

The core-shell structure of nanogels is composed of inner and outer layers that contribute to the functionality of the delivery system. The properties of the core or the inner compartment can be adjusted to protect a loaded substance from incompatible environments or provide a hydrophilic cavity for the hydrophilic therapeutic substances. Nanogels can be fabricated using two or more distinct polymers, and thus the ratio of hydrophilicity and hydrophobicity can be adjusted to meet the requirements of various types of substances. Moreover, the amphiphilicity of the nanogel can transport substances with a low aqueous solubility through the circulatory system. Thus, the inner layer is typically fabricated to enhance the capability of the nanogel to hold and/or stabilize the payload, while the outer compartment, the so-called shell, covers the core, and acts as a protective layer exposed to the surrounding environment. Furthermore, the shell can be modified with specialized functional groups for specific features, such as targetability, stimuli-responsiveness, colloidal stability, or increased retention time in the circulatory system.

In 2011, He F. et al. [28] developed multi-responsive semi-interpenetrating network (semi-IPN) hydrogels whose nanostructure comprised core-shell spherical nanoparticles. The stimuli-responsive semi-IPN hydrogels were modified by magnetic nanoparticles with Fe_3O_4 nanoparticles as a core structure together with the combination of poly-N-isopropylacrylamide (PNIPAM, thermo-responsive) and polyacrylic acid (PAA, pH-responsive). This study demonstrated the successful fabrication of multifunctional materials using core-shell structure nanoparticles [28]. Gonzalez-Urias A. et al. [29] introduced pH-sensitive core-shell nanoparticles using a poly (N,N-diethylaminoethyl methacrylate) (PDEAEMA; anionic polymer) or poly (2-methacryloyloxi benzoic acid) (P2MBA; cationic polymer) core shielded with polyethylene glycol (PEG) that can be used for treating cancer. These nanogels can specifically deliver cisplatin, a cancer drug, to a tumor by targeting its acidic environment. The results indicated that the charge of the inner layer influenced nanogel drug release, cell viability, and cell internalization. Additionally, core-shell nanogels composed of gold nanoparticles as a core and biodegradable chitosan as the outer segment were developed to enhance curcumin cytotoxicity to cancer cells [30]. The findings demonstrated an improved cell uptake of the nanogels in breast cancer cells, which resulted in increased toxicity to the cancer cells. This study revealed that polymer and metal materials could be combined to form core-shell nanoparticles for cancer therapy. The findings of these studies indicate that the core-shell structured nanoparticles can be developed from a wide variety of materials to modify the materials with characteristics specific to the required therapeutic properties of the nanogel. In this review, the nanogels systems were categorized into two groups based on the major components of the core or shell of the nanoparticles. Additionally, hollow sphere drug delivery systems with a functionalized shell and empty core are discussed. Illustrations of the three types of nanogels are presented in Figure 1A.

The homogeneity of the nanogel is critical for the stability of the system, and thus the nanogel macrostructure was frequently investigated using transmission electron microscope (TEM). Figure 1B illustrates hyaluronic acid-based nanogel's TEM images. Likewise, nanogel microstructure was also observed after freeze drying that demonstrated the polymer's porous structure and interconnected network [31]. The porosity of the microscale structure can increase the mass encapsulation of specific substances, such as oxygen or nutrients [32]. Moreover, the void space between the polymer network increases the diffusion of the nanogels compared with hydrogels [33].

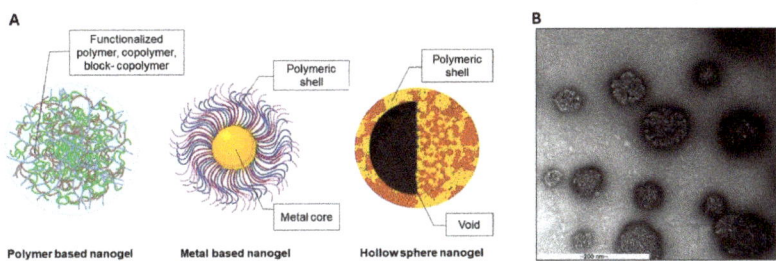

Figure 1. (**A**) Core-shell structures of nanogels (**B**) TEM image of hyaluronic acid-based polymeric nanogels (unpublished data).

2.1. Polymer Based Polymeric Nanogels

A nanogel is a colloidal system consisting of a crosslinked, water-swellable, 3-dimensional polymeric network whose size can reach ~1000 nm in a fully-swelled state [34]. The crosslinked network of a nanogel can swell or shrink based on external physical or chemical stimuli. This morphological reversibility leads to the use of nanogels as a smart drug carrier because the site-specific drug release is controllable. Furthermore, several materials possess unique characteristics that are beneficial for the functional design of nanogels. Hydrophobically modified polymers are used to cross the lipid bilayer structure of the cell membrane for an effective cell-targeted delivery system. Liechty WB. et al. [35] introduced the tunability of the hydrophobicity and polymer charge by incorporating hydrophobic moieties, tert-butyl methacrylate (TBMA) and 2-(tert-butylamino)ethyl methacrylate (TBAEMA), to the pH-responsive polymer, P(DEAEMA-g-PEGMA; PDET) [36].

The presence of cationic moieties can destabilize the integrity of the cell membrane, which possesses a negative charge surface, leading to increased cell internalization. In addition, amphiphilic polymers were shown to be internalized by fungal cells in Horvat S. et al. [37]. Thiol-functionalized poly (glycidol) nanogels were fabricated, and the amount of alkyl chain conjugated with the thiol groups was varied to optimize the amphiphilicity of the polymers. These nanogels demonstrated an enhanced antifungal effect of entrapped amphotericin B by lowering the minimal inhibitory concentration or MIC, and reduced biofilm formation after treatment.

A polymeric nanogel was used as a DDS for gene delivery. Costa D. et al. used a conjugated polyamine (spermine, protamine sulfate, or polyethylenimine) on an ethylene glycol diglycidyl ether (EGDE) backbone to encapsulate plasmid DNA. The polymeric nanogels facilitated the delivery of plasmid DNA and anti-cancer drugs, e.g., doxorubicin, epirubicin, and paclitaxel, by increasing the drug-loading capacity because the active substance chemically bonded with EGDE. Moreover, the EGDE crosslinking is broken by ultraviolet light, resulting in drug release [38].

A hollow sphere, a unique structure that contains an empty space surrounded by a polymer has also been reported as a smart delivery system, because the core can facilitate drug encapsulation and controlled release [39]. Together with the functionalized polymeric shell, hollow sphere nanogels can be site specific or stimuli-responsive. Many research groups have used the distinctive properties of a hollow sphere in biomedical applications because the inner void provides a large space for drug encapsulation and the shell thickness is tunable [40].

A combination of thermo-responsive polymers that were chosen to increase the advantages of the hollow sphere structure was described by Li G. et al. The thermosensitive polymeric hollow spheres were assembled using sodium alginate-graft-poly(N-isopropylacrylamide) (ALG-g-PNIPAM) and β-cyclodextrin (β-CD) for 5-fluorouracil

(5-FU) controlled release. The nanoparticles in the nanorods and coils were prepared using β-CD/PNIPAM and sodium alginate, respectively. The fabricated hollow sphere nanoparticles were expected to increase 5-FU loading. At temperatures above the lower critical solution temperature (LCST) of PNIPAM or in an acidic environment, loaded 5-FU release was enhanced due to the morphological change of the nanoparticles [39]. Furthermore, the characteristics of the hollow sphere can be optimized by modified shell polymers.

The fabrication of a hollow shell-shell nanocontainer composed of thermo-responsive PNIPAM and poly(N-isopropylmethacrylamide) or PNIPMAM as an inner and outer shell was reported by Schmid AJ. et al. [39]. An in silico study of the shrink-swell behavior of the inner shell demonstrated a controllable uptake and drug release, while the outer shell facilitated colloidal stability and maintains the hollow sphere's void. More recently, hollow sphere nanogels were used as a skin hydration and penetration enhancer. Osorio-Blanco ER. et al. fabricated a thermo-responsive hollow sphere nanocapsule using PNIPAM and PNIPMAM as shell polymers, which provided thermo-responsive behavior to the system. The ratio of the polymers was varied to optimize the volume phase transition temperature and size of the nanogels. After being triggered by heat, this nanocarrier demonstrated improved skin penetration because it could move through the stratum corneum to the viable epidermis. Interestingly, the nanocapsule exhibited the penetration of a high molecular weight Atto oxa12 (MW = 835 g/mol) into the viable epidermis, which indicated increased skin penetration compared with DMSO, which was used as the positive control [41].

Wang Z. et al. introduced gold hollow/nanoshells conjugated with small interference RNA (siRNA) designed to interfere with the expression of heat shock protein 70 (hsp70) that is produced during photothermal therapy (PTT) of a tumor and is associated with tumor resistance and tumor recurrence. In this study, siRNA was entrapped by the gold hollow/nanoshells that protected the siRNA from enzymatic degradation and effectively delivered it to the cell cytosol. Together with the light absorption and light-heat transformation potential of gold hollow/nanoshells, hsp70 was down-regulated, resulting in increased PTT efficacy [42]. An inorganic substance was also incorporated with hollow sphere nanogels. Han H. et al. used PS-PAA (polystyrene- poly(acrylic acid)) core-shell nanogels as a template for hollow silica nanoparticles. After removing the PS-PAA core, a pH-responsive silica hollow sphere was obtained and demonstrated successful pH-triggered drug release [43].

Functionalized polymers have also been fabricated using layer-by-layer techniques, the so-called nanofilms [44]. In these core-shell particles, nanofilms were used as a polymeric shell coated on particles or microspheres to facilitate multifunctionality [45,46], specificity [47], controlled release [48], or cytoprotective effects [49]. The various uses of the functionalized polymers represent their advantages as a smart drug delivery material, especially the tunability of the polymers that influence efficacy, specificity, or stimuli-responsiveness of the fabricated carriers.

2.2. Metal Based Polymeric Nanogels

Numerous studies have demonstrated wide applications of metal-based nanomaterials, such as catalysts or nanoconductors. The distinct features of metal-based materials, quantum dots and metallic hybrid nanogels, have resulted in their use in biomedical fields [50]. Several metallic substances are light-responsive or emit fluorescence when in an excited state. Hence, the nanogels loaded or conjugated with these materials are used as diagnostic devices, such as optical nanosensors, contrast imaging, and cellular tracking.

Wu W. et al. developed silver (Ag)/gold (Au) dual-metal nanoparticles coated with a hydrophobic polystyrene (PS) layer as an inner core and shieled with PEG. These bimetallic nanoparticles emitted intense fluorescence, which could be used for cell imaging. Moreover, near-infrared (NIR) light can be absorbed by these nanogels, leading to photothermal conversion that can be beneficial for drug release [51]. Au has photothermal conversion activity that can enhance the effectiveness of anti-cancer therapeutics [50], and the use of other metals were also reported.

Selenium (Se) was conjugated to sulfhydryl groups as Se-S to create reduction-responsive linkages. The linkages were designed to prepare a doxorubicin (DOX)-loaded superparamagnetic nanogel with reduction and pH-responsiveness for drug release. The release of DOX from the magnetic Se-S nanogels coated with alginate was enhanced in the high glutathione (GSH) tumor environment [52].

Nanoclusters embedded in nanogels were reported by Gao X. et al. [53]. Copper, as copper nanoclusters (CuNCs) with cysteine ligands, were incorporated into glycol chitosan to form nanocomposites. The modified nanoparticles were designed to detect physiological Zn^{2+}, because the level of Zn^{2+} in the human body impacts several biological mechanisms, such as growth and neurotransmission. These nanogels can be localized due to their increased photoluminescence intensity from the transition between the dispersed and aggregated state of CuNCs@GC in the presence of Zn^{2+}. This feature provides an interesting tool for live cell imaging.

Quantum dots or QDs exhibit outstanding properties, such as photo-luminescence with less dye fading compared with traditional dyes. This property makes QDs efficient tools for in vitro and in vivo imaging. Adipic acid dihydrazide-modified QDs were conjugated to the carboxyl groups of hyaluronic acid as an initial nanogel polymer. The obtained nanogels demonstrated an intense fluorescence signal after cell uptake, which was related to the affinity of the modified hyaluronic acid-based nanogels to CD44 receptors on the cell surface [54]. Colloidal semiconductor QDs display a unique physical property where the size of the particle is dependent on the optical properties owing to their electronic state limit [55].

In multiple QD studies, a QD-modified core–shell structure was designed to replace metals, which significantly reduced the cytotoxicity of the materials. QDs-chitosan core-shell nanogels as an "OFF"/"ON" biosensor were fabricated as a cancer cell targeted probe. This system was composed of Mn^{2+}-doped CdS/ZnS, which was a glutathione sensitive QD core, attached to dopamine as an intracellular simulation moiety and model drug. The complex core was coated with chitosan functionalized folic acid as a targeted ligand for cancer cells and fluorescein isothiocyanate (FITC). The interaction between this QD nanogel and cells that expressed different levels of folic acid was demonstrated along with being nontoxic up to 100 mg/L. [56]

Cancer has become a worldwide health issue because it exhibits complicated pathologies [57] and can metastasize through the blood to distal parts of the human body [58]. Various therapeutic substances and delivery systems were investigated to overcome the challenges of this disease. To estimate the disease severity and efficacy of cancer treatments, tracking nanogels were designed and evaluated in vitro and in vivo.

Tan L. et al. reported that ZnO nanoparticles decorated with 2,3-dimethylmaleic anhydride (DMMA) and loaded with DOX exhibited a synergistic cytotoxicity to cancer cells under acidic conditions. The ZnO QD system prolonged the retention time in the blood and an accumulated at the tumor that improved the efficacy of the loaded chemotherapeutics [59]. Notably, in Liu Y. et al., the tumor-suppressor gene p53 was effectively delivered using dextran-QD nanogel to elevate its stability and imaging property. The dextran-QD nanogels were functionalized with polycation PGEA (ethanolamine-functionalized poly (glycidyl methacrylate)), and this nanoparticle demonstrated improved gene transfection efficiency and real-time in vivo fluorescent imaging [60].

QDs can also be modified by a stimuli-responsive polymer as described in Gui R. et al. In this study, nanospheres of QDs-embedded mesoporous silica nanoparticles (Q-MS) coated with chitosan-functionalized PNIPAM or PNIPAM-g-CS shell were fabricated. The results suggested the potential of nanospheres as a cell imaging agent in Hep2 cells along with the thermo-responsive behavior of PNIPAM, which can further facilitate the specificity to particular target sites [61].

3. Stimuli-Responsive Nanogels

Over the past decade, various smart polymers have been introduced as drug carriers with increased effectiveness, specificity, and fewer side effects. Improving polymers by incorporating proteins or other polymers can enhance specific characteristics and overcome some limitations of the base polymers. Furthermore, polymer modification can increase site specificity or cell internalization that can result in increased therapeutic effects.

Arteche Pujana M. et al. demonstrated a biodegradable crosslinker, genipin, grafted on chitosan to modify the water solubility of a chitosan derivative by minimizing its hydrogen bonding. Compared with the unconjugated chitosan, the modified structure exhibited improved water solubility and was pH-responsive that can be used for pH-triggered drug release [62]. The elevated transport of tenofovir across vaginal epithelium using an increased ratio of the unionized/ionized forms of tenofovir after encapsulated in nanoparticles composed of a combination of poly (lactic-co-glycolic acid) and methacrylic acid was demonstrated by Zhang T. et al. [63] The combination of two or more polymers resulted in improved material properties with more efficient drug entrapment and increased site-specificity.

Stimuli-responsive behavior is an important feature for biomedical applications. As presented in Figure 2, various smart polymers respond to diverse stimuli such as pH, temperature, enzymes, light, and chemicals. Several studies illustrated the usefulness of modified polymers where their responsiveness provided improved functions and enhanced delivery system specificity.

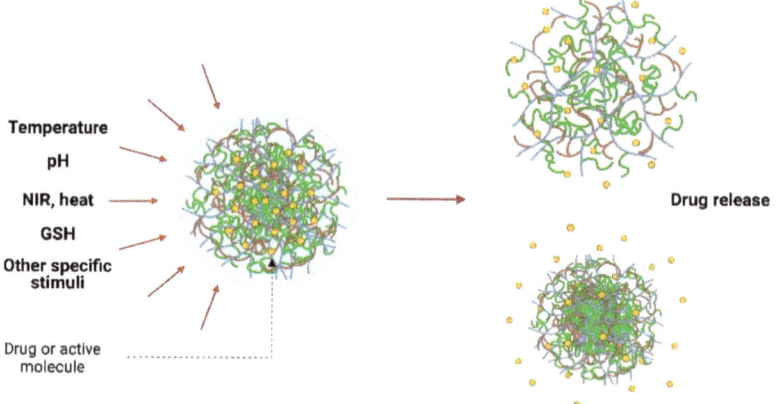

Figure 2. Shrink-swell behavior of nanogels after stimulation by specific stimuli resulting in targeted drug release.

3.1. Thermo-Responsive Nanogels

Temperature is a stimulus that has been investigated in stimuli-responsive drug carriers because most therapeutic substances have been designed to be activated at physiological temperature. Moreover, some diseases, e.g., cancers, have specific temperatures that can be exploited for targeted therapies.

Luckanagul J.A. et al. [64] reported the conjugation of a thermo-responsive polymer, PNIPAM, to chitosan that was used as a base material to form a drug carrier to enhance the aqueous solubility of a hydrophobic drug, curcumin. PNIPAM degrades too slowly, therefore conjugation with chitosan improves is biocompatibility due to improved biodegradability and reduced PNIPAM toxicity. This study also investigated the thermo-responsiveness of the modified polymer. The results demonstrated that the hydrodynamic size increased as the temperature increased and the thermo-responsiveness of the nanogels depended on the length and the grafting density. The cytotoxicity of the curcumin-loaded nanogels to cancer cells, MDA-MB-231, Caco-2, HepG2, and HT-29 cells, was evaluated and the results suggested the potential therapeutic effects of the nanogels in several cancer types [64].

The thermo-responsiveness of PNIPAM-modified nanogels is finely tunable using β-CD as reported by Yi P. et al. [65]. The result indicated increased thermo-sensitivity of the polymers, which resulted from linking β-CD and the isopropyl group of PNIPAM. The thermo-responsiveness of the nanogel was investigated based on the in vitro DOX release at 25, 37, and 40 °C, and faster release was observed at 40 °C. Moreover, the accelerated release of DOX in acidic pH and high reducing environment was reported, suggesting the use of the nanogels in cancer therapy.

In addition to PNIPAM, 2-(2-methoxyethoxy)ethyl methacrylate (MEO_2MA), a thermo-responsive monomer, was incorporated into silver nanoparticles (AgNPs) generating metal-based nanoparticles that were thermo-responsive. Soto-Quintero A. et al. [66] used a one-pot synthesis based on the reducing property of the phenolic moieties of curcumin to induce the formation of MEO_2MA-conjugated AgNPs and the interfacial energy reduction between MEO_2MA monomer and the metal surface. The curcumin-incorporated AgNPs exhibited antioxidant and thermo-responsive activities. The metallic core enhanced the capacity to encapsulate the hydrophobic curcumin and provided luminescence that can be beneficial for diagnostic and therapeutic purposes.

There has been increased attention paid to the diagnostic capability of stimuli-responsive nanogels. Thermo-responsive copolymers for high contrast magnetic resonance imaging (MRI) were initially reported by Kolouchova K. et al. [67]. Two block polymers, poly[N-(2-hydroxypropyl)-methacrylamide] (PHPMA) or poly (2-methyl-2-oxazoline) (PMeOx), and a thermo-responsive poly-[N(2,2difluoroethyl)acrylamide] (PDFEA) were synthesized as amphiphilic copolymers that can undergo self-assembly to form nanogels under physiological conditions. The self-assembled nanogels were used to encapsulate a magnetic-sensitive fluorine to use as high sensitivity MRI nanoparticles.

In addition to drug delivery and diagnostic devices, thermo-responsive polymers can be employed as an initial material in the self-assembly of polymeric nanoparticles. Poly (ethylene glycol)-methacrylate-based maleimide-bearing copolymer was used as the initial nanoaggregate in the reaction of a Clickable-nanogel due to its stability in an aqueous solution. Cell imaging and internalization of the nanogel were evaluated in the study [68].

Thermal triggering can occur by other sources, such as NIR light, which generates heat from the excitation of specific molecules. Moreover, NIR light can penetrate to deep tissues with less damage compared with direct heat or UV light. Shang T. et al. used the core-shell structure of Au_{rod}@PNIPAM-PEGMA nanogels (PNIPAM-PEGMA = cross-linking of PNIPAM and poly-(ethylene glycol)-methacrylate (PEGMA)), which responds to NIR light stimulation. Nanoparticles containing the Au rod absorbed the NIR light and converted it to heat. The elevated temperature caused the thermo-sensitive PNIPAM-PEGMA nanogels to shrink and release the payload. Moreover, the molar ratio of PNIPAM to PEGMA was varied and these nanogels demonstrated increased LCST at a reduced molar ratio, revealing the fine tunability of thermo-responsive behavior of the polymer that is useful for its controlled release of a material [69].

NIR light-sensitive nanogels with photo-sensing properties were designed by Wang H. et al. as bifunctional nanoparticles composed of a fluorescent carbon dot core coated with PNIPAM-co-acrylamide-based hydrogels. The photo-sensitivity of the carbon dots was used for in vivo imaging of tumor tissue together with the NIR light-triggered temperature increase, resulting in the thermo-responsive behavior of the nanogel for specific drug release [70].

Ma X. et al. [71] reported a nanogel system that was well-designed as a multi-stimuli-responsive carrier of a chemotherapeutic prodrug, 10-hydroxycamplothecin, and a photosensitizer, purpurin18 (P18). The nanogels demonstrated drug release behavior in the reducing tumor microenvironment along with ROS generation because the release of P18 was stimulated using 660-nm laser irradiation. Interestingly, P18 also responded to the

NIR excitation, resulting in a real-time monitor of the treatment, and can be further used for diagnosis.

3.2. pH-Responsive Nanogels

pH is a physiological factor that affects the ionization state of drug molecules, and influences the behavior of pH-sensitive drug delivery systems. Typically, the pH value at the target sites in several diseases differs from the physiological pH. Thus, designing a drug carrier with a specific pH-responsiveness can enhance the therapeutic efficacy and reduce side effects to the surrounding organs or tissues. Due to the different charge characteristics of each polymer, chemical modification or polymer blending could convey advantages to the distinct functionalities among other nano-carriers.

The morphology transition of anionic nanogels under different pH conditions was reported by Ashrafizadeh M. et al. [72]. They used a nano-system composed of ionizable acrylic acid, butyl acrylate, and ethylene glycol dimethacrylate as a crosslinker. The modified polymers were pH-responsive, where their swelling was increased at pH 7.5. The morphology of the modified nanogels transformed from a sphere structure at acidic pH to a core-shell structure as the pH increased, before forming a thin-walled sphere as pH further rose (pH >9). The maximum swelling indicated the feasibility of drug release under physiological conditions. Furthermore, Pal S. et al. [73] used a pH-responsive nanogel system composed of dextrin and poly (acrylic acid) using N,N′-methylene bis acrylamide (MBA) as a crosslinker. Protonation of the carboxyl groups of poly (acrylic acid) was responsible for controlling the drug release at an acidic pH (pH 5.5) that can be utilized for cancer therapy. Furthermore, the dextrin provided a greater biocompatibility to human mesenchymal stem cells [73].

pH-responsive hyaluronic acid-based nanogels have been fabricated to deliver a terephthalaldehyde (TPA) protein drug, cytochrome C (CC) and enhance cell uptake in the tumor environment [74]. Hyaluronic acid-grafted-methoxy PEG-diethylenetriamine copolymers was crosslinked with TPA by forming a benzoic imine bond. The benzoic imine facilitated the pH-responsiveness of the system because it can convert a negatively charged surface to a positively charged surface in the acidic tumor microenvironment, thereby releasing CC. This study illustrated using a polymeric nano-system to deliver macromolecules that widened the applications of the biologic drug delivery system.

Metallic nanoparticles can be pH-responsive, as demonstrated by Wu W. et al. Ni-Ag nanoparticle core/PEG-co-methacrylic acid shell nanogels were used as a targeted drug carrier for cancer treatment using 5-FU as a model drug. In this study, the metallic core was photoluminescent and its intensity depended on the pH value being 5.0–7.4. The nanogels also exhibited the rapid release of 5-FU at the tumor site due to acidic pH-triggered release. In addition, the drug loading capacity of the hybrid nanogels was high due to the intermolecular complex between the polymeric shell, the ether oxygens of PEG, and carboxyl groups of the methacrylic acid moiety [75].

3.3. Reduction-Responsive Nanogels

Tumor cells typically exhibit a higher amount of reducing compounds compared with normal cells. This distinct environment has led to the development of targeted cancer chemotherapy using a drug delivery system that is affected by a reducing environment [76]. GSH-responsive nanogels consisting of a poly (oligo (ethylene oxide) monomethyl ether methacrylate) crosslinked network and disulfide-functionalized dimethacrylate were synthesized in which the disulfide bond could be cleaved by a high GSH concentration. Indeed, a water-soluble fluorescent dye, Rhodamine 6G (R6G), was readily released when the R6G-loaded nanogels were incubated with C2C12 cancer cells, confirming a reduction-sensitive release. DOX, an anticancer drug, was also loaded in the nanogels and an enhanced cytotoxicity to HeLa cells was observed [77].

Reduction-responsive nanogels have been extensively reported as intracellular targeting delivery systems, because the intracellular concentration of GSH ranges from 1–10 mM, which is relatively high compared with the extracellular space [78]. This cell environment results in intracellular-specific payload release. Carboxyl-functionalized polyvinylpyrrolidone nanogels were developed with (3-[(2-aminoethyl)dithio]propionic acid), a redox-sensitive spacer that contains disulfide bridges. The highest release of DOX was achieved in the presence of a reductant, dithiothreitol, due to the reductant sensitivity of disulfide bonds that led to the cleavage and relaxed polymeric networks.

This reduction-triggered release system was also used for siRNA delivery, due to the instability of siRNA in the circulatory system and the low transfection efficiency of bare nucleotides. The siRNA was conjugated with polyvinylpyrrolidone to form nanogels with reduction-sensitive bonding without losing the inherent silencing properties [79]. Furthermore, plasmid DNA can be efficiently delivered using the nanogels as reported by Liwinska W. et al. [80]. Plasmid DNA was crosslinked to the base polymers containing disulfide bonds, thus, the disulfide cleavage occurred in the high GSH intracellular environment.

The use of metal-containing GSH-responsive nanogels, composed of poly [methacrylic acid-co-poly(ethylene glycol) methyl ether methacrylate, as a tumor-targeted drug carrier was reported [81]. N,N-bis(acryloyl)cystamine facilitated the nanogels to respond to the GSH gradient along the distance from the external part of tumors to the core. Along with the encapsulated DOX, in vitro studies demonstrated markedly reduced release at the tissue periphery based on the drug-leakage protection provided by the crosslinked metal and accelerated drug release at the tumors due to the highly concentrated GSH in the tissue.

Glucose is also a stimulus that can be used when designing a responsive drug delivery system, especially for diabetes mellitus [82] or infection [83]. Several glucose-sensitive hybrid nanogels were designed to respond to specific glucose levels and specifically release glucose-lowering agents. Phenylboronic acid is a polymer used for sensing glucose levels as a phenylborate ion or the dissociated (charged) form that can react with the 1,2-cis-diols of glucose in an aqueous solution [84]. Poly (4-vinylphenylboronic acid-co-2-(dimethylamino)ethylacrylate), a boronic acid derivative, was coated on an Ag core to form nanogels that are glucose-sensitive. The glucose binding by nanogels was investigated based on the swelling of the particles in higher glucose concentrations, enhancing the release of entrapped molecules.

Infected cells exhibit a high level of mannose receptors on their surface, and thus mannosyl ligand-functionalized nanogels were fabricated target infected tissues. The nanogels were then ingested by macrophages at the infection site. Additionally, the bacteria at the infected site can secrete vesicles containing phosphatase and/or phospholipase, which disassociated the PEG grafted-polyphosphoester. Nanogels made of this material showed an accelerated release of loaded vancomycin and cytotoxicity to methicillin-resistance *Staphylococcus aureus* (MRSA).

3.4. Multi-Responsive Nanogels

To provide additional functionality to nanogels, multi-responsive features, which are responsive to two or more stimuli, were introduced to deliver drugs with higher efficacy and safety. Temperature-, pH-, and NaCl-sensitive nanogels were demonstrated by Xu Y. et al. to be dual-function injectable nanocarriers that enhanced drug delivery and functioned as biological probes [85]. This multi-responsive delivery system, composed of PNIPAM-co-adenine (PNIPAM-co-A) polymer, increased high temperature-responsiveness from PNIPAM and fluorescent properties from nucleobase (adenine), suggesting its potential as a theranostic system.

Combined pH/temperature sensitive nanogels were designed to facilitate controlled release at a specific tumor site with an acidic pH and to improve the thermal stability and biocompatibility of the materials. Nita L.E. et al. fabricated this type of system using poly (itaconic anhydride-co-3,9-divinyl-2,4,8,10-tetraoxaspiro (5.5) undecane) copolymer crosslinked with 1,12-dodecandiol [86]. Changing the polymer charge through a pH range (pH 4–11) can change the shrink-swell behavior of a nanogel, which affected drug release, were investigated by varying the comonomer ratio in the copolymer chain. Moreover, temperature responsiveness was also affected by the different comonomer ratios that caused the hydrodynamic radius and shrink-swell behavior of the nanogels to change. In addition, an interesting host-guest interaction between dextran grafted benzimidazole and thiol-β-CD was fabricated and used as a nanogel polymer chain [87]. An oxidative hydrosulfide group was employed as crosslinking group and GSH sensitive marker. Consequently, the size of the nanogels increased after being treated with GSH at pH 7.4, which can be inferred to be its response in the reductive cancer environment. The system was also designed to respond to acidic pH, therefore drug release from the nanogels increased at pH 5.3 without GSH.

Lou S. et al. fabricated hepatocyte-specific multi-responsive nanogels consisting of poly (6-Ovinyladipoyl-D-galactose-ss-N-vinylcaprolactam-ss-methacrylic acid) or P(ODGal-VCL-MAA) [34]. β-D-galactose (GAL) or N-acetylgalactosamine that included asialogly co-protein receptor (ASGP-R) ligands, which are abundantly expressed in mammalian hepatocytes. The hepatocyte specificity of the ODGal ligands was identified using human hepatoma HepG2 cells. The cytotoxicity of the DOX-loaded P (ODGal-VCL-MAA) nanogels was significantly improved compared with those without ODGal ligands. The multi-responsiveness of the nanogels was observed in various temperature, pH, and reductase conditions. A higher phase transition temperature was observed at higher ODGal concentration. The nanogel's pH-dependent drug release behavior was also investigated, the release rate of the loaded DOX increased as the pH decreased from 7.4 to 6.0. In the presence of GSH, increased drug release in the elevated reductase condition was observed.

An effective delivery system for cancer therapy was demonstrated using crosslinked hyaluronic acid (HA) and keratin to fabricate multi-responsive nanogels [88]. Keratin is a hair protein that is biocompatible with no immunogenicity due to the abundant cysteines and sulfhydryl bonds in its structure that can be degraded by several enzymes, including trypsin. Additionally, keratin can increase the production of intracellular nitric oxide (NO) that attacks tumor cells, enhancing the cytotoxicity of chemotherapeutic substances. Furthermore, HA binding to CD44 promotes cell internalization. Due to HA's negative charge, a large amount of positively charged DOX can occupy the nanogels surface. These well-designed nanogels displayed high drug encapsulation, combining pH-, trypsin-, and GSH-responsive drug release, as well as increased efficacy promoted by increased NO levels at the targeted cancer cells.

Graphene oxide (GO), an inorganic compound, exhibits a distinct drug loading capacity. Due to its strong polarity, GO forms an electrostatic interaction with some cancer therapeutics, and it was combined with PNIPAM monomer to improve thermal responsiveness when exposed to NIR light, as well as enhancing its colloidal stability [89]. The results demonstrated the optimized DOX loading capacity together with improved colloidal stability and biocompatibility of the native GO. Furthermore, the multiple stimuli-response components provided targeted drug delivery under acidic and reducing conditions that were specific to the cancer cell microenvironment [90]. An in vitro cytotoxic study in HeLa cells indicated the local efficacy of the DOX-loaded nanogels after NIR light exposure. These nanogels demonstrated higher cellular uptake and bioimaging potential, confirming from the results of confocal laser scanning microscope analysis.

Moreover, GO-DOX coated with crosslinked hyaluronic acid nanogels were used as a theranostic system [91]. When incorporated with GO, the nanogels emitted light when exposed to red-light radiation that can be used as cancer-cell tracking. The loaded DOX can be specifically released at low pH and red light stimulation and the crosslinked hyaluronic acid specifically targets the CD44 receptors on A549 cells. The specificity of the nanogels decreases the side effects of the anti-cancer drugs and increases the accuracy of tumor/cancer identification.

4. Biomedical Applications

As smart drug carriers, the modified polymeric nanogel systems are expected to enhance the efficiency of drug delivery, which can reduce side effects by enhancing the specificity to target sites and improve the drug's pharmacokinetics that can prolong the circulation time, leading to a greater bioavailability and an elevated therapeutic efficacy. Moreover, with the stimuli-sensitiveness of polymers, polymeric nanogels can act as diagnostic tools for tracking specific organs or tissue with a higher precision. Furthermore, there are various modifications based on a combination of therapeutic and diagnosis, the so-called theranostic systems, which indicate the multiple functionalities of the modified nanogels.

4.1. Therapeutics

To enhance the therapeutic properties of the nanogels, nano-carriers were designed to release their payloads upon the application of a stimulus, known as stimuli-responsive nanogels. The nanogels formulated as a therapeutic eye drop were typically made of chitosan, which was known for its mucoadhesive properties [92]. A chitosan derivative, N-succinyl-chitosan, was synthesized with self-crosslinking properties that stabilize the nanogel system. The results suggested the potential of this nanogel as a mucoadhesive enhancer due to the higher affinity to mucous epithelium compared with native chitosan. Additionally, the nanogels displayed a double-step release; an initial burst that results in a high therapeutic concentration in a short period, followed by a consequent sustained release to maintain the therapeutic level of the drug in a prolonged fashion.

The development of biopharmaceutical products containing macromolecules, i.e., proteins, has increased. The major limitation of protein delivery is maintaining its stability during storage or exposure to physiological enzymes, which affects its structure. Hemoglobin (Hb), a functional protein that transports oxygen, carbon dioxide, and other gases to support normal body functions in mammals, was used as a model protein in Wei X. et al. [93] Hb was loaded into a functionalized nanogel containing dextran grafted with succinic anhydride to promote self-assembly, and dopamine (DA) to modify the hydrophobicity of the nanoparticles. A gas-binding capacity study revealed that the nanogels demonstrated a positive trend in binding and releasing CO, N_2, or O_2 without the loss of Hb's inherent bioactivities. Higher oxygen affinity compared with free Hb and increased stability of the protein in physiological buffers were observed. The nanoparticles also showed an enhanced biocompatibility and hemo-compatibility features because there were no signs of significant cell death or hemorrhagic abnormalities in the tested conditions.

Dandan Li et al. [94] used reducing-sensitive nanogels loaded with ovalbumin (OVA) as a model antigen for vaccine applications. Methacrylated dextran and trimethyl aminoethyl methacrylate (TMAEMA)-based nanogels were conjugated with pyridyldisulfide groups to facilitate the response to a presence of a reducing agent. In the circulatory system or extracellular matrices, which are considered to be non-reducing conditions, the OVA antigen was entrapped in the nanoparticles, enhancing its stability against physiological enzymes. The OVA release from nanogels occurred as a response to the reductive urokinase-type plasminogen activator (uPA) environment after cell internalization. Therefore, the antigen-presenting property can be enhanced by this reducing-responsive nanogel, because dendritic cells, a potent antigen presenting cell, can exploit this stimuli-responsiveness to promote MHC class I cascades.

Another intriguing example is multifunctional polymeric nanogels that were developed for improving the specificity of a rescue drug for stroke and decrease fatal side effects, including pulmonary edema and intracranial hemorrhage. PEG-conjugated glycol chitosan hollow nanogels were used to stimulate drug release after ultrasonic stimulation, which limited their specific target release and enhanced uPA thrombolysis activity and stability in circulatory system. The developed nanogels with ultrasonic stimulation showed a decreased risk of hemorrhage because the loaded uPA did not penetrate the blood-brain barrier [95,96]. In addition, PNIPAM core-acrylic acid shell nanogels were fabricated to deliver fibrin to enhance the efficacy of tissue-type plasminogen activator (tPA) for treating disseminated intravascular coagulation [97]. The findings indicated that the dual functionalities of the nanogels, the thrombolytic effect on microthrombi that can reduce risk of organ failure and the ability to ameliorate blood clots at the injured sites, provide a promising tool for targeted tPA treatment together with a restoration the hemostatic balance in patients.

The tumor microenvironment displays distinctive conditions that allows for wide functional design of drug delivery systems. A co-release system of DOX and cisplatin made of a tri-coblock polymer, poly (acrylic acid-b-PNIPAM-b-acrylic acid), or PNA, was fabricated [98]. DOX was loaded into the core and electrostatically bound to the carboxyl groups of PNA, and cisplatin was incorporated into the shell. Due to the different loading compartments of the drugs, the DOX and cisplatin exhibited different release behaviors, resulting in synergistic tumor suppression. The DOX and cisplatin ratio can be optimized to achieve different release patterns. Moreover, the thermo-responsiveness of PNIPAM contributed to the sol-gel transition of the nanogels that was beneficial for targeted release at the tumor site.

Topical nanogel formulations have also been widely investigated. Zhu J. et al., fabricated chlorhexidine diacetate-loaded nanogels enclosed in aminoethyl methacrylate hyaluronic acid and methacrylated methoxy polyethylene glycol crosslink system, referred to as Gel@CLN hydrogels as a wound dressing. The results demonstrated that Gel@CLN hydrogels reduced infection by *E. coli* and *S. aureus* with prolonged inhibition for up to 10 days. Gel@CLN hydrogels treatment also accelerated wound healing due to its high water content, which enhanced its moisturizing effects and supported rapid wound closure [99].

4.2. Diagnotics and Theranostics

Apart from using nanogels as drug delivery systems to improve their therapeutic effects, the nanogels also possess multi-functional features that can be used for diagnosis. QD-containing polymeric nanogels that emitted a specific light/fluorescent signal were used as biological dyes for live cells or in vivo imaging. Multi-responsive biosensing nanoparticles containing Cadmium-Selenium (CdSe) QDs shielded with chitosan-poly (methacrylic acid) were fabricated to combine the optical properties of QDs and the pH-responsiveness of the modified chitosan [82]. Therefore, the controlled release of loaded drugs occurred at a pH of 5.0–7.4, and the internalization of the nanogels can be visually observed due to the optical activities of CdSe QDs. Thermo-responsive copolymers were used as a diagnostic device for high contrast MRI, by Kolouchova K. et al. [67]. Two block co-polymers, PHPMA or PMeOx, and a thermo-responsive PDFEA, were fabricated, and the materials self-assembled into nanoparticles at human body temperature. The nanogels were then used to encapsulate fluorine atoms as a probe for high sensitivity MRI.

Various targeted diagnostic nanogel platforms have been evaluated for their potential biomedical application; however, for those containing metals, their toxicity is a concern. Hence, several core-shell structured nanogels were used to reduce the cytotoxicity by shielding the metals from being exposed to the cell surface. Gold nanoparticles (AuNPs)-based chitin-MnO_2 ternary composite nanogels (ACM-TNGs) were fabricated as a radio-assisted cancer therapy [100]. This system was designed to reduce the toxicity of MnO_2 nanorods that absorbed low frequency radio waves. To minimize the toxicity from the metals, chitin nanogels were used as a biocompatible shell that wrapped the AuNPs and MnO_2 nanorods.

A cytotoxicity study using L929 HDF, MG63, T47D, and A375 cell lines demonstrated the cytocompatibility of ACM-TNGs and that their cell internalization did not alter cell morphology. Moreover, this nano-system also killed breast cancer cells at a 100 watts/2 min radio frequency. The potential of the nanogel as a cell tracker was investigated using NIR light-triggered drug release [101].

Kimura A. et al. used ultra-small Galodium-gelatin nanogels as an MRI contrast agent because the coated gelatin shell minimizes the risk of nanoparticles leaking through the blood-brain barrier and blood-cerebrospinal fluid barrier [102]. The developed nanogels provided high contrast MRI images with low toxicity to cultured cells. The in vivo biodistribution in tumor-bearing mice demonstrated rapid renal clearance with no accumulation in the kidney and liver, including over long-term observation. Moreover, no nanoparticles were found in the blood stream and cerebrospinal fluid, which confirmed the high safety profile of the developed nanoparticles as a MRI contrast agent.

Theranostic is a term used to describe tools, devices, or systems that have combined diagnostic and therapeutic features. For a drug delivery system, these platforms were designed to improve the therapeutic potential and decrease side effects, along with being able to diagnose or indicate the drug's biodistribution after administration. Wu W. et al. used pH-responsive polymers, PNIPAM-co-acrylic acid with embedded Au nanoparticles, as an imaging probe [103]. The nanogels exhibited pH-sensitive behavior due to the protonation of their acrylic acid moieties, and increased hydrophobic drug loading capacity from the temperature-sensitive hydrophobic NIPAM groups. In another study, folate-terminated poly (ethylene glycol)-modified hyaluronic acid crosslinked with carbon dots was synthesized to target tumor cells that highly express folate receptors [90]. This system provided pH- and acidic-sensitive drug release that should be specific to tumor tissue conditions, and the carbon dots could be used for bioimaging.

A graphene-DOX-entrapped hyaluronic acid nanogel was evaluated as a theranostic system [91]. The nanogels emitted light from the entrapped graphene when exposed to red-light radiation, which is advantageous for cancer-cell imaging. Due to the high expression of CD44, hyaluronic acid receptors, DOX-targeted release via pH/red light-responsiveness was seen in cultured non-small cell lung cancer A549 cells. The specificity of the nanogels decreased the side effects of the cancer drug and increased the accuracy of the cancer identification.

Peng N. et al. formulated novel hybrid nanogels composed of a disulfide-modified alginate shield covering superparamagnetic iron oxide nanoparticles (SPIONs, an MRI probe, for theranostic applications [104]. The nanogels can be used as an imaging agent in acidic and reductive environments, which was beneficial for tumor-targeted release and diagnostic targetability. The combination of SPIONs and biocompatible alginate derivative as a delivery and cell imaging system also exhibited high DOX loading and high toxicity to tumor cells.

In conclusion, the biomedical uses of nanogels require the biocompatibility and specific designs of the nanogels to minimize the side effects to surrounding tissue and enhance the efficacy and stability of active drugs, respectively. The core-shell structure of the nanogels protects the encapsulated drugs and increases the specificity to the nanocarriers because its components influence the responsiveness to various stimuli (physiological and external environment). In addition, the adjustable design of polymers results in the stability of active compounds in the physiological environment and having feature-specific functions. Moreover, the stimuli-responsive behavior of nanogels allows them to be used as biocompatible diagnostic tools that provide precise results. The practical biomedical use of nanogels is being increasingly investigated to develop efficient and specific medical tools and advance the use of nanotechnology in drug delivery systems.

Author Contributions: Conceptualization, T.K. and J.A.L.; Data curation, T.K.; Formal analysis, T.K.; Funding acquisition, J.A.L.; Investigation, T.K.; Methodology, T.K.; Project administration, J.A.L.; Resources, J.A.L.; Supervision, J.A.L.; Writing—original draft, T.K.; Writing—review & editing, C.L., Y.P., J.A.L. All authors have read and agreed to the published version of the manuscript.

Funding: This work was supported by Thailand Science Research and Innovation (TSRI) Fund, grant number CU_FRB640001_01_32_1, and Second Century Fund (C2F), Chulalongkorn University. T.K. acknowledges the 100th Anniversary Chulalongkorn University Fund for Doctoral Scholarship and the 90th Anniversary Chulalongkorn University Fund (Ratchadaphiseksomphot Endowment Fund) for Ph.D. scholarship.

Institutional Review Board Statement: Not applicable.

Informed Consent Statement: Not applicable.

Data Availability Statement: Not applicable.

Conflicts of Interest: The authors declare no conflict of interest.

References

1. Son, S.J.; Bai, X.; Lee, S.B. Inorganic hollow nanoparticles and nanotubes in nanomedicine: Part Drug/gene delivery applications. *Drug Discov. Today* **2007**, *12*, 650–656. [CrossRef] [PubMed]
2. Lee, Y.; Thompson, D.H. Stimuli-responsive liposomes for drug delivery. *WIREs Nanomed. Nanobiotechnol.* **2017**, *9*, e1450. [CrossRef]
3. Jamkhande, P.G.; Ghule, N.W.; Bamer, A.H.; Kalaskar, M.G. Metal nanoparticles synthesis: An overview on methods of preparation, advantages and disadvantages, and applications. *J. Drug Deliv. Sci. Technol.* **2019**, *53*, 101174. [CrossRef]
4. George, A.; Shah, P.A.; Shrivastav, P.S. Natural biodegradable polymers based nano-formulations for drug delivery: A review. *Int. J. Pharmaceut.* **2019**, *561*, 244–264. [CrossRef] [PubMed]
5. Sunshine, J.C.; Akanda, M.I.; Li, D.; Kozielski, K.L.; Green, J.J. Effects of base polymer hydrophobicity and end-group modification on polymeric gene delivery. *Biomacromolecules* **2011**, *12*, 3592–3600. [CrossRef]
6. Wang, Y.; Li, J.; Chen, Y.; Oupický, D. Balancing polymer hydrophobicity for ligand presentation and siRNA delivery in dual function CXCR4 inhibiting polyplexes. *Biomater. Sci.* **2015**, *3*, 1114–1123. [CrossRef]
7. Wang, C.; Zhou, D.-D.; Gan, Y.-W.; Zhang, X.-W.; Ye, Z.-M.; Zhang, J.-P. A partially fluorinated ligand for two super-hydrophobic porous coordination polymers with classic structures and increased porosities. *Natl. Sci. Rev.* **2020**, *8*. [CrossRef]
8. Liu, D.; Ma, L.; An, Y.; Li, Y.; Liu, Y.; Wang, L.; Guo, J.; Wang, J.; Zhou, J. Thermoresponsive Nanogel-Encapsulated PEDOT and HSP70 Inhibitor for Improving the Depth of the Photothermal Therapeutic Effect. *Adv. Funct. Mater.* **2016**, *26*, 4749–4759. [CrossRef]
9. Ji, Y.; Winter, L.; Navarro, L.; Ku, M.-C.; Periquito, J.S.; Pham, M.; Hoffmann, W.; Theune, L.E.; Calderón, M.; Niendorf, T. Controlled Release of Therapeutics from Thermoresponsive Nanogels: A Thermal Magnetic Resonance Feasibility Study. *Cancers* **2020**, *12*, 1380. [CrossRef]
10. Shi, X.; Ma, X.; Hou, M.; Gao, Y.-E.; Bai, S.; Xiao, B.; Xue, P.; Kang, Y.; Xu, Z.; Li, C.M. pH-Responsive unimolecular micelles based on amphiphilic star-like copolymers with high drug loading for effective drug delivery and cellular imaging. *J. Mater. Chem. B* **2017**, *5*, 6847–6859. [CrossRef]
11. Liu, S.; Ono, R.J.; Yang, C.; Gao, S.; Ming Tan, J.Y.; Hedrick, J.L.; Yang, Y.Y. Dual pH-Responsive Shell-Cleavable Polycarbonate Micellar Nanoparticles for in Vivo Anticancer Drug Delivery. *ACS Appl. Mater. Interfaces* **2018**, *10*, 19355–19364. [CrossRef] [PubMed]
12. Cazotti, J.C.; Fritz, A.T.; Garcia-Valdez, O.; Smeets, N.M.B.; Dubé, M.A.; Cunningham, M.F. Graft modification of starch nanoparticles with pH-responsive polymers via nitroxide-mediated polymerization. *J. Polym. Sci.* **2020**, *58*, 2211–2220. [CrossRef]
13. Maiti, C.; Parida, S.; Kayal, S.; Maiti, S.; Mandal, M.; Dhara, D. Redox-responsive core-cross-linked block copolymer micelles for overcoming multidrug resistance in cancer cells. *ACS Appl. Mater. Interfaces* **2018**, *10*, 5318–5330. [CrossRef]
14. Aramoto, H.; Osaki, M.; Konishi, S.; Ueda, C.; Kobayashi, Y.; Takashima, Y.; Harada, A.; Yamaguchi, H. Redox-responsive supramolecular polymeric networks having double-threaded inclusion complexes. *Chem. Sci.* **2020**, *11*, 4322–4331. [CrossRef]
15. Vicario-de-la-Torre, M.; Forcada, J. The Potential of Stimuli-Responsive Nanogels in Drug and Active Molecule Delivery for Targeted Therapy. *Gels* **2017**, *3*, 16. [CrossRef] [PubMed]
16. Zhang, Y.; Dosta, P.; Conde, J.; Oliva, N.; Wang, M.; Artzi, N. Prolonged Local In Vivo Delivery of Stimuli-Responsive Nanogels That Rapidly Release Doxorubicin in Triple-Negative Breast Cancer Cells. *Adv. Healthc. Mater.* **2020**, *9*, 1901101. [CrossRef]
17. Li, S.; Zhang, T.; Xu, W.; Ding, J.; Yin, F.; Xu, J.; Sun, W.; Wang, H.; Sun, M.; Cai, Z. Sarcoma-targeting peptide-decorated polypeptide nanogel intracellularly delivers shikonin for upregulated osteosarcoma necroptosis and diminished pulmonary metastasis. *Theranostics* **2018**, *8*, 1361. [CrossRef]
18. Pan, G.; Guo, Q.; Cao, C.; Yang, H.; Li, B. Thermo-responsive molecularly imprinted nanogels for specific recognition and controlled release of proteins. *Soft Matter* **2013**, *9*, 3840–3850. [CrossRef]
19. Liu, P. 1—Redox—and pH-responsive polymeric nanocarriers. In *Stimuli Responsive Polymeric Nanocarriers for Drug Deliv. Applications*; Makhlouf, A.S.H., Abu-Thabit, N.Y., Eds.; Woodhead Publishing: Cambridge, UK, 2019; pp. 3–36.
20. Badeau, B.A.; DeForest, C.A. Programming Stimuli-Responsive Behavior into Biomaterials. *Annu. Rev. Biomed. Eng.* **2019**, *21*, 241–265. [CrossRef]
21. Caro, C.; García-Martín, M.L.; Pernia Leal, M. Manganese-Based Nanogels as pH Switches for Magnetic Resonance Imaging. *Biomacromolecules* **2017**, *18*, 1617–1623. [CrossRef]
22. Zhang, X.; Chen, X.; Wang, H.-Y.; Jia, H.-R.; Wu, F.-G. Supramolecular Nanogel-Based Universal Drug Carriers Formed by "Soft–Hard" Co-Assembly: Accurate Cancer Diagnosis and Hypoxia-Activated Cancer Therapy. *Adv. Ther.* **2019**, *2*, 1800140. [CrossRef]
23. Wang, Y.; Zu, M.; Ma, X.; Jia, D.; Lu, Y.; Zhang, T.; Xue, P.; Kang, Y.; Xu, Z. Glutathione-Responsive Multifunctional "Trojan Horse" Nanogel as a Nanotheranostic for Combined Chemotherapy and Photodynamic Anticancer Therapy. *ACS Appl. Mater. Interfaces* **2020**, *12*, 50896–50908. [CrossRef]

24. Zhang, H.; Ba, S.; Lee, J.Y.; Xie, J.; Loh, T.-P.; Li, T. Cancer Biomarker-Triggered Disintegrable DNA Nanogels for Intelligent Drug Delivery. *Nano Lett.* **2020**, *20*, 8399–8407. [CrossRef] [PubMed]
25. Cazares-Cortes, E.; Espinosa, A.; Guigner, J.-M.; Michel, A.; Griffete, N.; Wilhelm, C.; Ménager, C. Doxorubicin Intracellular Remote Release from Biocompatible Oligo(ethylene glycol) Methyl Ether Methacrylate-Based Magnetic Nanogels Triggered by Magnetic Hyperthermia. *ACS Appl. Mater. Interfaces* **2017**, *9*, 25775–25788. [CrossRef]
26. Khan, A.R.; Yang, X.; Du, X.; Yang, H.; Liu, Y.; Khan, A.Q.; Zhai, G. Chondroitin sulfate derived theranostic and therapeutic nanocarriers for tumor-targeted drug delivery. *Carbohydr. Polym.* **2020**, *233*, 115837. [CrossRef] [PubMed]
27. Fusco, L.; Gazzi, A.; Peng, G.; Shin, Y.; Vranic, S.; Bedognetti, D.; Vitale, F.; Yilmazer, A.; Feng, X.; Fadeel, B.; et al. Graphene and other 2D materials: A multidisciplinary analysis to uncover the hidden potential as cancer theranostics. *Theranostics* **2020**, *10*, 5435–5488. [CrossRef]
28. He, F.; Zhang, Y.; Li, J.; Liu, S.; Chi, Z.; Xu, J. Preparation and properties of multi-responsive semi-IPN hydrogel modified magnetic nanoparticles as drug carrier. *J. Control. Rel.* **2011**, *152*, e119–e121. [CrossRef]
29. Gonzalez-Urias, A.; Zapata-Gonzalez, I.; Licea-Claverie, A.; Licea-Navarro, A.F.; Bernaldez-Sarabia, J.; Cervantes-Luevano, K. Cationic versus anionic core-shell nanogels for transport of cisplatin to lung cancer cells. *Colloids Surf. B Biointerfaces* **2019**, *182*, 110365. [CrossRef]
30. Amanlou, N.; Parsa, M.; Rostamizadeh, K.; Sadighian, S.; Moghaddam, F. Enhanced cytotoxic activity of curcumin on cancer cell lines by incorporating into gold/chitosan nanogels. *Mater. Chem. Phys.* **2019**, *226*, 151–157. [CrossRef]
31. Qiu, M.; Wu, H.; Cao, L.; Shi, B.; He, X.; Geng, H.; Mao, X.; Yang, P.; Jiang, Z. Metal–Organic Nanogel with Sulfonated Three-Dimensional Continuous Channels as a Proton Conductor. *ACS Appl. Mater. Interfaces* **2020**, *12*, 19788–19796. [CrossRef] [PubMed]
32. Tang, J.; Cui, X.; Caranasos, T.G.; Hensley, M.T.; Vandergriff, A.C.; Hartanto, Y.; Shen, D.; Zhang, H.; Zhang, J.; Cheng, K. Heart Repair Using Nanogel-Encapsulated Human Cardiac Stem Cells in Mice and Pigs with Myocardial Infarction. *ACS Nano.* **2017**, *11*, 9738–9749. [CrossRef]
33. Shoueir, K.R.; Sarhan, A.A.; Atta, A.M.; Akl, M.A. Macrogel and nanogel networks based on crosslinked poly (vinyl alcohol) for adsorption of methylene blue from aqua system. *Environ. NanoTechnol. Monitor. Manag.* **2016**, *5*, 62–73. [CrossRef]
34. Lou, S.; Gao, S.; Wang, W.; Zhang, M.; Zhang, J.; Wang, C.; Li, C.; Kong, D.; Zhao, Q. Galactose-functionalized multi-responsive nanogels for hepatoma-targeted drug delivery. *Nanoscale* **2015**, *7*, 3137–3146. [CrossRef]
35. Liechty, W.B.; Scheuerle, R.L.; Vela Ramirez, J.E.; Peppas, N.A. Uptake and function of membrane-destabilizing cationic nanogels for intracellular drug delivery. *Bioeng. Transl. Med.* **2019**, *4*, 17–29. [CrossRef] [PubMed]
36. Liechty, W.B.; Scheuerle, R.L.; Peppas, N.A. Tunable, responsive nanogels containing t-butyl methacrylate and 2-(t-butylamino)ethyl methacrylate. *Polymer* **2013**, *54*, 3784–3795. [CrossRef]
37. Horvat, S.; Yu, Y.; Bojte, S.; Tessmer, I.; Lowman, D.W.; Ma, Z.; Williams, D.L.; Beilhack, A.; Albrecht, K.; Groll, J. Engineering Nanogels for Drug Delivery to Pathogenic Fungi Aspergillus fumigatus by Tuning Polymer Amphiphilicity. *Biomacromolecules* **2020**, *21*, 3112–3121. [CrossRef]
38. Costa, D.; Valente, A.J.; Queiroz, J. Plasmid DNA nanogels as photoresponsive materials for multifunctional bio-applications. *J. Biotechnol.* **2015**, *202*, 98–104. [CrossRef]
39. Schmid, A.J.; Dubbert, J.; Rudov, A.A.; Pedersen, J.S.; Lindner, P.; Karg, M.; Potemkin, I.I.; Richtering, W. Multi-Shell Hollow Nanogels with Responsive Shell Permeability. *Sci. Rep.* **2016**, *6*, 22736. [CrossRef]
40. Yasun, E.; Gandhi, S.; Choudhury, S.; Mohammadinejad, R.; Benyettou, F.; Gozubenli, N.; Arami, H. Hollow micro and nanostructures for therapeutic and imaging applications. *J. Drug Deliv. Sci. Technol.* **2020**, *60*. [CrossRef] [PubMed]
41. Osorio-Blanco, E.R.; Rancan, F.; Klossek, A.; Nissen, J.H.; Hoffmann, L.; Bergueiro, J.; Riedel, S.; Vogt, A.; Ruhl, E.; Calderon, M. Polyglycerol-Based Thermoresponsive Nanocapsules Induce Skin Hydration and Serve as a Skin Penetration Enhancer. *ACS Appl. Mater. Interfaces* **2020**, *12*, 30136–30144. [CrossRef]
42. Wang, Z.; Li, S.; Zhang, M.; Ma, Y.; Liu, Y.; Gao, W.; Zhang, J.; Gu, Y. Laser-Triggered Small Interfering RNA Releasing Gold Nanoshells against Heat Shock Protein for Sensitized Photothermal Therapy. *Adv. Sci.* **2017**, *4*, 1600327. [CrossRef] [PubMed]
43. Han, H.; Li, L.; Tian, Y.; Wang, Y.; Ye, Z.; Yang, Q.; Wang, Y.; von Klitzing, R.; Guo, X. Spherical polyelectrolyte nanogels as templates to prepare hollow silica nanocarriers: Observation by small angle X-ray scattering and TEM. *RSC Adv.* **2017**, *7*, 47877–47885. [CrossRef]
44. Ranzoni, A.; Cooper, M.A. Chapter One—The Growing Influence of Nanotechnology in Our Lives. In *Micro and NanoTechnol. in Vaccine Development*; Skwarczynski, M., Toth, I., Eds.; William Andrew Publishing: Norwich, NY, USA, 2017; pp. 1–20.
45. Roberts, J.R.; Ritter, D.W.; McShane, M.J. A design full of holes: Functional nanofilm-coated microdomains in alginate hydrogels. *J. Mater. Chem. B* **2013**, *1*, 3195–3201. [CrossRef]
46. Pan, H.M.; Yu, H.; Guigas, G.; Fery, A.; Weiss, M.; Patzel, V.; Trau, D. Engineering and Design of Polymeric Shells: Inwards Interweaving Polymers as Multilayer Nanofilm, Immobilization Matrix, or Chromatography Resins. *ACS Appl. Mater. Interfaces* **2017**, *9*, 5447–5456. [CrossRef]
47. Bornhoeft, L.R.; Biswas, A.; McShane, M.J. Composite hydrogels with engineered microdomains for optical glucose sensing at low oxygen conditions. *Biosensors* **2017**, *7*, 8. [CrossRef]
48. Choi, D.; Heo, J.; Park, J.H.; Jo, Y.; Jeong, H.; Chang, M.; Choi, J.; Hong, J. Nano-film coatings onto collagen hydrogels with desired drug release. *J. Ind. Eng. Chem.* **2016**, *36*, 326–333. [CrossRef]

49. Kim, M.; Kim, H.; Lee, Y.-s.; Lee, S.; Kim, S.-E.; Lee, U.-J.; Jung, S.; Park, C.-G.; Hong, J.; Doh, J.; et al. Novel enzymatic cross-linking–based hydrogel nanofilm caging system on pancreatic β cell spheroid for long-term blood glucose regulation. *Sci. Adv.* **2021**, *7*, eabf7832. [CrossRef]
50. Lu, S.; Neoh, K.G.; Huang, C.; Shi, Z.; Kang, E.T. Polyacrylamide hybrid nanogels for targeted cancer chemotherapy via co-delivery of gold nanoparticles and MTX. *J. Colloid Interface Sci.* **2013**, *412*, 46–55. [CrossRef] [PubMed]
51. Wu, W.; Shen, J.; Banerjee, P.; Zhou, S. Water-dispersible multifunctional hybrid nanogels for combined curcumin and photothermal therapy. *BioMaterials* **2011**, *32*, 598–609. [CrossRef]
52. Xue, Y.; Xia, X.; Yu, B.; Tao, L.; Wang, Q.; Huang, S.W.; Yu, F. Selenylsulfide Bond-Launched Reduction-Responsive Superparamagnetic Nanogel Combined of Acid-Responsiveness for Achievement of Efficient Therapy with Low Side Effect. *ACS Appl. Mater. Interfaces* **2017**, *9*, 30253–30257. [CrossRef]
53. Gao, X.; Zhuang, X.; Tian, C.; Liu, H.; Lai, W.-F.; Wang, Z.; Yang, X.; Chen, L.; Rogach, A.L. A copper nanocluster incorporated nanogel: Confinement-assisted emission enhancement for zinc ion detection in living cells. *Sens. Actuat. B Chem.* **2020**, *307*, 127626. [CrossRef]
54. Jiang, G.; Park, K.; Kim, J.; Kim, K.S.; Hahn, S.K. Target Specific Intracellular Delivery of siRNA/PEI−HA Complex by Receptor Mediated Endocytosis. *Molec. Pharmaceut.* **2009**, *6*, 727–737. [CrossRef] [PubMed]
55. Nagahama, K.; Sano, Y.; Kumano, T. Anticancer drug-based multifunctional nanogels through self-assembly of dextran-curcumin conjugates toward cancer theranostics. *Bioorg. Med. Chem. Lett.* **2015**, *25*, 2519–2522. [CrossRef]
56. Maxwell, T.; Banu, T.; Price, E.; Tharkur, J.; Campos, M.G.; Gesquiere, A.; Santra, S. Non-Cytotoxic Quantum Dot-Chitosan Nanogel Biosensing Probe for Potential Cancer Targeting Agent. *NanoMater.* **2015**, *5*, 2359–2379. [CrossRef]
57. The global challenge of cancer. *Nat. Cancer* **2020**, *1*, 1–2. [CrossRef]
58. Martin, T.A.; Ye, L.; Sanders, A.J.; Lane, J.; Jiang, W.G. Cancer invasion and metastasis: Molecular and cellular perspective. In *Madame Curie BioSci. Database [Internet]*; Landes Bioscience: Austin, TX, USA, 2013.
59. Tan, L.; He, C.; Chu, X.; Chu, Y.; Ding, Y. Charge-reversal ZnO-based nanospheres for stimuli-responsive release of multiple agents towards synergistic cancer therapy. *Chem. Eng. J.* **2020**, *395*. [CrossRef]
60. Liu, Y.; Zhao, N.; Xu, F.J. pH-Responsive Degradable Dextran-Quantum Dot Nanohybrids for Enhanced Gene Delivery. *ACS Appl. Mater. Interfaces* **2019**, *11*, 34707–34716. [CrossRef]
61. Gui, R.; Wang, Y.; Sun, J. Embedding fluorescent mesoporous silica nanoparticles into biocompatible nanogels for tumor cell imaging and thermo/pH-sensitive in vitro drug release. *Colloids Surf. B Biointerfaces* **2014**, *116*, 518–525. [CrossRef]
62. Arteche Pujana, M.; Perez-Alvarez, L.; Cesteros Iturbe, L.C.; Katime, I. Biodegradable chitosan nanogels crosslinked with genipin. *Carbohydr. Polym.* **2013**, *94*, 836–842. [CrossRef]
63. Zhang, T.; Sturgis, T.F.; Youan, B.B. pH-responsive nanoparticles releasing tenofovir intended for the prevention of HIV transmission. *Eur. J. Pharm. Biopharm.* **2011**, *79*, 526–536. [CrossRef]
64. Luckanagul, J.A.; Pitakchatwong, C.; Ratnatilaka Na Bhuket, P.; Muangnoi, C.; Rojsitthisak, P.; Chirachanchai, S.; Wang, Q.; Rojsitthisak, P. Chitosan-based polymer hybrids for thermo-responsive nanogel delivery of curcumin. *Carbohydr. Polym.* **2018**, *181*, 1119–1127. [CrossRef]
65. Yi, P.; Wang, Y.; Zhang, S.; Zhan, Y.; Zhang, Y.; Sun, Z.; Li, Y.; He, P. Stimulative nanogels with enhanced thermosensitivity for therapeutic delivery via beta-cyclodextrin-induced formation of inclusion complexes. *Carbohydr. Polym.* **2017**, *166*, 219–227. [CrossRef]
66. Soto-Quintero, A.; Guarrotxena, N.; Garcia, O.; Quijada-Garrido, I. Curcumin to Promote the Synthesis of Silver NPs and their Self-Assembly with a Thermoresponsive Polymer in Core-Shell Nanohybrids. *Sci. Rep.* **2019**, *9*, 18187. [CrossRef]
67. Kolouchova, K.; Sedlacek, O.; Jirak, D.; Babuka, D.; Blahut, J.; Kotek, J.; Vit, M.; Trousil, J.; Konefal, R.; Janouskova, O.; et al. Self-Assembled Thermoresponsive Polymeric Nanogels for (19)F MR Imaging. *Biomacromolecules* **2018**, *19*, 3515–3524. [CrossRef]
68. Aktan, B.; Chambre, L.; Sanyal, R.; Sanyal, A. "Clickable" Nanogels via Thermally Driven Self-Assembly of Polymers: Facile Access to Targeted Imaging Platforms using Thiol-Maleimide Conjugation. *Biomacromolecules* **2017**, *18*, 490–497. [CrossRef]
69. Shang, T.; Wang, C.D.; Ren, L.; Tian, X.H.; Li, D.H.; Ke, X.B.; Chen, M.; Yang, A.Q. Synthesis and characterization of NIR-responsive Aurod@pNIPAAm-PEGMA nanogels as vehicles for delivery of photodynamic therapy agents. *Nanoscale Res. Lett.* **2013**, *8*, 4. [CrossRef]
70. Wang, H.; Yi, J.; Mukherjee, S.; Banerjee, P.; Zhou, S. Magnetic/NIR-thermally responsive hybrid nanogels for optical temperature sensing, tumor cell imaging and triggered drug release. *Nanoscale* **2014**, *6*, 13001–13011. [CrossRef] [PubMed]
71. Ma, X.; Zhang, T.; Qiu, W.; Liang, M.; Gao, Y.; Xue, P.; Kang, Y.; Xu, Z. Bioresponsive prodrug nanogel-based polycondensate strategy deepens tumor penetration and potentiates oxidative stress. *Chem. Eng. J.* **2020**. [CrossRef]
72. Ashrafizadeh, M.; Tam, K.C.; Javadi, A.; Abdollahi, M.; Sadeghnejad, S.; Bahramian, A. Synthesis and physicochemical properties of dual-responsive acrylic acid/butyl acrylate cross-linked nanogel systems. *J. Colloid Interface Sci.* **2019**, *556*, 313–323. [CrossRef] [PubMed]
73. Das, D.; Rameshbabu, A.P.; Ghosh, P.; Patra, P.; Dhara, S.; Pal, S. Biocompatible nanogel derived from functionalized dextrin for targeted delivery of doxorubicin hydrochloride to MG 63 cancer cells. *Carbohydr. Polym.* **2017**, *171*, 27–38. [CrossRef]
74. Yang, H.Y.; Li, Y.; Jang, M.-S.; Fu, Y.; Wu, T.; Lee, J.H.; Lee, D.S. Green preparation of pH-responsive and dual targeting hyaluronic acid nanogels for efficient protein delivery. *Eur. Polym. J.* **2019**, *121*, 109342. [CrossRef]

75. Wu, W.; Shen, J.; Gai, Z.; Hong, K.; Banerjee, P.; Zhou, S. Multi-functional core-shell hybrid nanogels for pH-dependent magnetic manipulation, fluorescent pH-sensing, and drug delivery. *BioMaterials* **2011**, *32*, 9876–9887. [CrossRef] [PubMed]
76. Maciel, D.; Figueira, P.; Xiao, S.; Hu, D.; Shi, X.; Rodrigues, J.; Tomas, H.; Li, Y. Redox-responsive alginate nanogels with enhanced anticancer cytotoxicity. *Biomacromolecules* **2013**, *14*, 3140–3146. [CrossRef]
77. Oh, J.K.; Siegwart, D.J.; Lee, H.I.; Sherwood, G.; Peteanu, L.; Hollinger, J.O.; Kataoka, K.; Matyjaszewski, K. Biodegradable nanogels prepared by atom transfer radical polymerization as potential drug delivery carriers: Synthesis, biodegradation, in vitro release, and bioconjugation. *J. Am. Chem. Soc.* **2007**, *129*, 5939–5945. [CrossRef]
78. Forman, H.J.; Zhang, H.; Rinna, A. Glutathione: Overview of its protective roles, measurement, and biosynthesis. *Molec. Aspects Med.* **2009**, *30*, 1–12. [CrossRef] [PubMed]
79. Adamo, G.; Grimaldi, N.; Campora, S.; Bulone, D.; Bondi, M.L.; Al-Sheikhly, M.; Sabatino, M.A.; Dispenza, C.; Ghersi, G. Multi-Functional Nanogels for Tumor Targeting and Redox-Sensitive Drug and siRNA Delivery. *Molecules* **2016**, *21*, 1594. [CrossRef]
80. Liwinska, W.; Stanislawska, I.; Lyp, M.; Stojek, Z.; Zabost, E. Switchable conformational changes of DNA nanogel shells containing disulfide–DNA hybrids for controlled drug release and efficient anticancer action. *RSC Adv.* **2019**, *9*, 13736–13748. [CrossRef]
81. Yang, W.; Zhao, X. Glutathione-Induced Structural Transform of Double-Cross-Linked PEGylated Nanogel for Efficient Intracellular Anticancer Drug Delivery. *Mol. Pharm.* **2019**, *16*, 2826–2837. [CrossRef] [PubMed]
82. Wu, W.; Mitra, N.; Yan, E.C.; Zhou, S. Multifunctional hybrid nanogel for integration of optical glucose sensing and self-regulated insulin release at physiological pH. *Am. Chem. Soc.* **2010**, *4*, 9. [CrossRef]
83. Xiong, M.H.; Li, Y.J.; Bao, Y.; Yang, X.Z.; Hu, B.; Wang, J. Bacteria-responsive multifunctional nanogel for targeted antibiotic delivery. *Adv. Mater.* **2012**, *24*, 6175–6180. [CrossRef]
84. Zhang, Y.; Guan, Y.; Zhou, S. Synthesis and volume phase transitions of glucose-sensitive microgels. *Biomacromolecules* **2006**, *7*, 3196–3201. [CrossRef] [PubMed]
85. Xu, Y.; Yang, M.; Ma, Q.; Di, X.; Wu, G. A bio-inspired fluorescent nano-injectable hydrogel as a synergistic drug delivery system. *New J. Chem.* **2021**, *45*, 3079–3087. [CrossRef]
86. Nita, L.E.; Chiriac, A.P.; Diaconu, A.; Tudorachi, N.; Mititelu-Tartau, L. Multifunctional nanogels with dual temperature and pH responsiveness. *Int. J. Pharm.* **2016**, *515*, 165–175. [CrossRef]
87. Chen, X.; Chen, L.; Yao, X.; Zhang, Z.; He, C.; Zhang, J.; Chen, X. Dual responsive supramolecular nanogels for intracellular drug delivery. *Chem. Commun.* **2014**, *50*, 3789–3791. [CrossRef]
88. Sun, Z.; Yi, Z.; Cui, X.; Chen, X.; Su, W.; Ren, X.; Li, X. Tumor-targeted and nitric oxide-generated nanogels of keratin and hyaluronan for enhanced cancer therapy. *Nanoscale* **2018**, *10*, 12109–12122. [CrossRef] [PubMed]
89. Zhang, W.; Ai, S.; Ji, P.; Liu, J.; Li, Y.; Zhang, Y.; He, P. Photothermally Enhanced Chemotherapy Delivered by Graphene Oxide-Based Multiresponsive Nanogels. *ACS Appl. Bio. Mater.* **2018**, *2*, 330–338. [CrossRef]
90. Jia, X.; Han, Y.; Pei, M.; Zhao, X.; Tian, K.; Zhou, T.; Liu, P. Multi-functionalized hyaluronic acid nanogels crosslinked with carbon dots as dual receptor-mediated targeting tumor theranostics. *Carbohydr. Polym.* **2016**, *152*, 391–397. [CrossRef]
91. Khatun, Z.; Nurunnabi, M.; Nafiujjaman, M.; Reeck, G.R.; Khan, H.A.; Cho, K.J.; Lee, Y.K. A hyaluronic acid nanogel for photo-chemo theranostics of lung cancer with simultaneous light-responsive controlled release of doxorubicin. *Nanoscale* **2015**, *7*, 10680–10689. [CrossRef]
92. Nasr, F.H.; Khoee, S. Design, characterization and in vitro evaluation of novel shell crosslinked poly(butylene adipate)-co-N-succinyl chitosan nanogels containing loteprednol etabonate: A new system for therapeutic effect enhancement via controlled drug delivery. *Eur. J. Med. Chem.* **2015**, *102*, 132–142. [CrossRef] [PubMed]
93. Wei, X.; Xiong, H.; He, S.; Wang, Y.; Zhou, D.; Jing, X.; Huang, Y. A facile way to prepare functionalized dextran nanogels for conjugation of hemoglobin. *Colloids Surf. B Biointerfaces* **2017**, *155*, 440–448. [CrossRef]
94. Li, D.; Chen, Y.; Mastrobattista, E.; van Nostrum, C.F.; Hennink, W.E.; Vermonden, T. Reduction-Sensitive Polymer-Shell-Coated Nanogels for Intracellular Delivery of Antigens. *ACS Biomater. Sci. Eng.* **2017**, *3*, 42–48. [CrossRef]
95. Jin, H.; Tan, H.; Zhao, L.; Sun, W.; Zhu, L.; Sun, Y.; Hao, H.; Xing, H.; Liu, L.; Qu, X.; et al. Ultrasound-triggered thrombolysis using urokinase-loaded nanogels. *Int. J. Pharm.* **2012**, *434*, 384–390. [CrossRef]
96. Teng, Y.; Jin, H.; Nan, D.; Li, M.; Fan, C.; Liu, Y.; Lv, P.; Cui, W.; Sun, Y.; Hao, H.; et al. In vivo evaluation of urokinase-loaded hollow nanogels for sonothrombolysis on suture embolization-induced acute ischemic stroke rat model. *Bioact. Mater.* **2018**, *3*, 102–109. [CrossRef] [PubMed]
97. Mihalko, E.P.; Sandry, M.; Mininni, N.; Nellenbach, K.; Deal, H.; Daniele, M.; Ghadimi, K.; Levy, J.H.; Brown, A.C. Fibrin-modulating nanogels for treatment of disseminated intravascular coagulation. *Blood Adv.* **2021**, *5*, 613–627. [CrossRef] [PubMed]
98. Zhao, H.; Zhao, Y.; Xu, J.; Feng, X.; Liu, G.; Zhao, Y.; Yang, X. Programmable co-assembly of various drugs with temperature sensitive nanogels for optimizing combination chemotherapy. *Chem. Eng. J.* **2020**, *398*. [CrossRef]
99. Zhu, J.; Li, F.; Wang, X.; Yu, J.; Wu, D. Hyaluronic Acid and Polyethylene Glycol Hybrid Hydrogel Encapsulating Nanogel with Hemostasis and Sustainable Antibacterial Property for Wound Healing. *ACS Appl Mater. Interfaces* **2018**. [CrossRef]
100. Rejinold, N.S.; Ranjusha, R.; Balakrishnan, A.; Mohammed, N.; Jayakumar, R. Gold–chitin–manganese dioxide ternary composite nanogels for radio frequency assisted cancer therapy. *RSC Adv.* **2014**, *4*. [CrossRef]
101. Zan, M.; Li, J.; Huang, M.; Lin, S.; Luo, D.; Luo, S.; Ge, Z. Near-infrared light-triggered drug release nanogels for combined photothermal-chemotherapy of cancer. *Biomater. Sci.* **2015**, *3*, 1147–1156. [CrossRef]

102. Kimura, A.; Jo, J.I.; Yoshida, F.; Hong, Z.; Tabata, Y.; Sumiyoshi, A.; Taguchi, M.; Aoki, I. Ultra-small size gelatin nanogel as a blood brain barrier impermeable contrast agent for magnetic resonance imaging. *Acta Biomater.* **2021**. [CrossRef]
103. Wu, W.; Zhou, T.; Berliner, A.; Banerjee, P.; Zhou, S. Smart Core−Shell Hybrid Nanogels with Ag Nanoparticle Core for Cancer Cell Imaging and Gel Shell for pH-Regulated Drug Delivery. *Chem. Mater.* **2010**, *22*, 1966–1976. [CrossRef]
104. Peng, N.; Ding, X.; Wang, Z.; Cheng, Y.; Gong, Z.; Xu, X.; Gao, X.; Cai, Q.; Huang, S.; Liu, Y. Novel dual responsive alginate-based magnetic nanogels for onco-theranostics. *Carbohydr. Polym.* **2019**, *204*, 32–41. [CrossRef] [PubMed]

Article

Development, Characterization, and Evaluation of α-Mangostin-Loaded Polymeric Nanoparticle Gel for Topical Therapy in Skin Cancer

Shadab Md [1,2,3,*], Nabil A. Alhakamy [1,2,3], Thikryat Neamatallah [4], Samah Alshehri [5], Md Ali Mujtaba [6], Yassine Riadi [7], Ammu K. Radhakrishnan [8], Habibullah Khalilullah [9], Manish Gupta [10] and Md Habban Akhter [11,*]

[1] Department of Pharmaceutics, Faculty of Pharmacy, King Abdulaziz University, Jeddah 21589, Saudi Arabia; nalhakamy@kau.edu.sa

[2] Center of Excellence for Drug Research & Pharmaceutical Industries, King Abdulaziz University, Jeddah 21589, Saudi Arabia

[3] Mohamed Saeed Tamer Chair for Pharmaceutical Industries, King Abdulaziz University, Jeddah 21589, Saudi Arabia

[4] Department of Pharmacology & Toxicology, Faculty of Pharmacy, King Abdulaziz University, Jeddah 21589, Saudi Arabia; taneamatallah@kau.edu.sa

[5] Department of Pharmacy Practice, Faculty of Pharmacy, King Abdulaziz University, Jeddah 21589, Saudi Arabia; Salshehri1@kau.edu.sa

[6] Department of Pharmaceutics, Faculty of Pharmacy, Northern Border University, Rafha 91911, Saudi Arabia; mmujtaba@nbu.edu.sa

[7] Department of Pharmaceutical Chemistry, College of Pharmacy, Prince Sattam Bin Abdulaziz University, Al-Kharj 11942, Saudi Arabia; y.riadi@psau.edu.sa

[8] Jeffrey Cheah School of Medicine and Health Sciences, Monash University, Subang Jaya 47500, Malaysia; ammu.radhakrishnan@monash.edu

[9] Department of Pharmaceutical Chemistry and Pharmacognosy, Unaizah College of Pharmacy, Qassim University, Unaizah 51911, Saudi Arabia; h.abdulaziz@qu.edu.sa

[10] Department of Pharmaceutical Sciences, School of Health Sciences, University of Petroleum and Energy Studies (UPES), Dehradun 248007, India; manish.gupta@ddn.upes.ac.in

[11] School of Pharmaceutical and population Health Informatics (SoPPHI), DIT University, Dehradun 248009, India

* Correspondence: shaque@kau.edu.sa (S.M.); habban.akhter@dituniversity.edu.in (M.H.A.)

Citation: Md, S.; Alhakamy, N.A.; Neamatallah, T.; Alshehri, S.; Mujtaba, M.A.; Riadi, Y.; Radhakrishnan, A.K.; Khalilullah, H.; Gupta, M.; Akhter, M.H. Development, Characterization, and Evaluation of α-Mangostin-Loaded Polymeric Nanoparticle Gel for Topical Therapy in Skin Cancer. *Gels* **2021**, *7*, 230. https://doi.org/10.3390/gels7040230

Academic Editors: Emanuele Mauri, Chien-Chi Lin and Filippo Rossi

Received: 22 September 2021
Accepted: 19 November 2021
Published: 24 November 2021

Publisher's Note: MDPI stays neutral with regard to jurisdictional claims in published maps and institutional affiliations.

Copyright: © 2021 by the authors. Licensee MDPI, Basel, Switzerland. This article is an open access article distributed under the terms and conditions of the Creative Commons Attribution (CC BY) license (https:// creativecommons.org/licenses/by/ 4.0/).

Abstract: The aim of this study was to prepare and evaluate α-mangostin-loaded polymeric nanoparticle gel (α-MNG-PLGA) formulation to enhance α-mangostin delivery in an epidermal carcinoma. The poly (D, L-lactic-co-glycolic acid) (PLGA) nanoparticles (NPs) were developed using the emulsion–diffusion–evaporation technique with a 3-level 3-factor Box–Behnken design. The NPs were characterized and evaluated for particle size distribution, zeta potential (mV), drug release, and skin permeation. The formulated PLGA NPs were converted into a preformed carbopol gel base and were further evaluated for texture analysis, the cytotoxic effect of PLGA NPs against B16-F10 melanoma cells, and in vitro radical scavenging activity. The nanoscale particles were spherical, consistent, and average in size (168.06 ± 17.02 nm), with an entrapment efficiency (EE) of 84.26 ± 8.23% and a zeta potential of −25.3 ± 7.1 mV. Their drug release percentages in phosphate-buffered solution (PBS) at pH 7.4 and pH 6.5 were 87.07 ± 6.95% and 89.50 ± 9.50%, respectively. The release of α-MNG from NPs in vitro demonstrated that the biphasic release system, namely, immediate release in the initial phase, was accompanied by sustained drug release. The texture study of the developed α-MNG-PLGA NPs gel revealed its characteristics, including viscosity, hardness, consistency, and cohesiveness. The drug flux from α-MNG-PLGA NPs gel and α-MNG gel was 79.32 ± 7.91 and 16.88 ± 7.18 µg/cm^2/h in 24 h, respectively. The confocal study showed that α-MNG-PLGA NPs penetrated up to 230.02 µm deep into the skin layer compared to 15.21 µm by dye solution. MTT assay and radical scavenging potential indicated that α-MNG-PLGA NPs gel had a significant cytotoxic effect and antioxidant effect compared to α-MNG gel ($p < 0.05$). Thus, using the developed α-MNG-PLGA in treating skin cancer could be a promising approach.

Keywords: α-Mangostin; polymeric nanoparticle; skin cancer; gel; Behnken design; MTT assay; antioxidant assay

1. Introduction

Skin cancer is the most common malignancy affecting a large percentage of people in many countries worldwide [1]. The skin has a large surface area, subjected to several harmful environmental factors such as chemicals, toxic substances, and ultraviolet rays, which apparently can cause abnormal growth of cells and eventually cancers [2]. The skin cancer type is identified based on the cellular origin, either melanoma skin cancer (MSC) or nonmelanoma skin cancer (NMSC) [3]. Melanoma is the least common type, but it is responsible for a high number of deaths among patients with skin cancer [4]. However, the incidence of NMSC is higher, and it includes two types: basal cell carcinoma (BCC) and squamous cell carcinoma (SCC) [5]. The cancerous cells may travel to a distant part of the body and metastasize to develop melanoma, which may remain undiagnosed and subsequently not be treated timely [6].

Chemotherapy is considered the standard approach for destroying skin cancerous cells. However, it has several undesirable outcomes in treating skin cancer, including reduced systemic effects and bioavailability [7]. In addition, chemotherapeutic agents cannot differentiate between normal and cancerous cells. Thus, nanotechnology emerges as a remarkable solution for drug delivery [8]. This technology utilizes polymer nanoparticles (NPs) that allow the primary drug to reach the targeted region using a delivery vehicle that is versatile, biocompatible, biodegradable, and nanosized. In cancer, prolonged circulation of NPs enriches drug availability to the tumor tissues through direct permeation and diffusion to the cells or enhanced permeation and retention effects [9–12]. The transport of NPs across distinguished skin layers may face difficulty in permeation and penetration, probably due to the hard and horny multilayer of stratum corneum, followed by dermis barriers. The NPs may transport the drug through the skin's appendages, such as hair follicles and sebaceous and sweat glands. Hair follicles may act as reservoirs for polymeric NPs due to their deeper penetration and release of drugs into different skin layers, including the stratum corneum. Particles smaller than 10 nm are easily excreted out and phagocytized. NPs are internalized in the target cells via a series of biological events, including early and late endosome, thereby fusing with the lysosome, and diffusing into the cytoplasm, and finally destined in the nucleus of the cells. They may also internalize into cells through pinocytosis or phagocytosis process [13]. The transport of NPs into the tumor region mainly relies on physicochemical features, including size, surface, topography, and nano-biointeraction in the biological milieu [14].

A gel is a three-dimensional swellable polymeric network with hydrophilic groups cross-linked by a strong force of interaction that can incorporate both hydrophilic and lipophilic drugs. It has excellent biocompatibility, biodegradability, flexibility, drug stability, and controlled drug release. Moreover, it has a tendency to retain a large quantity of aqueous fluid without dissolving onto them and is widely explored in drug delivery and biomedical engineering. The polymeric network in the gel system plays a significant role in determining physicochemical properties and hydrophilicity, and controlling particle size; small pores in the polymeric network enables the accommodation of small molecules [15].

In the current era, bioactive agents from natural sources have been utilized enormously in pharmaceutical and nutraceutical preparations due to their wide range of biological activity. Interestingly, incorporating nanotechnology and phytomedicine may provide alternative treatment strategies for cancer as these bioactive agents normally cause no harm and most of them only exhibit minimal side-effects and toxicity at appropriate doses [16]. The activity of the phytoconstituents can be enhanced by encapsulating them in polymeric nanocarriers. For this purpose, several phyto-based NPs are designed and developed for site-specific targeting to improve drug therapeutic efficacy and potency.

Polymeric NP/gel is composed of poly D,L-lactic-co-glycolic acid (PLGA) a biodegradable linear copolymer produced by polymerization of monomeric units of lactic and glycolic acid that leads to nontoxic degraded products, i.e., water and carbon dioxide. The degradation rate of PLGA depends on the lactic acid/glycolic acid ratio and the molecular weight of the polymer. Polyglycolic acid has low water solubility due to its crystalline hydrophilic nature. On the other hand, the polylactic acid counterpart is somehow a stiff, hydrophobic polymer with low mechanical strength [17,18].

Alpha-Mangostin (α-Mangostin) is a natural xanthin found in the bark and fruit of the tropical plant *Garciniamangostana* Linn. The fruit is dark and reddish, with a soft, juicy, sweet, acidic pulp. The pericarp of the fruit is used in traditional medicine to treat various disorders, including skin infections. The main constituent of α-mangostin is xanthone glycoside, 1,3,6,7-tetrahydroxyxanthone-C2-β-d glucoside (Figure 1) [19,20]. It has several biological activities such as antioxidant, anticancer, anti-inflammatory, antiaging, antiviral, and antidiabetic effects [21,22]. The antioxidant properties are attributed to polyphenols, which combat free radicals (ROS) generation, thereby preventing DNA and cell/tissue damage. The generated free radicals require detoxification and should be eliminated from the body to avoid the occurrence of cancers.

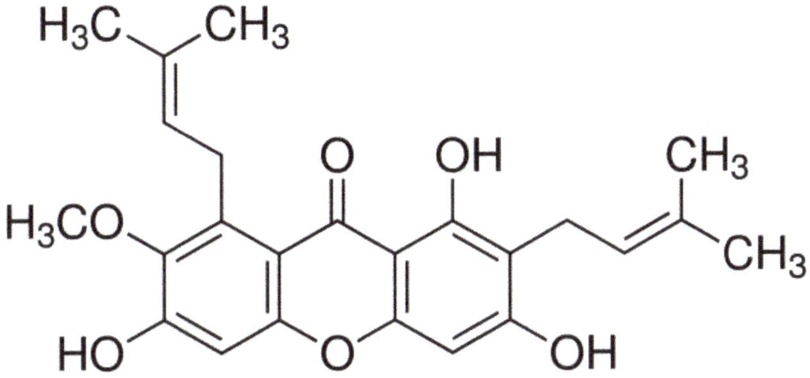

Figure 1. Chemical structure of α-MNG.

The current study aimed to design, optimize, and develop α-MNG-PLGA NPs and analyze the skin permeation and retention of the drug from nanocarriers within the distinguished layer of animal skin [23,24]. The formulation was tested in a melanoma cell line for its anticancer and antioxidant properties. Moreover, a stability study of the developed formulation was evaluated.

2. Results and Discussion

2.1. Optimization

A Box–Behnken experimental design was utilized for optimizing the formulation. It is a commonly used design as it costs less to run the experiment within the stipulated period, and it is a more efficient optimization technique than the conventional optimization procedure. The independent variables were polymer concentration, surfactant concentration, and sonication time (ST) and their impact effects were studied on critical quality attributes, such as particle size (PS), PDI, and entrapment efficiency (EE) of PLGA NPs. The dependent variables were used in three levels: low (-1), medium (0), and high ($+1$); accordingly, at these levels, the polymer concentrations (X1) were 2.5 mg (low); 3.00 mg (medium), and 3.50 mg (high); surfactant concentrations (X2) were 1% w/v (low), 1.75% w/v (medium), and 2.50% w/v (high); ST (X3) were 4 min (low), 8.00 min (medium), and 12.00 min (high), respectively. The low (-1) and high ($+1$) ranges of the independent variables were selected based on a preliminary examination and were found to be suitable for robust and consistent formulation. At the concentration of $X_1 > 3.00\%$ w/v, the PS was too large; below 2.5% w/v, the EE and drug loading were too low. Furthermore, ST > 12.00 min caused a further reduction in PS, which led to surface charge generation and ultimately caused the reaggregation of particles, which further impeded the stability of the formulation. ST < 4.00 min did not appropriately reduce PS to the desired size. The independent variables levels (low, medium, and high) and corresponding responses (minimize or maximize) are given in Table 1. The experimental design generated 17 distinguished formulations of varying compositions and responses (Table 2). The five replicas as shown in Table 2 were software-generated and an in vitro investigation revealed that they had varied particle size (nm) (Y_1); PDI (Y_2); and entrapment efficiency (%) (Y_3) but variation was insignificant. Similar results were observed in a previous study [25–30].

Table 1. The independent variables and their levels low (0), medium (−1), and high (+1) used in optimization α-MNG-PLGA NPs using experimental design.

Independent Variables	Level Used (Coded)		
	Low (−1)	Medium (0)	High (+1)
X1: Polymer concentration (% w/v)	2.5	3.00	3.50
X2: Surfactant concentration (% w/v)	1.00	1.75	2.50
X3: Sonication time (min)	4	8	12
Dependent variables			
Y_1: Particle size (nm)		Minimize	
Y_2: PDI		Minimize	
Y_3: Entrapment efficiency (%)		Maximize	

Table 2. The analyzed responses in experimental design for the preparation of α-MNG-PLGA NPs.

Formulation Code	Independent Variables			Responses		
	X1 (% w/v)	X2 (% w/v)	X3 (min)	Y_1 (nm)	Y_2	Y_3 (%)
* NP1	3.00	1.75	8.00	176 ± 19.04	0.205 ± 0.02	78 ± 8.67
NP2	2.50	1.00	8.00	187 ± 24.12	0.302 ± 0.03	68 ± 6.12
* NP3	3.00	1.75	8.00	171 ± 15.67	0.210 ± 0.04	79 ± 9.34
NP4	3.50	1.75	12.00	167 ± 13.45	0.233 ± 0.02	83 ± 7.98
* NP5	3.00	1.75	8.00	172 ± 21.81	0.206 ± 0.01	74.9 ± 6.80
NP6	3.00	2.50	12.00	125 ± 10.01	0.276 ± 0.05	75 ± 10.07
NP7	2.50	1.75	4.00	230 ± 33.76	0.281 ± 0.01	51 ± 9.23
NP8	3.50	1.00	8.00	172 ± 11.93	0.215 ± 0.03	83 ± 9.65
NP9	3.00	2.50	4.00	198 ± 19.62	0.306 ± 0.04	72 ± 11.92
NP10	2.50	2.50	8.00	167 ± 17.12	0.321 ± 0.02	53 ± 4.47
* NP11	3.00	1.75	8.00	173 ± 16.02	0.199 ± 0.01	76 ± 11.21
NP12	3.50	2.50	8.00	162 ± 14.03	0.268 ± 0.02	90 ± 12.76
NP13	3.00	1.00	4.00	194 ± 24.98	0.203 ± 0.07	74 ± 9.67
NP14	3.50	1.75	4.00	174 ± 17.54	0.222 ± 0.09	81 ± 8.78
NP15	3.00	1.00	12.00	166 ± 19.32	0.312 ± 0.06	84 ± 7.43
* NP16	3.00	1.75	8.00	172 ± 21.78	0.212 ± 0.03	77 ± 9.56
NP17	2.50	1.75	12.00	132 ± 12.67	0.321 ± 0.05	61 ± 7.12

X1: Polymer concentration (% w/v); X2: Surfactant concentration (% w/v); X3: Sonication time (min); Y_1: Particle size (nm); Y_2: PDI; Y_3: Entrapment efficiency (%). * Replicas.

The linear correlation plots between predicted vs. actual, perturbation chart, and interaction plot have shown a good fit for the PS, PDI, and EE responses, as shown in Figure 2a–c. The 3D response surface plot suggesting the comparative impact of factors on PS, PDI, and EE is shown in Figure 3A–C. The best-fitted model with a coefficient of correlation (R^2)~1 was selected for further study after the data fitted to a different model. Models with a good fit for every response were taken with ANOVA by estimating F values. Table 3 shows the results of the regression analysis for the selected responses, PS (Y_1), PDI (Y_2), and EE (Y_2), for adjustment into the polynomial equation. The optimum values of different responses, PS, PDI, and EE, were obtained using the point prediction method on desirability criteria.

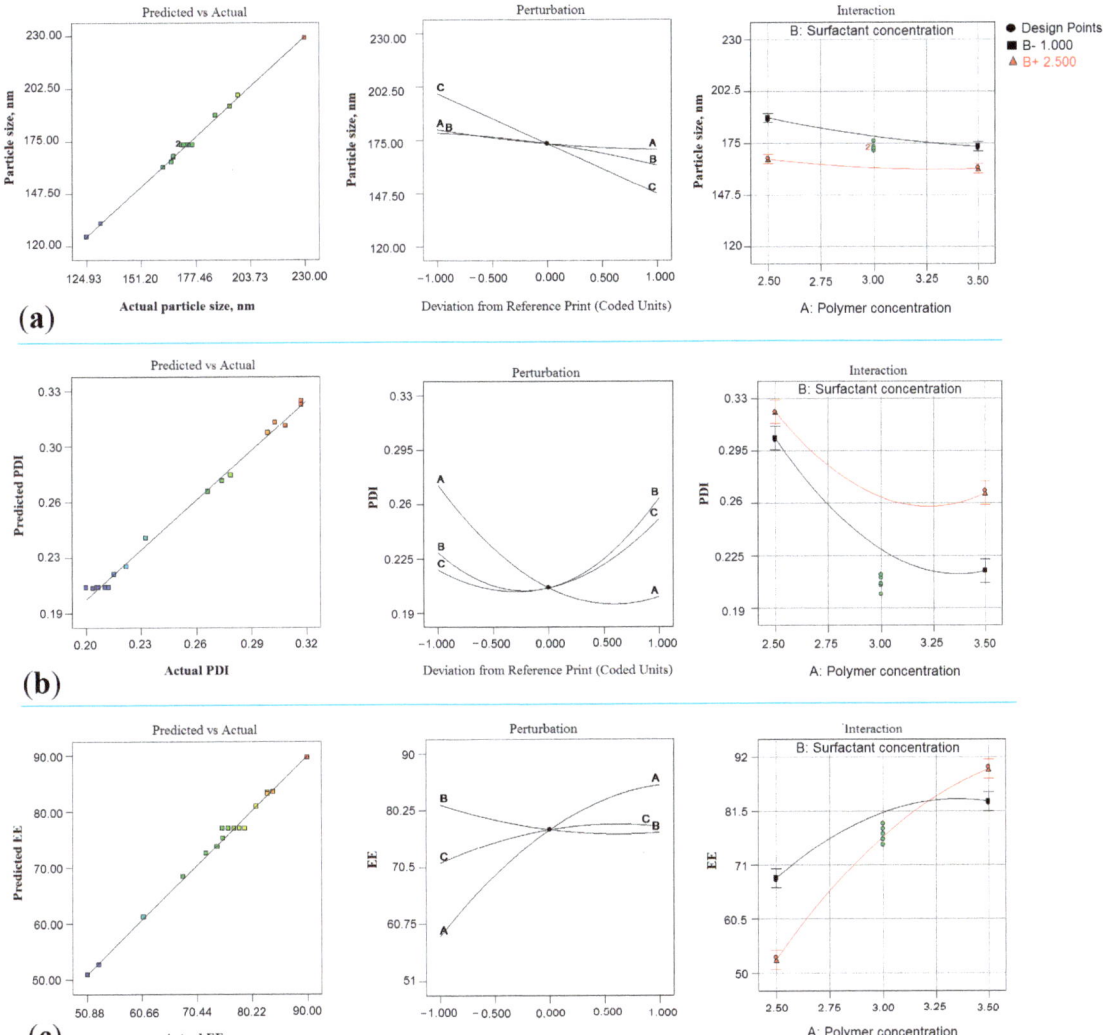

Figure 2. Model diagnostic plot representing linear correlation plot between predicted vs. actual, perturbation chart and interaction plot for responses particle size, Y_1 (**a**); entrapment efficiency, Y_2 (**b**) and PDI, Y_3 (**c**). (A = X_1 = Polymer concentration; B = X_2 = Surfactant concentration; and C = X_3 = Sonication time).

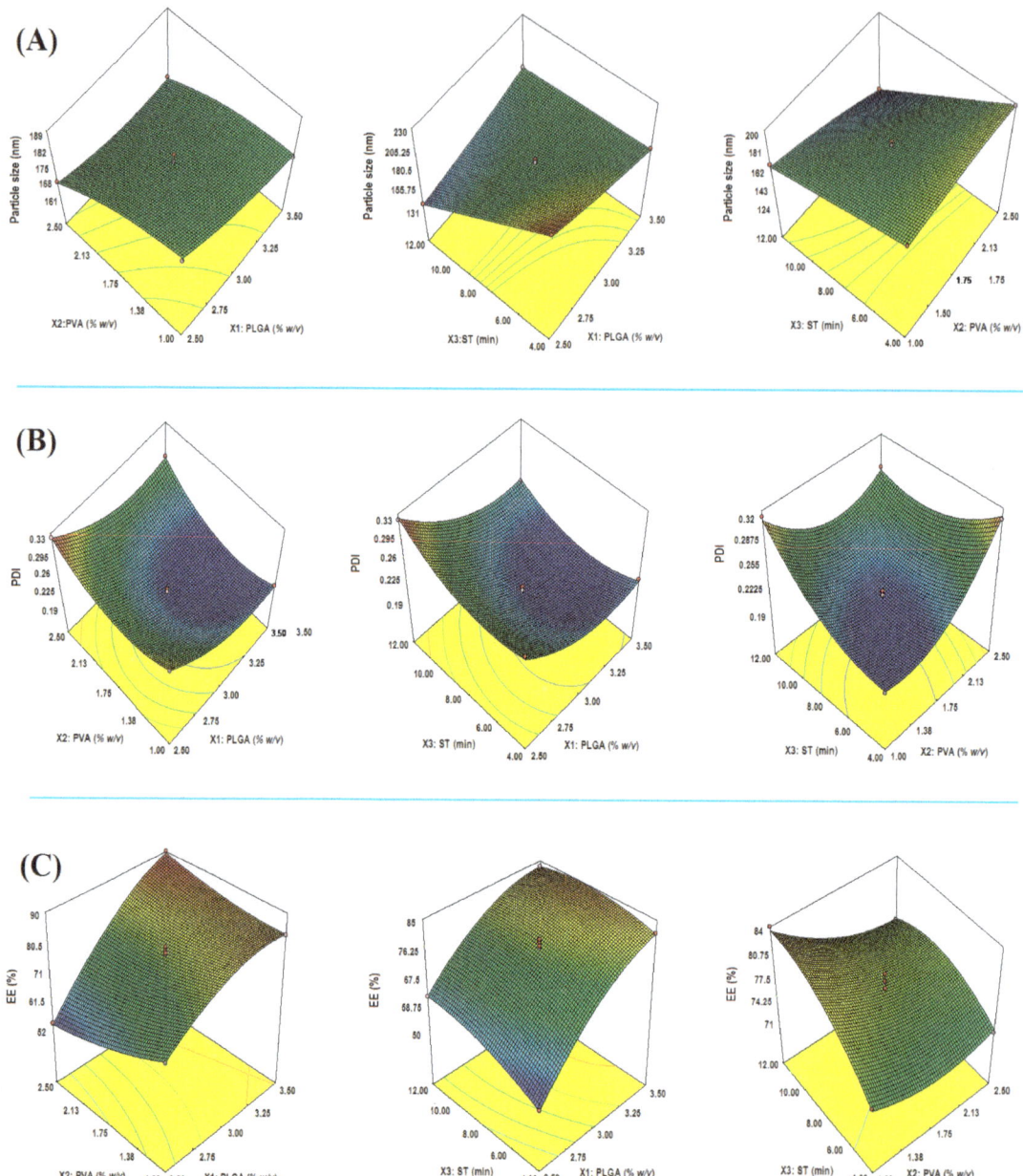

Figure 3. 3-D response surface morphology (**A**–**C**) exemplifying comparative impact of input attributes; PLGA (% *w/v*), PVA (% *w/v*), and ST (min) on critical quality attributes; particle size (**A**), PDI (**B**) and entrapment efficiency (**C**).

Table 3. Regression analysis summary for responses particle size (PS; Y_1), polydispersity index (PDI; Y_2), and entrapment efficiency (EE; Y_3) for fitting into quadratic equation.

Response Surface Quadratic Model	R-Squared	Adj R-Squared	Pred R-Squared	Adeq Precision	PRESS	% CV	Mean	SD
Response 1 (Y_1)	0.9970	0.9940	0.9853	79.146	128.59	1.04	172.82	1.81
Response 2 (Y_2)	0.9942	0.9867	0.9476	28.134	1.874×10^{-3}	2.16	0.25	5.457×10^{-3}
Response 3 (Y_3)	0.9935	0.9852	0.9799	39.150	36.26	1.74	74.11	1.29

PS = +173.16 − 5.12 × X1 − 8.37 × X2 − 25.75 × X3 + 2.50 × X1 × X2 + 22.75 × X1 × X3 − 11.25 × X2 × X3 + 2.14 × X1² − 2.86 × X2²

PDI = +0.21 − 0.036 × X1 + 0.017 × X2 + 0.016 × X3 + 8.500E − 03 × X1 × X2 − 7.250 − 0.03 × X1 × X3 − 0.035 × X2 × X3 + 0.03 × X1² + 0.040 × X2² + 0.028 × X3²

EE = +76.98 + 13.00 × X1 − 2.37 × X2 + 3.13 × X3 + 5.50 × X1 × X2 − 2.00 × X1 × X3 − 1.75 × X2 × X3 − 5.37 × X1² + 1.88 × X2² − 2.61 × X3²

2.2. Effect on Particle Size (Y_1)

The effects of PLGA, PVA, and ST on size are explicated by the quadratic equation as follows:

$$PS = +173.16 - 5.12 \times X1 - 8.37 \times X2 - 25.75 \times X3 + 2.50 \times X1 \times X2 + 22.75 \times X1 \times X3 - 11.25 \times X2 \times X3 + 2.14 \times X1^2 - 2.86 \times X2^2 \quad (1)$$

Model F-value of 334.52 entailed that the model was significant. Herein, X1, X2, X2, X1X2, X1X3, X2X3, X1², and X2² are significant model terms. The lack of fit (F-value) of 0.76 implies that the model was insignificant. "Pred R²" of 0.9970 is in reasonable agreement with "Adj R²" of 0.9940, as shown in Table 3. The adequate precision (signal-to-noise ratio) of 79.146 indicates an adequate signal. Therefore, the model was qualified for navigating the design space. In the quadratic equation, positive sign (+) indicates the synergistic effect and negative sign expressed antagonistic effect, explaining the individual, combined, and quadratic effects [28,29,31]. The effects of all the input attributes (X1, X2, and X3) on PS are shown in the 3D plot (Figure 3A).

2.2.1. Impact of PLGA

The PS significantly influences the fate of NPs in a biological fluid, such as the processes of dissolution, absorption, biodistribution, and cellular uptake. The PLGA concentration had a lesser negative impact on the PS of NPs. The ranges of PS were 125.00 ± 10.01 to 230.00 ± 33.76 nm. The PS at low PLGA concentration (2.5% *w/v*) was observed at 132 ± 12.67 nm (NP17) and 167 ± 17.12 nm (NP10). By increasing the PLGA concentration to 3.00% *w/v*, the maximum increase in PS was observed to be 176 ± 19.04 nm, as seen in formulation NP1 and 194 ± 24.98 nm in the case of formulation NP13. Furthermore, an increase in PLGA concentration reduced PS to 162 ± 14.03 nm (NP12), probably due to the interaction effect with a higher concentration of PVA (2.5% *w/v*) in this formulation composition. The results are consistent with those of the previously reported method [32].

2.2.2. Impact of PVA

The surfactant plays a significant role in controlling NPs size and generally has a negative impact. The increasing PVA (1.75% *w/v*) concentration reduced the PS to 132 ± 12.67 nm in the NP17 formulation. Furthermore, an increase in PVA concentration (2.5% *w/v*) led to a blending effect on the PS, namely, reduced PS to the maximum value of 125 ± 10.01 nm as seen in the NP6 formulation, whereas another formulation NP9 has shown increased PS up to 198 ± 19.62 nm. At a higher surfactant concentration (2.5% *w/v*), the PS increased to 198 ± 19.62 nm due to a significant reduction in the surface area, which led to the reagglomeration of nanoparticles. The surfactant was incorporated into the formulation to reduce the PS, prevent agglomeration, coalesceparticles, and stabilize the nanodroplets during the processing of the emulsion system [30,33]. The surfactant molecules stabilize the nanosystem by aligning themselves at the nanodroplet surface and reduce interfacial tension by lowering free energy [34].

2.2.3. Impact of Sonication Time (ST)

The ST had a strong negative impact on PS. At low ST (4 min), bigger PSs were 198 ± 19.62 nm and 230 ± 33.76 nm in formulations NP9 and NP7, respectively. Further increasing ST (8 min) led to a maximum PS reduction to 162 ± 14.03 nm, as seen in the formulation NP12. Increasing the ST to 12 min yielded reduced PS of 132 ± 12.67 nm and 125 ± 10.01 nm in formulations NP17 and NP6, respectively. The subjecting probe sonication to NPs dispersion for optimal time increases the collision frequency, which disintegrates the large particles and agglomerates into monodisperse and homogeneous dispersion [35].

2.3. Effect on PDI (Y_2)

The derived quadratic equation for PDI is given as follows.

$$PDI = +0.21 - 0.036 \times X1 + 0.017 \times X2 + 0.016 \times X3 + 8.500E-003 \times X1 \times X2 - 7.250\text{-}003 \times X1 \times X3 - 0.035 \times X2 \times X3 + 0.03 \times X1^2 + 0.040 \times X2^2 + 0.028 \times X3^2 \quad (2)$$

The model F-value of 132.56 entails that the model was significant. Herein, X1, X2, X3, X1X3, X2X3, and $X3^2$ are significant model terms. The lack of fit F-value of 1.41 entails that it was insignificant (p = 0.3622). "Pred R^2" of 0.9942 was in reasonable agreement with "Adj R^2" of 0.9967, as shown in Table 3. The model diagnostic plot representing a linear correlation between predicted vs. actual, perturbation chart, and interaction plot for response PDI is shown in Figure 2b. The impact of all the input attributes (X1, X2, and X3) on PDI is shown in the 3D plot (Figure 3B).

2.3.1. Impact of PLGA

As illustrated in the response surface curve, the increasing polymer has a negative influence on PDI. At low PLGA concentration (2.50% w/v) PDI was recorded as 0.302 ± 0.03 (NP2), 0.321 ± 0.02 (NP10), and 0.321 ± 0.05 (NP17), respectively. Further, by increasing PLGA concentration to (3.00% w/v), PDI decreased as to 0.276 ± 0.05 (NP6); 0.212 ± 0.03 (NP16); and 0.206 ± 0.01 (NP5), respectively. However, at a higher concentration of PLGA (3.5% w/v), PDI in some formulations was raised to 0.268 ± 0.02 (NP12), may be due to the large polymer concentration. The result is consistent with that of a previous report [36].

2.3.2. Impact of PVA

At a low PVA concentration (1.00 w/v), PDI was 0.302 ± 0.03 (NP2) and 0.203 ± 0.07 (NP13), respectively. Furthermore, by increasing PVA concentration to 1.75% w/v, PDI was dropped to 0.199 ± 0.01 as seen in formulation NP11. However, at a higher concentration of the surfactant (2.5 % w/v), the PDI slightly increased, probably due to the combined effect of PLGA. The results agree with those of previous work [37]. As per the polynomial equation, PVA has a less positive impact on PDI. The low PDI with narrow size distribution is considered effective and ideal for NPs in drug delivery. The low value of PDI indicates the homogeneity and uniform distribution of PS in the developed formulation. The international standard organization has stated that the PDI value <0.05 indicates monodisperse size distribution, whereas value >0.7 is classified as a polydisperse size distribution [38].

2.3.3. Impact of ST

ST has a less positive impact on the PDI of the formulation. The PDI of drug-loaded nanoparticles ranges between 0.199 ± 0.01 and 0.312 ± 0.06. The impact of independent variables on the NPs' colloidal properties, especially the PS and PDI, is important as it determines the fate of NPs penetrating ability across cutaneous barriers [31]. The polydispersity measures the heterogeneity and size distribution of the sample as the process of agglomeration is high in nanoparticles. At ST of 4 min, the PDI was observed 0.306 ± 0.04

(NP9); at 8 min, further increasing of the blend effect was observed and PDI was reduced to 0.205 ± 0.02 in NP1 and 0.210 ± 0.04 in NP3 and increased to 0.302 ± 0.03 in NP2 and 0.321 ± 0.02 in NP10. At ST of 12 min, the PDI was 0.312 ± 0.06 in formulation NP15. The excess reduction in particles leads to reaggregation of particles, resulting in a larger particle diameter, which may be attributed to the increased PDI of the formulation. Despite the optimized formulation reported, the PDI value of 0.201 ± 0.01 (Table 4) indicates that the distribution of particle population has changed from a bimodal to unimodal shape [39,40].

Table 4. Composition, experimental vs. predicted value with percentage error of optimized α-MNG-PLGA NPs.

Variables	Optimum Composition	Response	Observed Value of Response	Predicted Value of Response	Percentage Error
X1	3.39 % w/v	Y_1	168.06 ± 17.02	150.87	11.39
X2	1.82 % w/v	Y_2	0.201 ± 0.01	0.214	−6.07
X3	8.79 min	Y_3	84.26 ± 8.23	79.16	6.44

Predicted error (%) = (observed value − predicted value)/predicted value × 100%. X1: Polymer concentration (% w/v); X2, Surfactant concentration (% w/v), X3: Sonication time; Y_1: Particle size (nm), Y_2: PDI, Y_3: Entrapment efficiency (%).

2.4. Y3: Effect of on EE

The generated quadratic equation for EE is as follows:

$$EE = +76.98 + 13.00 \times X1 − 2.37 \times X2 + 3.13 \times X3 + 5.50 \times X1 \times X2 - 2.00 \times X1 \times X3 − 1.75 \times X2 \times X3 − 5.37 \times X1^2 + 1.88 \times X2^2 − 2.61 \times X3^2 \quad (3)$$

The model F-value of 119.42 entails that the model was significant. In this case, X1, X3, X1X2, X1X3, and X2X3 are significant model terms. The lack of fit of 0.16 implies that it was insignificant, while "Pred R^2" of 0.9799 was in reasonable agreement with "Adj R^2" of 0.9852 (Table 3). The adequate precision (signal-to-noise ratio), 39.150, indicates an adequate signal. Thus, the model is suitable for navigating the design space. The impact of all the input attributes (X1, X2, and X3) on EE is shown in the 3D plot (Figure 3C).

2.4.1. Impact of PLGA

The quadratic equation revealed that PLGA had a significant positive impact on EE of α-MNG in NPs. The EE of developed nanoparticles ranges from 51 ± 9.23% (NP7) to 90 ± 12.76% (NP12). Increasing the polymer increases the EE of a drug in NPs. The highest EE percentage reported for the formulation NP12 was 3.5% w/v of PLGA concentration. The EE mainly depends on the drug miscibility in the organic phase and drug-polymer interaction. Moreover, drugs solubilize in the polymeric solution due to enhanced emulsification properties [41]. The EE results agree with the findings regarding PLGA NPs reported in previous work [42].

2.4.2. Impact of PVA

The above quadratic equation expressed the negative effect of PVA on EE. At a lower surfactant concentration (PVA, 1% w/v), the EEs were estimated to be 68 ± 6.12%, 74 ± 9.67%, and 84 ± 7.43%, respectively. Furthermore, when the PVA concentration is increased to 1.75% w/v, EE declines to 77 ± 9.56% in NP16 and 61 ± 7.12% in NP17, respectively. When the PVA concentration increased to 2.5% w/v, the least EE was recorded: 72 ± 11.92% in NP9 and 53 ± 4.47% in NP10. The main objective of adding PVA in the formulation is to reduce the PS and obtain the optimum size range. The PS reduction decreases the core volume of the polymeric shell, thus reducing the amount of drug entrapped. The concentration and type of surfactant are essential in the formulation of NPs, which plays a crucial role in stabilizing and protecting the nanodispersion system from coalescence and maintaining the homogeneity of the formulation [43].

2.4.3. Impact of ST

ST has a less positive impact on the EE. The sonicating PLGA NPs for 4 min yielded EEs of 51 ± 9.23% in NP7 and 72 ± 11.92% in NP9. The maximum entrapment was 90 ± 12.76% by sonicating the formulation for 8 min (NP12), probably due to the combined effect of PVA and PLGA.

2.5. Numerical Optimization

The numerical optimization technique was used to select the formulation with criteria of small PS, PDI, and high encapsulation efficiency. The selected optimized preparation indicates desirability closer to ~1, representing that the optimized method is vital. The selection of minimum PS provides more surface area for drug solubility and dissolution and, in the end, improves the bioavailability of the formulation [44,45]. The statistically optimized PLGA NPs are comprised of PLGA of 3.39% w/v and PVA of 1.82% w/v under probe ST of 8.79 min. The average observed PS was recorded at 168.06 ± 17.02 nm vs. the predicted PS of 150.87 nm. The average observed PDI value of PLGA NPs was 0.201 ± 0.01 vs. the predicted value of PDI 0.214. Similarly, the average EE of drugs was 84.26 ± 8.23% vs. predicted EE at 79.16%. The percentage errors in PS, PDI, and ST were 11.39, −6.07, and 6.44, respectively (Table 4).

2.6. Characterization of α-MNG-PLGA NPs

The transmission electron micrograph has shown that the optimized formulation of nanoparticles was consistent, spherical, well-dispersed, uniform in size, and deaggregated (Figure 4). The developed PLGA NPs possess high drug EE with reproducible size. The sonication of the polymeric system significantly controlled the PS using the preparation technique. The low value of PDI clearly indicates monodispersed nanoparticulate system and the negative surface charge of the nanoparticle was produced, which provides the colloidal stability and predicts the fate of NPs in vivo.

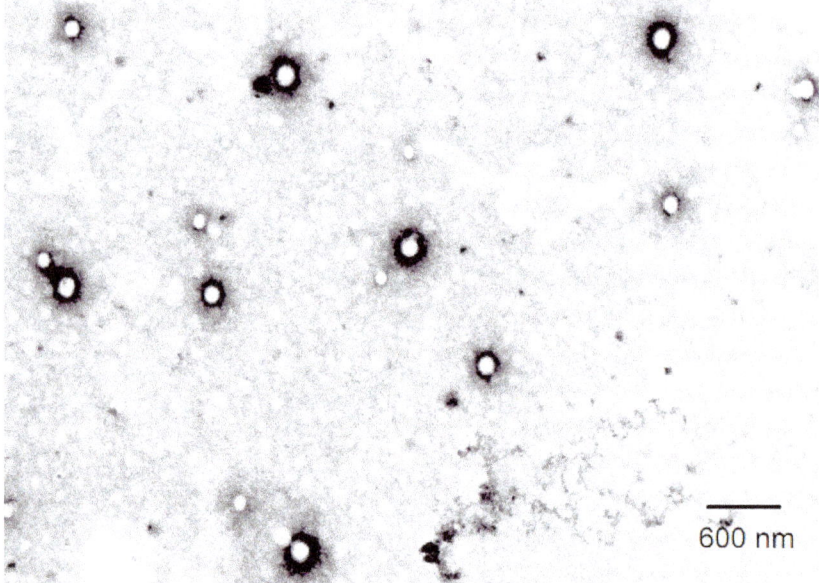

Figure 4. Transmission Electron microscopy image of Particle size α-MNG-PLGA NPs.

2.6.1. Differential Scanning Calorimetry of α-MNG NPs

The DSC thermogram is a vital tool to identify the nature and alteration in the melting range of excipients in the physical mixture and PLGA NPs formulation. The thermal analysis results of plain α-MNG and α-MNG-PLGA NPs are shown in Figure 5A,B. The DSC thermogram revealed the strong endothermic peak of plain α-MNG at a melting point of 183.04 °C, indicating the crystalline structure of α-MNG alone. However, due to some interaction between polymer and α-MNG, the crystallinity of the drug is reduced, which is evident from the lowering of the melting point [46].

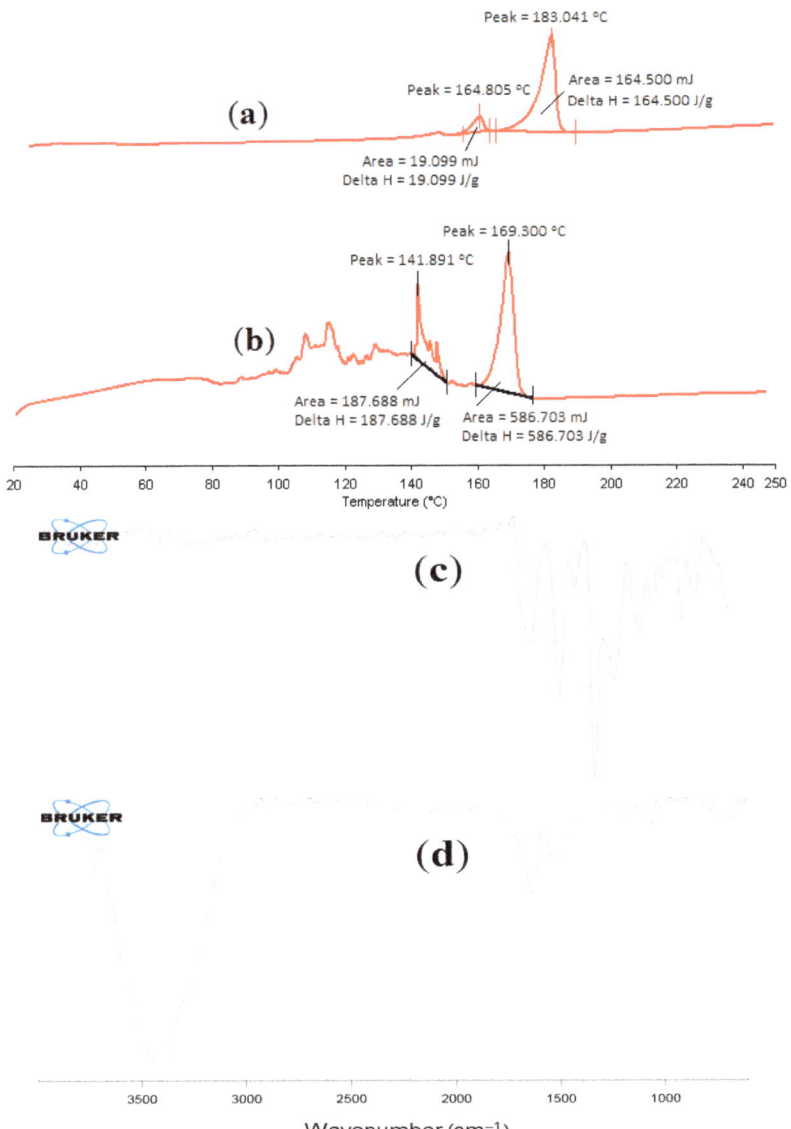

Figure 5. DSC thermogram showing melting point of plain α-MNG (**a**); and α-MNG-PLGA NPs (**b**). FT-IR of plain α-MNG (**c**) and α-MNG-PLGA NPs (**d**).

2.6.2. FT-IR Spectral Analysis of α-MNG NPs

IR spectroscopic technique reveals the chemical stability of the entrapment in the NPs. The FT-IR spectra of α-MNG and α-MNG-PLGA NPs are shown in Figure 5c,d. These absorption bands of α-mangostin were also observed in the α-MNG-PLGA NPs. However, they appear to diminish to a flat level, suggesting that α-MNG is not interacting with other excipients in the formulation and is considerably entrapped within the NPs. Thus, the chemical stability of α-MNG in the NP is corroborated [46,47].

2.7. Drug Release and Kinetic Study

The drug release profile of NPs at fixed time intervals is illustrated in Figure 6a,b. The results showed a biphasic release pattern; that is, drug release seems to burst in the early phase within 4 h (45.34%) accompanied by a sustained release profile over 24 h. The highest release of drug release from NPs at the end of the study (24 h) was 87.07 ± 6.95% in PBS at pH 7.4 compared to 17.43 ± 6.75% released from α-MNG dispersion. The minor drug release from drug dispersion may be due to the crystalline nature, poor solubility, and drug dissolution in PBS at a pH value of 7.4. Adversely, the drug release from polymeric NPs at pH 6.5 was 89.50 ± 9.50% vs. 21.48 ± 6.50% drug release from α-MNG dispersion. Moreover, drug release from polymeric NPs is significantly higher than plain α-MNG dispersion in both mediums ($p < 0.05$). However, the difference in the release under two different pH values was statistically nonsignificant ($p > 0.05$). The immediate drug release in contact with the dissolution medium from the polymeric surface or adsorbed drug and the controlled release from polymeric matrix comply with the biphasic drug release behavior from PLGA NPs. The drug release from PLGA NPs agreed with that reported in previous work [48]. The release profile of α-MNG from α-MNG-PLGA NPs was fitted to the different release kinetic models, for example, zero-order, Higuchi, first-order, and Hixson–Crowell models, to find the best suitable model. The regression value (R^2) obtained from various models indicated that the best-fitted model for α-MNG-PLGA NPs was the Higuchi model ($R^2 = 0.9793$). Furthermore, the underlying mechanism for drug release of α-MNG-PLGA NPs was assessed by applying the Korsmeyer–Peppas model, and the obtained n exponent valuewas 0.780 (between 0.5 and 0.89), which indicated that α-MNG release from NPs complied with non-Fickian diffusion [49]. The results displayed that α-MNG release from α-MNG-PLGA NPs is governed by a constant diffusion from the polymer matrix with the penetration of fluid, resulting in surface degradation and bulk erosion accompanied by coherent and ordered drug dissolution and succeeding release through diffusion in the acceptor compartment [50–52]. The drug release from α-MNG-PLGA NPs agrees with that reported in various studies on PLGA NPs and the research domain [53].

2.8. Gel Characteristics

The pH of the developed gel was slightly acidic, pH 6.72 ± 0.40, which did not cause skin irritation or erythema. The developed gel was of good appearance, consistency, and spreadability of 4.45 ± 0.45 cm within the acceptable range. The gel viscosity refers to the thickness of gel to flow over the skin. The viscosity of the α-MNG-PLGA NPs gel and the α-MNG gel was 3804.05 ± 102 and 3678.32 ± 302 cps, respectively. The gel is a stable colloidal particle with an excellent tunable and flexible surface, which can be easily manipulated with therapeutics. The soft and flexible architectureof the polymeric network makes penetration easy and improves the circulation time, thereby improving therapeutic drug delivery in vivo. The polymeric network can protect drug molecules and could release them intelligently in vivo. The gel characteristics in the current study were consistent with those of the previously reported carbopol gel [54].

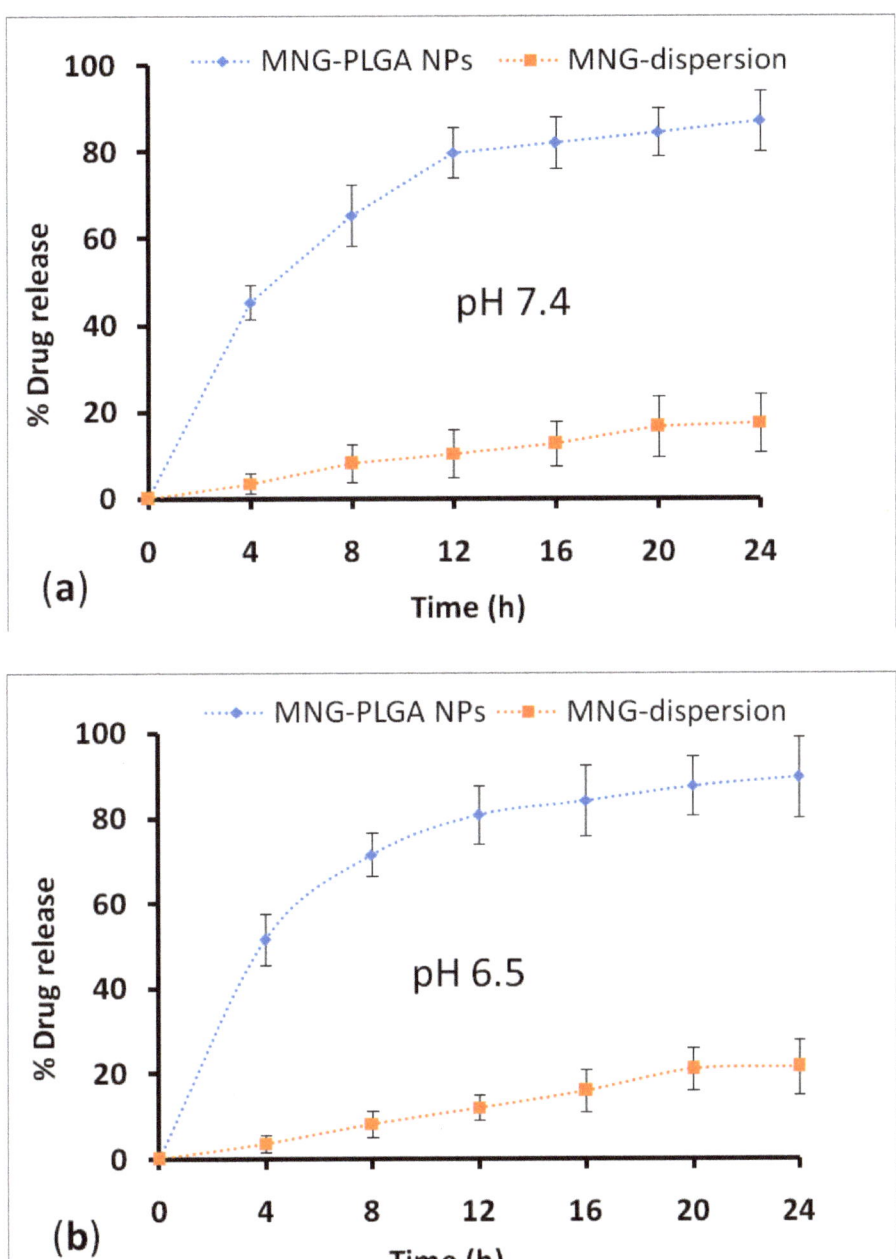

Figure 6. Percentage α-MNG release from α-MNG-PLGA NPs, and α-MNG-dispersion in PBS, pH 7.4 (**a**); and 6.5 (**b**), respectively.

2.9. Ex Vivo Skin Permeation

The drug diffusion and penetration study of α-MNG-PLGA NPs gel was carried out on rat skin using Logan's assembly (Model FDC-6, Effem Technologies, New Delhi, India). The drug permeation was assessed as steady-state flux (*fss*) in 24 h from the slope of the linear part of the graph. The processed animal skin was incorporated between donor and receptor chambers at a physiological temperature (37 ± 0.5°C). The maximum is determined at the end of the study on α-MNG-PLGA NPs gel, and the α-MNG gel was 79.32 ± 7.91 and 16.88 ± 7.18 µg/cm^2/h, respectively, as shown in Figure 7A. The *fss* obtained from α-MNG-PLGA NPs gel was significantly higher than the flux obtained from the α-MNG gel ($p < 0.05$). The drug delivery from NPs resulted in higher drug deposition in the target loci by enhancing stability, thereby enhancing the controlled and sustained release of therapeutic delivery. Furthermore, passively transported NPs after crossing epithelial skin barriers shows an enhanced permeation and retention (EPR) effect nearby the tumor microenvironment due to leaky vasculature, causing more apoptosis [55–57]. However, low flux from the α-MNG gel was probably due to less drug solubility, dissolution, and permeation through the cutaneous barrier. The carbopol gel increases in-depth penetration of therapeutic by improving contact time and circumventing loss of aqueous phase from the skin surface [58].

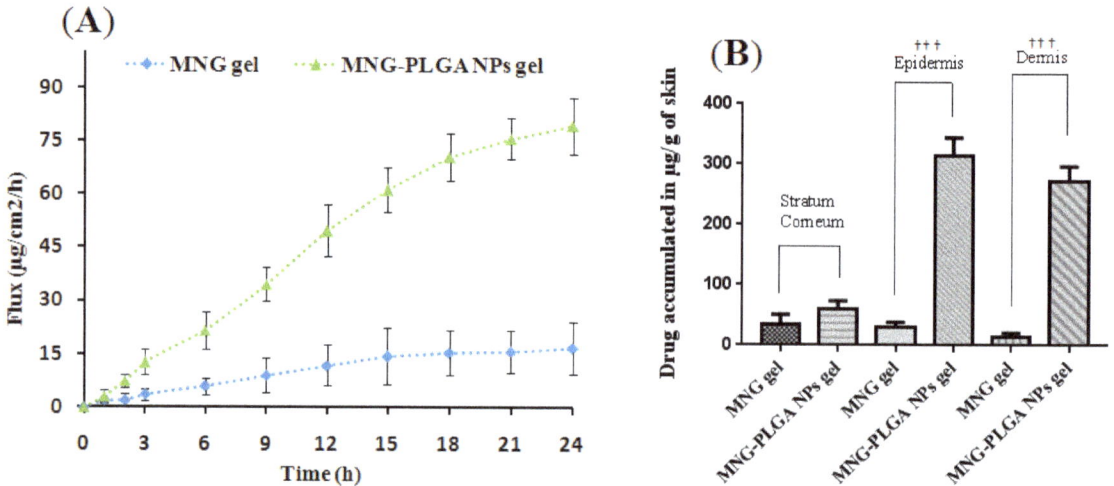

Figure 7. The steady state flux of α-MNG-PLGA NPs in skin permeation studies compared with flux of α-MNG gel (**A**). Concentration of drug retained in various layer of skin viz. stratum corneum, epidermis and dermis from α-MNG-PLGA NPs gel, α-MNG gel (Y), respectively (**B**). Data expressed as mean ± SD ($n = 3$) (††† $p \leq 0.01$).

Drug Estimation across Skin Layer

The concentrations of drugs from α-MNG-PLGA NPs gel retained in various strata of the skin such as stratum corneum (SC), epidermis layer (EL), and dermal layer (DL) were examined by HPLC and indicated as µg/g of skin tissues. The quantity of α-MNG deposited over SC from the α-MNG-PLGA NPs gel and α-MNG gel was 68.02 µg/g and 29.67 µg/g of skin tissue, respectively. Moreover, in EL, the quantities of drug from α-MNG-PLGA NPs gel and α-MNG gel deposited were 302 µg/g, and 51.21 µg/g of skin. Moreover, the DL had α-MNG-PLGA NPs gel and α-MNG gel concentrations of 267.78 and 21.03 µg/g of skin tissue, as shown in Figure 7B. The drug from α-MNG-PLGA NPs gel was significantly deposited in EL and DL compared to that from α-MNG gel ($p < 0.05$). The higher drug deposition from NPs may be attributed to the nanosize of particles, <200 nm, that circumvent the cutaneous barrier and improve physicochemical stability and compatibility with skin tissue.

2.10. Confocal Laser Microscopy

The fluorescent RhB dye incursion transversely to the skin layer was investigated using RhB solution, compared with RhB encapsulated PLGA NPs. The confocal micrograph (Figure 8A,B) revealed that the RhB solution penetrated up to the limited depth inside skin, 15.21 µm, whereas RhB-PLGA NPs (Figure 8C,D) achieved a depth of 230.02 µm inside the layer of skin. The variation in fluorescence intensity in every region of the tissue section indicates a time-dependent process; EL and DL showed more intensity for PLGA NPs than RhB solution. However, the fluorescence intensity of RhB-PLGA NPs and RhB solution at a predetermined time after topical administration showed a time-dependent enrichment of RhB-PLGA NPs in the in different strata of skin and hair follicles. The result indicated higher penetration of RhB from PLGA NPs due to the nanosize of particles (<200 nm) and better skin compatibility of the polymer.

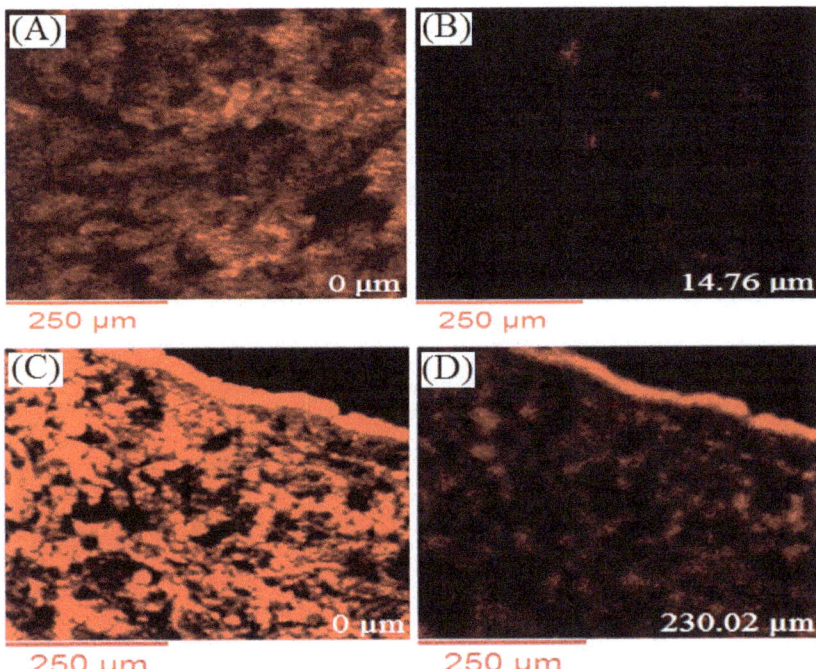

Figure 8. Confocal images of RhB solution (**A**,**B**); and RhB-PLGA NPs (**C**,**D**). Scale bar = 250 µm.

2.11. Cell Viability Assay

The MTT assay was performed using α-MNG gel and α-MNG-PLGA NPs gel, both containing an equivalent concentration of the drug, and blank PLGA NPs gel and blank gel to assess the cytotoxic effect of the formulation in comparison to α-MNG gel. The results showed that α-MNG-PLGA NPs gel decreased the viability of the cells in a time- and dose-dependent manner, as shown in Figure 9A,B. At the end of 24 h, the maximum cell viability percentages of treated cells with α-MNG gel and α-MNG-PLGA NPs gel were 89.67% and 41.56%, respectively. After 48 h, the cell viability was 18.50% for α-MNG-PLGA NPs gel and 80.87% for α-MNG gel. The comparative IC_{50} of α-MNG-PLGA NPs gel vs. α-MNG gel was estimated to be 10.00 ± 6.70 and 55.53 ± 6.70 µM after 24 h; 32.65 ± 3.45 and 5.63 ± 0.45 µM after 48 h of incubation in cell line. The observation demonstrated that α-MNG-PLGA NPs gel was more effective in inhibiting carcinoma cells proliferation and thus significantly enhanced apoptosis compared to α-MNG gel ($p < 0.05$). This significant cytotoxicity can be attributed to the high concentration of α-MNG released from PLGA NPs and passively transported inside the cells, thus enhancing the cytotoxic effect. The passive targeting strategy-mediated release of phyto active from the polymeric core could stabilize the therapeutics from protein fraction in blood component and, therefore, enhance the antiproliferation efficacy against melanoma cells [59]. Furthermore, the polymer releases drugs intracellularly through distinguished transporters located in the extracellular matrix of the cell membrane. The cell viability results were consistent with those in previous studies [60,61].

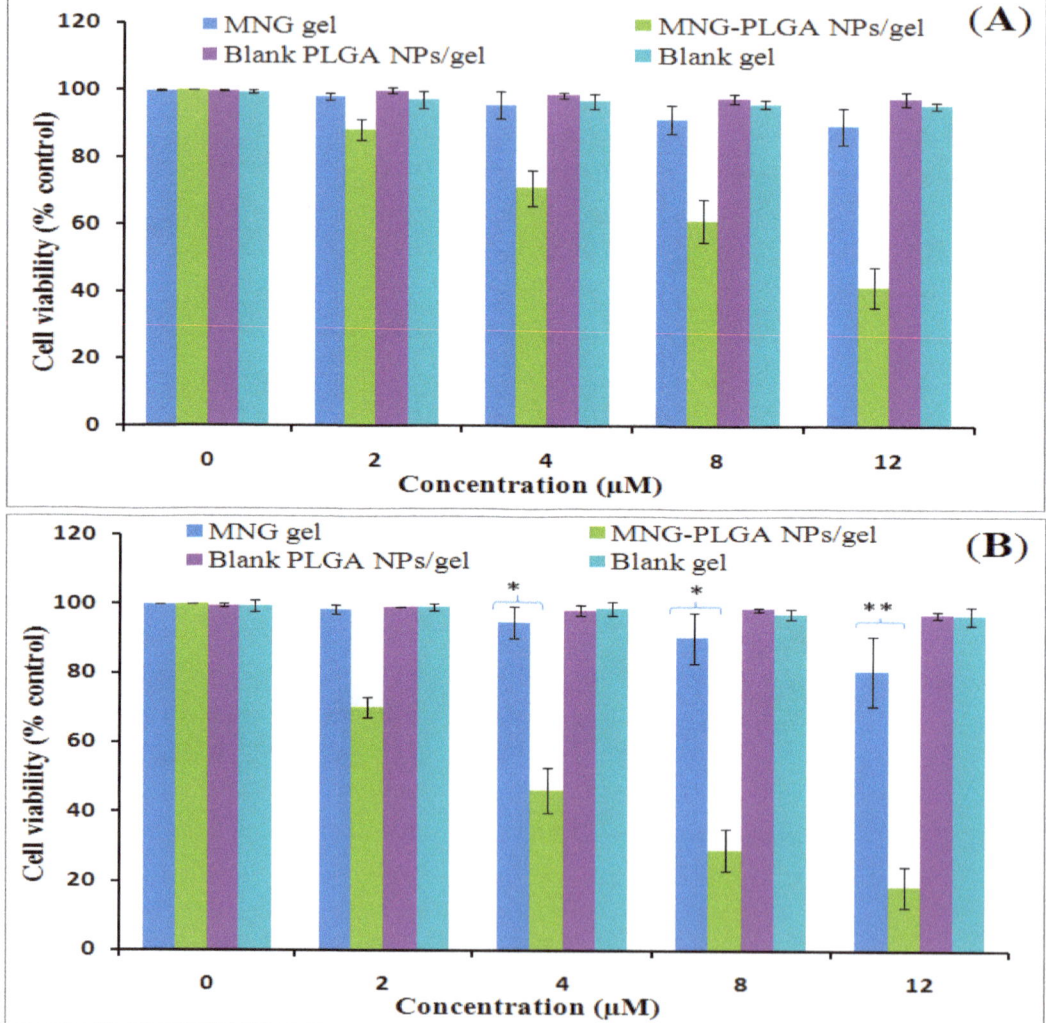

Figure 9. Cell viabilitystudy of α-MNG gel, α-MNG-PLGA NPs gel, blank PLGA NPs gel and blank gel at concentrations, (0, 2, 4, 8, and 12 μM) after incubation times at 24 h (**A**); and after 48 h (**B**) in skin cancer cell line. Data indicated as mean ± SD (n = 3); (* p < 0.05), (** p < 0.01) compared with α-MNG gel.

2.12. In Vitro Antioxidant Activity

DPPH assay was used to estimate the antiradical scavenging activity of α-MNG gel and α-MNG-PLGA NPs gel. Trolox equivalent antioxidant capacity (TEAC) was calculated from a standard curve of Trolox. The TEAC values of optimized α-MNG-PLGA NPs gel were 17, 31, 47, 72, and 87 µg Trolox equivalent per 10, 20, 40, 80, and 160 µg/mL of Trolox in DPPH assay, respectively (Figure 10A,B). The IC_{50} values reported were 32.44, 48.83, and 100.29 for α-MNG-PLGA NPs gel, standard Trolox, and α-MNG gel, respectively. The antioxidant activity of α-MNG-PLGA NPs gel was significantly higher than that of α-MNG gel ($p < 0.05$).

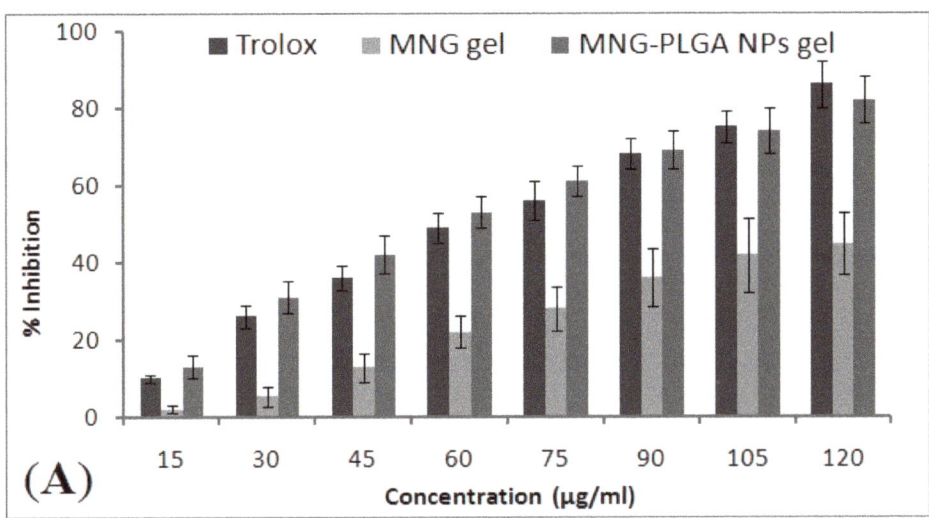

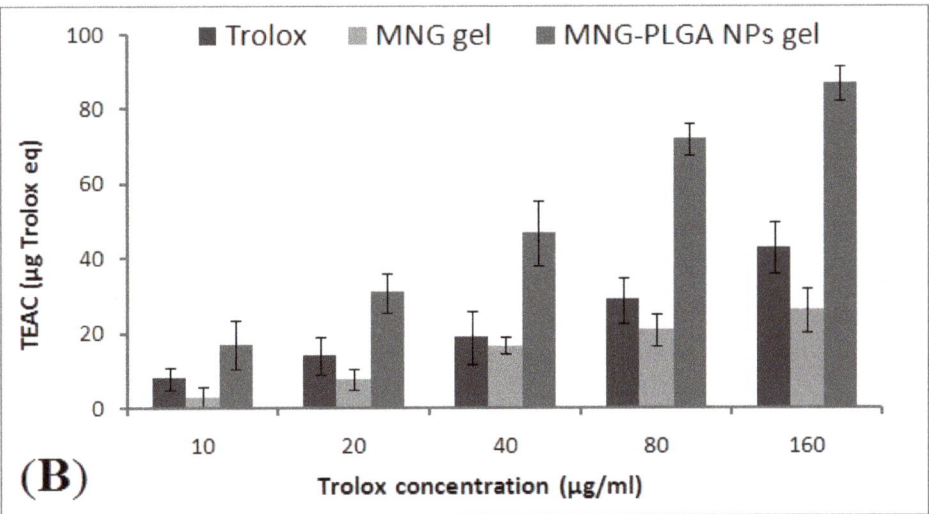

Figure 10. The DPPH antioxidant assay of α-MNG gel and α-MNG-PLGA gel. Antioxidant inhibition effect in percentage (**A**), Trolox equivalent antioxidant activity (TEAC) of α-MNG-PLGA gel compared with trolox and α-MNG gel (**B**). Values indicated as means ± SD ($n = 3$) analyzed by one way ANOVA with Bartlett's test for statistical significance ($p < 0.05$).

2.13. Stability Studies

A stability experiment studied for three months and a regular assessment at an interval of 30 days concluded that there is an insignificant alteration in the average PS, PDI, zeta potential, % drug entrapment, and % drug retained, which further indicates that the optimized formulation was stable within the specified period. The graphical presentation of the stability assessment profile of α-MNG-PLGA NPs gel under different experimental conditions is expressed in Figure 11.

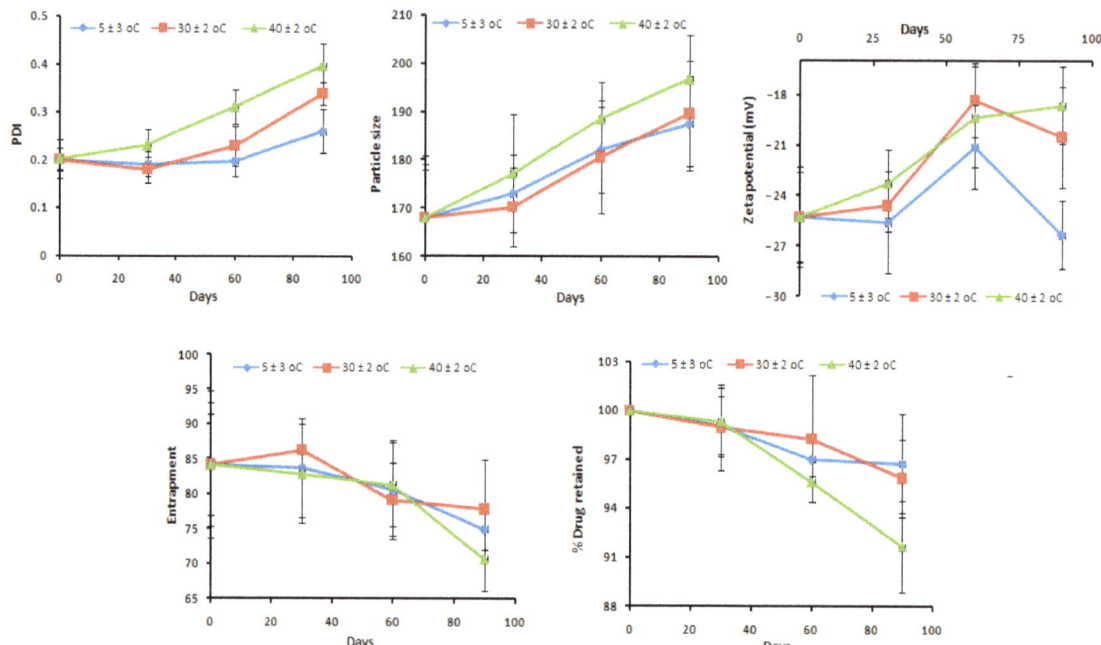

Figure 11. Three-month stability assessment profile of α-MNG-PLGA NPs gel in varying temperature (5 ± 3 °C, 30 ± 2 °C, and 40 ± 2 °C) condition.

3. Conclusions

The α-MNG-PLGA NPs were formulated using the emulsion–diffusion–evaporation technique and optimized successfully by applying a 3-level 3-factors Box–Behnken design. The α-MNG-PLGA NPs formulations compared well with the α-MNG gel, where the characterization parameters were consistent, noninteracting, and substantially suited to the study. The optimized formulation showed uniform particle distribution, good EE, and stability of the formulation. The selected factors successfully resulted in a positive impact on dependent variables by quadratic equation and well defined the in vitro characterizing fate of α-MNG-PLGA NPs. The analytical studies DSC and FT-IR corroborated that the distinct peak of a drug was also observed in polymeric NPs, which further validated that α-MNG is physicochemically stable and remains encapsulated inwardly to the polymer core of the nanoparticle. The drug release study of the formulation showed initial burst releases abiding by sustained drug release, and the drug release mechanism followed was non-Fickian diffusion of the Higuchi model. The developed formulation showed excellent flux across the skin layer in the skin permeation study, and skin probe fluorescent dye in confocal microscopy revealed significant penetration of NPs into the skin. MTT assay showed significant cytotoxicity of α-MNG-PLGA NPs gel in B16-F10 cell compared to α-MNG gel. Moreover, the radical scavenging activity of α-MNG in α-MNG-PLGA NPs gel showed a marked antioxidant effect compared to that in the α-MNG gel. Thus, α-MNG-PLGA NPs gel is a promising candidate in treating skin cancer and other skin disorders.

4. Materials and Methods

4.1. Materials

α-MNG (MW 410.46 g/mol; purity = 98%) was obtained from Sigma-Aldrich. PLGA [poly(D, L-lactide-co-glycolide)] lactide: glycolide (50:50) (Mw = 44,628 g/mol; viscosity, 0.71–1.0 dl/g) (Merck, India). Polyvinyl alcohol (PVA) was obtained from Sisco Laboratory (Mumbai, India) and Rhodamine B (RhB) from Sigma-Aldrich (St. Louis, MO, USA). The laboratory solvents, namely acetone, dichloromethane, acetonitrile, and deionized water, were procured from SD Fine Chemicals (Mumbai, India). LC-MS grade reagents, namely acetonitrile, methanol, ammonium acetate, ethyl acetate, glacial acetic acid, and water, were purchased from Sigma-Aldrich (St. Louis, MO, USA). The analytical grades of buffer reagents, i.e., phosphate-buffered saline (PBS), sodium dihydrogen phosphate, sodium hydroxide, disodium hydrogen phosphate, potassium dihydrogen phosphate, and ethanol, were procured from Central Drug House (New Delhi, India). For HPLC analysis, double distilled water was filtered through a micron-sized membrane filter (pore size, 0.45µ) and membrane filter (Durapore pore size, 0.21 µ) using an injectable needle.

4.2. Fabrication of α-Mangostin-Loaded PLGA Nanoparticle

As per our previously reported work, the sonication-tailored α-MNG NPs was constructed using the emulsion–diffusion–evaporation technique [30]. α-MNG (10 mg) and PLGA (2.5% *w/v*) were incorporated in organic solution, dichloromethane (2 mL), with sonication (one cycle, 30 kHz power, 60 W) to form the polymer-encapsulated α-MNG core. Then, the α-MNG polymeric solution was added to the surfactant solution (PVA, 1.5% *w/v*) at a slow flow rate (0.5 mL/min) and emulsified using a sonicator (Probe Sonicator, Hielscher Ultrasonics, Berlin, Germany) underneath an ice bath (one cycle, 30 kHz power, 80 W) to form nanodroplets. These nanodroplets were subjected to continuous magnetic stirring for 3 h, allowing evaporation of the organic phase, and the nanoparticulate suspension was dried overnight to harden the formed NPs. The NPs were separated using a cooling centrifuge (Optima™ LE-80K Ultracentrifuge) at 10,000× *g* for 30 min. Furthermore, the particles were washed thrice and separated from the unentrapped drug and stored in a freezer [39]. The preparation of α-MNG-PLGA NPs is outlined in Figure 12.

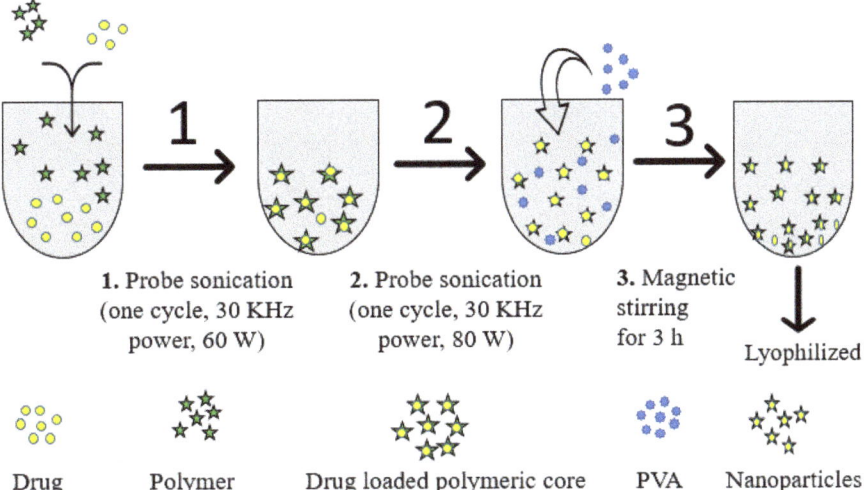

Figure 12. The schematic illustration of preparation of α-MNG-PLGA NPs.

4.3. Analytical Method

α-MNG content was quantified using the liquid chromatography-mass spectrometry (LC-MS) method. The column configuration was Agilent Zorbax® SB C_{18} (3.5μm; 150 × 3 mm) with guard cartridge C_{18} (4.2 × 2.2 mm). The analyte was separated in isocratic mode using a mobile phase combination of water, ammonium acetate, formic acid (0.1% v/v), and methanol at a 300 μL/min solvent flow rate.

4.4. Formulation Optimization

The Box–Behnken design was employed with three factors, three levels, and a total of 17 runs of the experiment using the Design-Expert software (Version 10; Stat-Ease Inc., Minneapolis, Minnesota) [28,29]. The levels of variables employed in the study are indicated in Table 1. The effects of concentrations of independent variables, that is, polymer (X1), surfactant (X2), and ST (X3), were examined on the responses, that is, PS (Y_1), PDI (Y_2) and % EE (Y_3), for the development of NP, as shown in Table 2. The second-order quadratic equations deduced from the design of the experiment for various responses are shown in Table 3. The multivariate linear regression equation generated, which details the relationship among independent and dependent variables, is as follows:

$$Y = A_0 + A_1Z_1 + A_2Z_2 + A_3Z_3 + A_{12}Z_1Z_2 + A_{23}Z_2Z_3 + A_{13}Z_1Z_3 + A_{11}Z_{12} + A_{22}Z_{22} + \alpha_{33}Z_{32}, \quad (4)$$

where Y indicates the measured responses of factors conflated with the level of each factor. A_0 indicates the intercept and A_1 to A_{33} are regression coefficients of the respective variables, while Z_1, Z_2, and Z_3 are the coded levels of independent variables. Z_1Z_2, Z_2X_3, Z_1Z_3, and Z_{12} (l = 1, 2, 3) show the interaction and quadratic impacts. The levels used in the experimental (base) design are low, medium, high, axis, and central point. After statistical analysis, the best-fitting model within various models (linear, quadratic, cubic, and 2FI) was taken. The F-value showed the best-fitted model; model fitting is shown by PRESS-value, and the lack of fit for the proposed model indicates that it is insignificant.

4.5. Nanoparticle Characterization

4.5.1. Particle Sizes and Their Distribution

The size of α-MNG-PLGA NPs and their distribution were analyzed using Zetasizer Nano ZSP (Malvern Instruments, Worcestershire, UK) based on a dynamic light scattering procedure. In brief, α-MNG-PLGA NPs (~2 mg) were diluted with deionized water (50 mL); and after sonication, 10 μL of sample volume was applied over the carbon-surfaced copper grid and then was negatively stained with 1% phosphotungstic acid. The studies were performed in triplicate (n = 3).

4.5.2. Drug Entrapment and Loading

The entrapped α-MNG and RbB in the polymeric dispersion were estimated before lyophilization, as previously reported, with minor modification [62]. The NPs were ultracentrifuged at high speed (10,000× g at 4 °C for 30 min) (REMI Cooling Centrifuge, India). The aliquots of both α-MNG-PLGA NPs dispersion and RhB-PLGA NPs were separated and subjected to estimation of α-MNG and RbB content using RP-HPLC. The drug entrapment efficiency (% EE) and drug loading were determined as follows:

$$\% \; EE = \frac{\text{cumulative quantity of } \alpha - \text{mangostin} \; - \; \text{quantity of } \alpha - \text{mangostin in the supernatant}}{\text{cumulative quantity of } \alpha - \text{mangostin}} \times 100$$

$$\% \; drug \; loading = \frac{\text{cumulative quantity of } \alpha - \text{mangostin} \; - \; \text{quantity of } \alpha - \text{mangostin on surface layer}}{\text{cumulative weight of } \alpha - \text{mangostin NPs}} \times 100$$

4.5.3. Transmission Electron Microscopy

The PS of the α-MNG-PLGA NPs was studied using transmission electron microscopy (TEM) (Techni TEM 200 Kv, Fei, Electron optics). The α-MNG NPs (1 mg/mL) sample were dispersed in distilled water and bath sonicated. The diluted NPs were disseminated onto the porous film grid and then dried for 10 min, and microscopy was executed at 100 kV, and the image was captured.

4.5.4. Infrared Spectroscopy

Fourier transform-infrared spectroscopy (FT-IR) of plain α-mangostin and α-MNG-PLGA NPs was assessed utilizing an FT-IR spectrometer (BRUKER Corporation, Billerica, MA, USA). The weighed quantity of plain drug and drug-loaded PLGA NPs ~ 5 mg was incorporated in direct contact with a light beam, and the spectrum was recorded in the scanning range of 4000–400 cm^{-1}.

4.5.5. Differential Scanning Calorimetry

The changes in the enthalpy of plain α-mangostin and α-MNG-PLGA NPs were described by differential scanning calorimetry (Pyris 4 DSC, Perkin Elmer, Waltham, MA, USA). The samples (5 mg) were transferred in aluminum pans, and on the other pan, the reference standard was placed. Both pans were heated simultaneously at a scanning rate of 20 °C/min (20–250 °C) using dry nitrogen gas as effluent gas.

4.5.6. Drug Release and Kinetics Studies

The in vitro drug release was executed for α-MNG-PLGA NPs and compared to α-MNG dispersion. The weighed quantities of NPs, 10 mg of α-MNG, and an equivalent dose α-MNG dispersion were enclosed in a dialysis bag (MW; 8–12 kDa; (Repligen, Waltham, MA, USA) with its ends fixed with a clip. Then, they were transferred to a beaker containing 100 mL of preheated (37 ± 0.5 °C) medium of PBS of pH values of 7.4 and 6.5, which was kept under gentle agitation at 100 rpm [63]. The dialysis bag was activated before the release study under treatment with PBS solution. At predetermined time intervals (0, 4, 8, 12, 16, 20, and 24), 0.5 mL of supernatant was taken from the dissolution medium and filtered using a Durapore® membrane filter of pore size 0.21 µ, using an injectable syringe. The withdrawn medium volume was supplanted by the new medium, so the sink condition is ascertained. The drug concentration was analyzed using RP-HPLC. Moreover, the optimized α-mangostin formulation was fitted to kinetic release models, such as Higuchi, Korsmeyer–Peppas, Hixson–Crowell, first-order, and zero-order model, using the graphical method. The model with good data fit was predicted.

4.5.7. Preparation of Gel

The previously reported technique was employed to prepare the gel base (Kausar et al., 2019). The carbopol 934 (1.0% *w/w*) was magnetically stirred in distilled water (10 mL) for 2 h. In addition, propylene glycol, methyl- and propyl paraben (qs), and triethanolamine (qs) were incorporated with undisturbed stirring until a transparent clear gel was formed. Furthermore, α-MNG-PLGA NPs and α-MNG dispersions were uniformly thrown into the preformed base of gel with uninterrupted stirring. Moreover, the characteristics of the developed gel were analyzed using a gel texture analyzer (TA.XT Plus Texture Analyzer, Stable Micro Systems Ltd., Surrey, UK).

Gel Characterization

The pH of the developed gel was measured using a pH meter, and the viscosity was assessed using a Brookfield viscometer. Gel (50 mL) was transferred into a container therein; a spindle 64 was inserted and rotated at a predetermined time; the speed and viscosity of the gel were recorded. The spreadability of the gel was evaluated by placing 500 mg of both α-MNG-PLGA NPs gel and α-MNG gel separately between the glass slides up to a diameter of 2 cm. After that, the gels (500 mg) were placed on the upper glass slide for 5 min and the gel spreading capacities were determined.

4.5.8. Ex Vivo Skin Permeation Studies

The skin penetration capability was investigated using Franz diffusion cell apparatus, having an effective surface area of 0.750 cm^2 using Logan's assembly. The rat's skin was excised, and the fatty layers were surgically removed, rinsed with ethanol, and refrigerated at −80 °C. The skin specimen was mounted safely on the Franz diffusion cells between the donor and receptor compartments with SC cladding towards the donor compartment. Firstly, 7.5 mL of PBS was added to the receptor compartment and agitated at every point using a small magnetic bead at approximately 500 rpm, with the temperature being maintained at 37 ± 0.5 °C. Then, 1 mL of the formulation [α-MNG-PLGA NPs gel or α-MNG gel] was placed on the donor compartment over the skin surface, and samples were taken at fixed intervals for 24 h from the receiver compartment, and the same quantity was replaced with fresh medium. Analysis of the drug content was conducted using RP-HPLC analysis.

4.5.9. Drug Concentration Estimation in Skin Strata

The mounted skin collected from the Franz diffusion cell was washed and cleaned with PBS to remove any adhered particles and was used to estimate the amount of drug present in the various skin strata. The outer SC was removed from the dermis using the tape-stripping method with a scotch crystal tape and sterile surgical scalpel [64,65]. The solution of tissue protein extracting reagent (T-PER) with a tissue: T-PER of 1:10 *w/v* ratio was used to improve the extraction from the skin and probe sonicated for 5 min. Furthermore, the tap strips, epidermis, and dermis were transferred into alcohol, then sonicated to extract the drug, and estimated by UV-spectrophotometry.

4.5.10. Confocal Microscopy

Confocal microscopy was performed to examine the α-MNG distribution and penetration from NPs formulation into various skin layers. In this study, a fluorescent dye, RhB, has to be entrapped into PLGA NPs in place of the drug. During the preparation of NPs using the emulsion–diffusion–evaporation technique, 0.02% *w/v* of fluorescent dye solution was loaded into the formulation, labelled RhB-loaded PLGA NPs. The processed animal skin was mounted onto the Franz diffusion cell on the donor compartment, where the skin SC confronted the donor compartment. A 1 mL of RhB-loaded PLGA NPs was placed on the donor compartment and probe dye release behavior from NPs in the receptor compartment was investigated for 24 h, as shown in the skin permeation study. The temperature of the receptor compartment was maintained at 32 ± 0.5 °C after being filled with 6 mL of PBS, pH 7.4. After the complete study, the skin specimen was removed, gently wiped with deionized water, and prepared on a glass slide. Therefore, it was studied under the confocal microscope set at 540 nm and 630 nm for excitation (λex) and emission wavelength (λem) using an argon laser beam and 65× objective lens (EC-Plan Neofluar 65×/01.40 Oil DICM27). The fluorescent dye, RhB release, and distribution pattern of NPs into distinguished layer of skin was compared with those of plain RhB solution. The confocal microscope optically scanned through z-axis and analyzed the fluorescent distribution and permeation in various layers of skin.

4.5.11. Cell Viability Study

MTT assay was used to study the cytotoxicity of the α-MNG-PLGA NPs gel and α-MNG gel, which was carried out in B16-F10 melanoma cells. The assay shows the transformation of tetrazolium (a yellow color salt) into an insoluble purple color formazan crystal inside the viable cells by a mitochondria enzyme, and the succinate dehydrogenase. About 1×10^6 cells/well were seeded in each well of the 96-well plate and left overnight. Then, a series α-MNG gel and α-MNG-PLGA NPs gel concentrations (0, 2, 4, 8, and 12 µM) were added to the cells in each well and cultured at 37 °C for 24 and 48 h in a humidified 5% CO$_2$ incubator. At the end of each culture period, the culture medium was removed, and the MTT reagent (5 µL) was added to each well for 3 h. After that, the plates were returned to the CO_2 incubator for 3 h. Then, the insoluble crystals formed were solubilized with dimethyl sulphoxide (DMSO) (130 µL/well), and the optical density (OD) at 570 nm was measured using a microplate reader (Bio-Rad, Des Plaines, IL, USA) [66,67]. Untreated cells or cells exposed to blank PLGA NPs/ gel or blank gel was used as controls (100 % viability). The IC_{50}, which is the concentration of the drug that kills 50% cell proliferation compared to control cells, was determined from the cell viability graph using the methods listed in [68]. The cell viability (%) was estimated as mean viability (%) ± standard deviation (SD) (n = 3) and calculated by the following:

$$\% \text{ Cell viability} = \text{Absorbance}_{treated} / \text{Absorbance}_{controlled} \times 100. \tag{5}$$

4.5.12. Free Radical Scavenging Activity

The radical scavenging power of α-MNG in optimized α-MNG-PLGA NPs gel and α-MNG gel was estimated using 2, 2-diphenyl-2-picrylhydrazyl (DPPH) assay as previously reported with minor modification [69]. A 40 µL of α-MNG-PLGA NPs gel and α-MNG gel was mixed with freshly prepared DPPH reagents (220 µL, 0.1 mM), followed by vigorous shaking and incubation for 30 min at 28 °C. OD was then measured using UV-spectrophotometry at λ_{max} of 517 nm. The DPPH antioxidant capacity of α-MNG in the optimized α-MNG-PLGA NPs gel and α-MNG gel was measured as % inhibition in DPPH and estimated as milligrams of Trolox equivalent (TEAC) per gram of sample. The quashed concentration of DPPH was calculated from the standard curve of standard Trolox, and the assay was performed in triplicate (n = 3).

The % radical scavenging activity is determined using the following equation:

$$\% \text{ Inhibition by DPPH assay} = Ao - A1/Ao \times 100, \tag{6}$$

where Ao is the blank; A1 is the sample absorbance calculated using milligram of Trolox equivalent per gram dry extract (mg TE/g extract). IC_{50} (µg/mL) was estimated from the dose–response curve of linear regression analysis, and the concentration calculated enabled achieving 50% of maximum scavenging effect by DPPH radical [70].

4.5.13. Stability Study

The stability study was performed on the optimized α-MNG-PLGA NPs gel as per the International Conference on Harmonization (ICH) guidelines protocol for three months. During the study period, the formulation was monitored for NPs size distribution, PDI, surface charge, drug entrapment, and % drug retained. For stability study, α-MNG-PLGA NPs gel was transferred in a stability chamber at a refrigerated temperature (5 ± 3 °C), at an ambient temperature (30 ± 2 °C/65 ± 5 % RH), and at an increased temperature of 40 ± 2 °C/75 ± 5% RH for three months; the sample was analyzed at intervals of 0, 30, 60 and 90 days [45]. The experiments were performed in triplicate (n = 3).

4.5.14. Statistical Analysis

The analysis was performed using ANOVA and Duncan's multiple range test (DMRT) using the Statistical Package of Social Sciences (SPSS), version 11.0, for Windows. The values were expressed as mean (n = 3) ± SD for at least three samples. Level of significance was $p < 0.05$.

Author Contributions: Conceptualization, S.M., N.A.A. and A.K.R.; Data curation, S.M., N.A.A., Y.R., A.K.R., M.G. and M.H.A.; Formal analysis, S.M., N.A.A., S.A., M.A.M., H.K., M.G. and M.H.A.; Funding acquisition, S.M., N.A.A. and A.K.R.; Investigation, T.N. and M.A.M.; Methodology, T.N., S.A., M.A.M., Y.R., H.K., M.G. and M.H.A.; Software, M.A.M., Y.R. and H.K.; Writing—original draft, S.A., M.G. and M.H.A.; Writing—review and editing, T.N. and A.K.R. All authors have read and agreed to the published version of the manuscript.

Funding: The Deanship of Scientific Research (DSR) at King Abdulaziz University, Jeddah, Saudi Arabia has funded this project, under grant no. (RG-12-166-42).

Institutional Review Board Statement: The animal protocol was approved by the animal ethical committee of the Faculty of Pharmacy, King Abdulaziz University, Jeddah, Saudi Arabia (PH-1443-06) dated 05.09.2021).

Informed Consent Statement: Not applicable.

Data Availability Statement: The data presented in this study are available in article.

Acknowledgments: The authors, therefore, acknowledge with thanks DSR for technical and financial support.

Conflicts of Interest: The authors declare that they have no competing interests.

References

1. Dianzani, C.; Zara, G.P.; Maina, G.; Pettazzoni, P.; Pizzimenti, S.; Rossi, F.; Gigliotti, C.L.; Ciamporcero, E.S.; Daga, M.; Barrera, G. Drug Delivery Nanoparticles in Skin Cancers. *BioMed Res. Int.* **2014**, *2014*, 895986. [CrossRef]
2. Omran, A.R. The Epidemiologic Transition: A Theory of the Epidemiology of Population Change. *Milbank Q.* **2005**, *83*, 731–757. [CrossRef]
3. Armstrong, B.K.; Kricker, A. The epidemiology of UV induced skin cancer. *J. Photochem. Photobiol. B Biol.* **2001**, *63*, 8–18. [CrossRef]
4. Esteva, A.; Kuprel, B.; Novoa, R.A.; Ko, J.; Swetter, S.M.; Blau, H.M.; Thrun, S. Dermatologist-level classification of skin cancer with deep neural networks. *Nature* **2017**, *542*, 115–118. [CrossRef]
5. Narayanan, D.L.; Saladi, R.N.; Fox, J.L. Review: Ultraviolet radiation and skin cancer. *Int. J. Dermatol.* **2010**, *49*, 978–986. [CrossRef]
6. Akhter, M.H.; Amin, S. An Investigative Approach to Treatment Modalities for Squamous Cell Carcinoma of Skin. *Curr. Drug Deliv.* **2017**, *14*, 597–612. [CrossRef]
7. Curran, S.; Vantangoli, M.M.; Boekelheide, K.; Morgan, J.R. Architecture of Chimeric Spheroids Controls Drug Transport. *Cancer Microenviron.* **2015**, *8*, 101–109. [CrossRef] [PubMed]
8. Zhao, M.; Li, H.; Ma, Y.; Gong, H.; Yang, S.; Fang, Q.; Hu, Z. Nanoparticle abraxane possesses impaired proliferation in A549 cells due to the underexpression of glucosamine 6-phosphate N-acetyltransferase 1 (GNPNAT1/GNA1). *Int. J. Nanomed.* **2017**, *12*, 1685–1697. [CrossRef]
9. Sahibzada, M.U.K.; Sadiq, A.; Faidah, H.S.; Khurram, M.; Amin, M.U.; Haseeb, A.; Kakar, M. Berberine nanoparticles with enhanced in vitro bioavailability: Characterization and antimicrobial activity. *Drug Des. Dev. Ther.* **2018**, *12*, 303–312. [CrossRef]
10. Gokhale, J.P.; Mahajan, H.S.; Surana, S.J. Quercetin loaded nanoemulsion-based gel for rheumatoid arthritis: In vivo and in vitro studies. *Biomed. Pharmacother.* **2019**, *112*, 108622. [CrossRef]
11. Akhter, M.H.; Rizwanullah, M.; Ahmad, J.; Amin, S.; Ahmad, M.Z.; Minhaj, A.; Mujtaba, A.; Ali, J. Molecular Targets and Nanoparticulate Systems Designed for the Improved Therapeutic Intervention in Glioblastoma Multiforme. *Drug Res.* **2021**, *71*, 122–137. [CrossRef] [PubMed]
12. Akhter, M.H.; Beg, S.; Tarique, M.; Malik, A.; Afaq, S.; Choudhry, H.; Hosawi, S. Receptor-based targeting of engineered nanocarrier against solid tumors: Recent progress and challenges ahead. *Biochim. Biophys. ActaBBA-Gen. Subj.* **2021**, *1865*, 129777. [CrossRef] [PubMed]
13. Barua, S.; Mitragotri, S. Challenges associated with penetration of nanoparticles across cell and tissue barriers: A review of current status and future prospects. *Nano Today* **2014**, *9*, 223–243. [CrossRef]
14. Akhter, M.H.; Khalilullah, H.; Gupta, M.; Alfaleh, M.A.; Alhakamy, N.A.; Riadi, Y.; Shadab, M. Impact of Protein Corona on the Biological Identity of Nanomedicine: Understanding the Fate of Nanomaterials in the Biological Milieu. *Biomedicines* **2021**, *9*, 1496. [CrossRef]
15. Li, C.; Obireddy, S.R.; Lai, W.-F. Preparation and use of nanogels as carriers of drugs. *Drug Deliv.* **2021**, *28*, 1594–1602. [CrossRef]
16. Boinpelly, V.C.; Verma, R.K.; Srivastav, S.; Srivastava, R.K.; Shankar, S. α-Mangostin-encapsulated PLGA na-noparticles inhibit colorectal cancer growth by inhibiting Notch pathway. *J. Cell. Mol. Med.* **2020**, *24*, 11343–11354. [CrossRef]
17. Rezvantalab, S.; Drude, N.; Moraveji, M.K.; Güvener, N.; Koons, E.K.; Shi, Y.; Lammers, T.; Kiessling, F. PLGA-Based Nanoparticles in Cancer Treatment. *Front. Pharmacol.* **2018**, *9*, 1260. [CrossRef] [PubMed]
18. Silva, A.T.C.R.; Cardoso, B.C.O.; e Silva, M.E.S.R.; Freitas, R.F.S.; Sousa, R.G. Synthesis, Characterization, and Study of PLGA Copolymer in Vitro Degradation. *J. Biomater. Nanobiotechnol.* **2015**, *6*, 8–19. [CrossRef]
19. Rodeiro, I.; Delgado, R.; Garrido, G. Effects of a *Mangiferaindica* L. stem bark extract and mangiferin on radiation-induced DNA damage in human lymphocytes and lymphoblastoid cells. *Cell Prolif.* **2013**, *47*, 48–55. [CrossRef]
20. Xiao, J.; Liu, L.; Zhong, Z.; Xiao, C.; Zhang, J. Mangiferin regulates proliferation and apoptosis in glioma cells by induction of microRNA-15b and inhibition of MMP-9 expression. *Oncol. Rep.* **2015**, *33*, 2815–2820. [CrossRef]
21. Peng, Z.; Yao, Y.; Yang, J.; Tang, Y.; Huang, X. Mangiferin induces cell cycle arrest at G2/M phase through ATR-Chk1 pathway in HL-60 leukemia cells. *Genet. Mol. Res.* **2015**, *14*, 4989–5002. [CrossRef] [PubMed]
22. Quan, X.; Wang, Y.; Ma, X.; Liang, Y.; Tian, W.; Ma, Q.; Jiang, H.; Zhao, Y. α-Mangostin Induces Apoptosis and Suppresses Differentiation of 3T3-L1 Cells via Inhibiting Fatty Acid Synthase. *PLoS ONE* **2012**, *7*, e33376. [CrossRef] [PubMed]

23. Akhter, M.H.; Rizwanullah, M.; Ahmad, J.; Ahsan, M.J.; Mujtaba, A.; Amin, S. Nanocarriers in advanced drug targeting: Setting novel paradigm in cancer therapeutics. *Artif. Cells Nanomed. Biotechnol.* **2018**, *46*, 873–884. [CrossRef] [PubMed]
24. Akhter, M.H.; Madhav, N.S.; Ahmad, J. Epidermal growth factor receptor based active targeting: A paradigm shift towards advance tumor therapy. *Artif. Cells Nanomed. Biotechnol.* **2018**, *46*, 1188–1198. [CrossRef]
25. Pongsumpun, P.; Iwamoto, S.; Siripatrawan, U. Response surface methodology for optimization of cinnamon essential oil nanoemulsion with improved stability and antifungal activity. *Ultrason. Sonochem.* **2020**, *60*, 104604. [CrossRef] [PubMed]
26. Kumbhar, S.A.; Kokare, C.R.; Shrivastava, B.; Gorain, B.; Choudhury, H. Preparation, characterization, and optimization of asenapine maleate mucoadhesivenanoemulsion using Box-Behnken design: In vitro and in vivo studies for brain targeting. *Int. J. Pharm.* **2020**, *586*, 119499. [CrossRef]
27. Venugopal, V.; Kumar, K.J.; Muralidharan, S.; Parasuraman, S.; Raj, P.V. Optimization and in-vivo evaluation of isradipine nanoparticles using Box-Behnken design surface response methodology. *OpenNano* **2016**, *1*, 1–15. [CrossRef]
28. Kausar, H.; Mujeeb, M.; Ahad, A.; Moolakkadath, T.; Aqil, M.; Ahmad, A.; Akhter, M.H. Optimization of ethosomes for topical thymoquinone delivery for the treatment of skin acne. *J. Drug Deliv. Sci. Technol.* **2019**, *49*, 177–187. [CrossRef]
29. Md, S.; Alhakamy, N.; Aldawsari, H.; Husain, M.; Khan, N.; Alfaleh, M.; Asfour, H.; Riadi, Y.; Bilgrami, A.; Akhter, M.H. Plumbagin-Loaded Glycerosome Gel as Topical Delivery System for Skin Cancer Therapy. *Polymers* **2021**, *13*, 923. [CrossRef]
30. Akhter, M.H.; Kumar, S.; Nomani, S. Sonication tailored enhance cytotoxicity of naringenin nanoparticle in pancreatic cancer: Design, optimization, and in vitro studies. *Drug Dev. Ind. Pharm.* **2020**, *46*, 659–672. [CrossRef]
31. Soni, K.; Mujtaba, A.; Akhter, H.; Zafar, A.; Kohli, K. Optimisation of ethosomalnanogel for topical nano-CUR and sulphoraphane delivery in effective skin cancer therapy. *J. Microencapsul.* **2019**, *37*, 91–108. [CrossRef] [PubMed]
32. Shkodra, B.; Grune, C.; Traeger, A.; Vollrath, A.; Schubert, S.; Fischer, D. Effect of surfactant on the size and stability of PLGA nanoparticles encapsulating a protein kinase C inhibitor. *Int. J. Pharm.* **2019**, *566*, 756–764. [CrossRef] [PubMed]
33. Rizwan, K.; Muhammad, A.; Sarfaraz, K.; Jiménez, A.N.; Park, D.R.; Yeom, I.T. The influence of ionic and nonionic surfactants on the colloidal stability and removal of CuO nanoparticles from water by chemical coagulation. *Int. J. Environ. Res. Pub. Health* **2019**, *16*, 1260. [CrossRef]
34. Kim, Q.; Jhe, W. Interfacial thermodynamics of spherical nanodroplets: Molecular understanding of surface tension via a hydrogen bond network. *Nanoscale* **2020**, *12*, 18701–18709. [CrossRef] [PubMed]
35. Pradhan, S.; Hedberg, J.; Blomberg, E.; Wold, S.; Wallinder, I.O. Effect of sonication on particle dispersion, administered dose and metal release of non-functionalized, non-inert metal nanoparticles. *J. Nanoparticle Res.* **2016**, *18*, 285. [CrossRef]
36. Hernández-Giottonini, K.Y.; Rodríguez-Córdova, R.J.; Gutiérrez-Valenzuela, C.A.; Peñuñuri-Miranda, O.; Zavala-Rivera, P.; Guerrero-Germán, P.; Lucero-Acuña, A. PLGA nanoparticle preparations by emulsification and nanoprecipitation techniques: Effects of formulation parameters. *RSC Adv.* **2020**, *10*, 4218–4231. [CrossRef]
37. Bhatt, P.C.; Verma, A.; Al-Abassi, F.A.; Anwar, F.; Kumar, V.; Panda, B.P. Development of surface-engineered PLGA nanoparticulate-delivery system of Tet-1-conjugated nattokinase enzyme for inhibition of $A\beta_{40}$ plaques in Alzheimer's disease. *Int. J. Nanomed.* **2017**, *12*, 8749–8768. [CrossRef]
38. Mudalige, T.; Qu, H.; Van Haute, D.; Ansar, S.M.; Paredes, A.; Ingle, T. Characterization of Nanomaterials. In *Nanomaterials for Food Applications*; Elsevier: Amsterdam, The Netherlands, 2019; pp. 313–353. [CrossRef]
39. Sharma, N.; Madan, P.; Lin, S. Effect of process and formulation variables on the preparation of parenteral paclitaxel-loaded biodegradable polymeric nanoparticles: A co-surfactant study. *Asian J. Pharm. Sci.* **2016**, *11*, 404–416. [CrossRef]
40. Budhian, A.; Siegel, S.J.; Winey, K.I. Haloperidol-loaded PLGA nanoparticles: Systematic study of particle size and drug content. *Int. J. Pharm.* **2007**, *336*, 367–375. [CrossRef]
41. Español, L.; Larrea, A.; Andreu, V.; Mendoza, G.; Arruebo, M.; Sebastian, V.; Aurora-Prado, M.S.; Kedor-Hackmann, E.R.M.; Santoro, M.I.R.M.; Santamaria, J. Dual encapsulation of hydrophobic and hydrophilic drugs in PLGA nanoparticles by a single-step method: Drug delivery and cytotoxicity assays. *RSC Adv.* **2016**, *6*, 111060–111069. [CrossRef]
42. Kızılbey, K. Optimization of Rutin-Loaded PLGA Nanoparticles Synthesized by Single-Emulsion Solvent Evaporation Method. *ACS Omega* **2019**, *4*, 555–562. [CrossRef]
43. Raya, S.; Mishrab, A.; Mandal, T.K.; Sac, B.; Chakraborty, J. Optimization of the process parameters for fabrication of polymer coated layered double hydroxide-methotrexate nanohybrid for possible treatment of osteosarcoma. *RSC Adv.* **2015**, *5*, 102574–102592. [CrossRef]
44. Liu, D.; Pan, H.; He, F.; Wang, X.; Li, J.; Yang, X.; Pan, W. Effect of particle size on oral absorption of carvedilolnanosuspensions: In vitro and invivoevaluation. *Int. J. Nanomed.* **2015**, *10*, 6425–6434. [CrossRef]
45. Akhter, M.H.; Ahmad, A.; Ali, J.; Mohan, G. Formulation and Development of CoQ10-Loaded s-SNEDDS for Enhancement of Oral Bioavailability. *J. Pharm. Innov.* **2014**, *9*, 121–131. [CrossRef]
46. Wathoni, N.; Rusdin, A.; Febriani, E.; Purnama, D.; Daulay, W.; Azhary, S.Y.; Panatarani, C.; Joni, I.M.; Lesmana, R.; Motoyama, K.; et al. Formulation and characterization of α-mangostin in chitosan nanoparticles coated by sodium alginate, sodium silicate, and polyethylene glycol. *J. Pharm. Bioallied Sci.* **2019**, *11*, S619–S627. [CrossRef]
47. Tejamukti, E.P.; Setyaningsih, W.; Yasir, B.; Alam, G.; Rohman, A. Application of FTIR Spectroscopy and HPLC Combined with Multivariate Calibration for Analysis of Xanthones in Mangosteen Extracts. *Sci. Pharm.* **2020**, *88*, 35. [CrossRef]
48. Cui, Y.-N.; Xu, Q.-X.; Davoodi, P.; Wang, D.-P.; Wang, C.-H. Enhanced intracellular delivery and controlled drug release of magnetic PLGA nanoparticles modified with transferrin. *Acta Pharmacol. Sin.* **2017**, *38*, 943–953. [CrossRef] [PubMed]

49. Azadi, A.; Rouini, M.-R.; Hamidi, M. Neuropharmacokinetic evaluation of methotrexate-loaded chitosan nanogels. *Int. J. Biol. Macromol.* **2015**, *79*, 326–335. [CrossRef] [PubMed]
50. Pandit, J.; Sultana, Y.; Aqil, M. Chitosan-coated PLGA nanoparticles of bevacizumab as novel drug delivery to target retina: Optimization, characterization, and in vitro toxicity evaluation. *Artif. Cells Nanomed. Biotechnol.* **2017**, *45*, 1397–1407. [CrossRef]
51. Körber, M. PLGA Erosion: Solubility- or Diffusion-Controlled? *Pharm. Res.* **2010**, *27*, 2414–2420. [CrossRef]
52. Siegel, S.; Kahn, J.B.; Metzger, K.; Winey, K.I.; Werner, K.; Dan, N. Effect of drug type on the degradation rate of PLGA matrices. *Eur. J. Pharm. Biopharm.* **2006**, *64*, 287–293. [CrossRef]
53. Yostawonkul, J.; Surassmo, S.; Namdee, K.; Khongkow, M.; Boonthum, C.; Pagseesing, S.; Saengkrit, N.; Ruktanonchai, U.R.; Chatdarong, K.; Ponglowhapan, S.; et al. Nanocarrier-mediated delivery of α-mangostin for non-surgical castration of male animals. *Sci. Rep.* **2017**, *7*, 16341. [CrossRef] [PubMed]
54. Nurman, S.; Yulia, R.; Noor, E.; Sunarti, T.C. IrmayantiThe Optimization of Gel Preparations Using the Active Compounds of Arabica Coffee Ground Nanoparticles. *Sci. Pharm.* **2019**, *87*, 32. [CrossRef]
55. Ahmad, J.; Ahmad, M.Z.; Akhter, H. Surface-Engineered Cancer Nanomedicine: Rational Design and Recent Progress. *Curr. Pharm. Des.* **2020**, *26*, 1181–1190. [CrossRef]
56. Akhter, M.H.; Ahsan, M.J.; Rahman, M.; Anwar, S.; Rizwanullah, M. Advancement in Nanotheranostics for Effective Skin Cancer Therapy: State of the Art. *Curr. Nanomed.* **2020**, *10*, 90–104. [CrossRef]
57. Ahmad, M.Z.; Rizwanullah, M.; Ahmad, J.; Alasmary, M.Y.; Akhter, M.H.; Abdel-Wahab, B.A.; Warsi, M.H.; Haque, A. Progress in nanomedicine-based drug delivery in designing of chitosan nanoparticles for cancer therapy. *Int. J. Polym. Mater.* **2021**, 1–22. [CrossRef]
58. Zheng, Y.; Ouyang, W.-Q.; Wei, Y.-P.; Syed, S.F.; Hao, C.-S.; Wang, B.-Z.; Shang, Y.-H. Effects of Carbopol® 934 proportion on nanoemulsion gel for topical and transdermal drug delivery: A skin permeation study. *Int. J. Nanomed.* **2016**, *11*, 5971–5987. [CrossRef]
59. Mohanty, C.; Sahoo, S.K. The in vitro stability and in vivo pharmacokinetics of curcumin prepared as an aqueous nanoparticulate formulation. *Biomaterials* **2010**, *31*, 6597–6611. [CrossRef] [PubMed]
60. Wang, J.J.; Sanderson, B.J.; Zhang, W. Cytotoxic effect of xanthones from pericarp of the tropical fruit mangosteen (Garciniamangostana Linn.) on human melanoma cells. *Food Chem. Toxicol.* **2011**, *49*, 2385–2391. [CrossRef]
61. Wang, J.J.; Sanderson, B.J.; Zhang, W. Significant anti-invasive activities of α-mangostin from the mangosteenperi-carp on two human skin cancer cell lines. *Anticancer. Res.* **2012**, *32*, 3805–3816.
62. Gan, Q.; Wang, T. Chitosan nanoparticle as protein delivery carrier—Systematic examination of fabrication conditions for efficient loading and release. *Colloids Surf. B Biointerfaces* **2007**, *59*, 24–34. [CrossRef] [PubMed]
63. Hafeez, A.; Kazmi, I. Dacarbazine nanoparticle topical delivery system for the treatment of melanoma. *Sci. Rep.* **2017**, *7*, 16517. [CrossRef] [PubMed]
64. Loffler, H.; Dreher, F.; Maibach, H. Stratum corneum adhesive tape stripping: Influence of anatomical site, application pressure, duration and removal. *Br. J. Dermatol.* **2004**, *151*, 746–752. [CrossRef]
65. Olesen, C.M.; Fuchs, C.S.K.; Philipsen, P.A.; Hædersdal, M.; Agner, T.; Clausen, M.-L. Advancement through epidermis using tape stripping technique and Reflectance Confocal Microscopy. *Sci. Rep.* **2019**, *9*, 12217. [CrossRef]
66. Lee, H.N.; Jang, H.Y.; Kim, H.J.; Shin, S.A.; Choo, G.S.; Park, Y.S.; Kim, S.K.; Jung, J.Y. Antitumor and apoptosis-inducing effects of α-mangostin extracted from the pericarp of the mangosteen fruit (*Garciniamangostana* L.)in YD-15 tongue mucoepidermoid carcinoma cells. *Int. J. Mol. Med.* **2016**, *37*, 939–948. [CrossRef]
67. Krishnakumar, N.; Sulfikkarali, N.; Prasad, R.; Karthikeyan, S. Enhanced anticancer activity of naringenin-loaded nanoparticles in human cervical (HeLa) cancer cells. *Biomed. Prev. Nutr.* **2011**, *1*, 223–231. [CrossRef]
68. Zeng, X.; Luo, M.; Liu, G.; Wang, X.; Tao, W.; Lin, Y.; Ji, X.; Nie, L.; Mei, L. Polydopamine-Modified Black Phosphorous Nanocapsule with Enhanced Stability and Photothermal Performance for Tumor Multimodal Treatments. *Adv. Sci.* **2018**, *5*, 1800510. [CrossRef]
69. Mohammad, N.A.; Zaidel, D.N.A.; Muhamad, I.I.; Hamid, M.A.; Yaakob, H.; Jusoh, Y.M.M. Optimization of the antioxidant-rich xanthone extract from mangosteen (*Garciniamangostana* L.) pericarp via microwave-assisted extraction. *Heliyon* **2019**, *5*, e02571. [CrossRef]
70. Phan, T.K.T.; Tran, T.Q.; Pham, D.T.N.; Nguyen, D.T. Characterization, Release Pattern, and Cytotoxicity of Liposomes Loaded With α-Mangostin Isolated From Pericarp of Mangosteen (*Garciniamangostana* L.). *Nat. Prod. Commun.* **2020**, *15*, 1–8. [CrossRef]

Article

Formulation Development and Evaluation of Pravastatin-Loaded Nanogel for Hyperlipidemia Management

Gaurav Kant Saraogi [1], Siddharth Tholiya [2], Yachana Mishra [3], Vijay Mishra [4,*], Aqel Albutti [5,*], Pallavi Nayak [4] and Murtaza M. Tambuwala [6]

1. Sri Aurobindo Institute of Pharmacy, Indore 453111, Madhya Pradesh, India; gauravsaraogi13@gmail.com
2. SVKM's NMIMS School of Pharmacy & Technology Management, Shirpur 425405, Maharashtra, India; siddhart619@gmail.com
3. Department of Zoology, Shri Shakti Degree College, Sankhahari, Ghatampur, Kanpur Nagar 209206, Uttar Pradesh, India; yachanamishra@gmail.com
4. School of Pharmaceutical Sciences, Lovely Professional University, Phagwara 144411, Punjab, India; pallavinayak97@gmail.com
5. Department of Medical Biotechnology, College of Applied Medical Sciences, Qassim University, Buraydah 52571, Saudi Arabia
6. School of Pharmacy and Pharmaceutical Sciences, Ulster University, Coleraine, Londonderry BT52 1SA, UK; m.tambuwala@ulster.ac.uk
* Correspondence: vijaymishra2@gmail.com (V.M.); as.albutti@qu.edu.sa (A.A.)

Abstract: Hyperlipidemia is a crucial risk factor for the initiation and progression of atherosclerosis, ultimately leading to cardiovascular disease. The nanogel-based nanoplatform has emerged as an extremely promising drug delivery technology. Pravastatin Sodium (PS) is a cholesterol-lowering drug used to treat hyperlipidemia. This study aimed to fabricate Pravastatin-loaded nanogel for evaluation of its effect in hyperlipidemia treatment. Pravastatin-loaded chitosan nanoparticles (PS-CS-NPs) were prepared by the ionic gelation method; then, these prepared NPs were converted to nanogel by adding a specified amount of 5% poloxamer solution. Various parameters, including drug entrapment efficacy, in vitro drug release, and hemolytic activity of the developed and optimized formulation, were evaluated. The in vitro drug release of the nanogel formulation revealed the sustained release (59.63% in 24 h) of the drug. The drug excipients compatibility studies revealed no interaction between the drug and the screened excipients. Higher drug entrapment efficacy was observed. The hemolytic activity showed lesser toxicity in nanoformulation than the pure drug solution. These findings support the prospective use of orally administered pravastatin-loaded nanogel as an effective and safe nano delivery system in hyperlipidemia treatment.

Keywords: nanogel; hyperlipidemia; pravastatin; polymer

1. Introduction

The term 'nanogel' is defined as the nanosized particles formed by a cross-linked polymer physically or chemically. It was first introduced by cross-linked bifunctional networks of a polyion and a non-ionic polymer for the delivery of polynucleotides [1]. Nanogels are typical formulations with a size range of 20–200 nm [2]. The volume fraction can be altered to maintain the three-dimensional (3D) structure by varying solvent quality and branching. Since gene delivery has become possible within cellular organelles for gene silencing therapy, nanogels have transformed the field of gene therapy [1]. As a result of their size, they can evade renal clearance and have a longer serum half-life. Nanogels tend to absorb a lot of water or physiological fluid without affecting their internal network structure. For targeted drug delivery, stimulus-responsive drug release, or the development of composite materials, chemical changes can be made to integrate a large number of ligands [3]. Nanogels have several characteristics that contribute to their effectiveness as a delivery system. They have exceptional thermodynamic stability, high solubilization capacity, low viscosity, and the ability to withstand rigorous sterilization techniques [4,5]. The majority of nanogel systems are made up of cross-linked synthetic polymers or natural biopolymers. Small molecules or biomacromolecules can be incorporated into the pores of the 3D network in nanogels. Polymeric nanogels as drug carriers have the advantages of artificially controlling drug dosage via external stimuli, concealing disagreeable odor of drug, enhancing therapeutic efficacy, and decreasing the drug's side effects. For example, anticancer drugs and proteins are excellent for administration via chemically cross-linked or physically constructed nanogel systems because they have severe side effects, a short circulation half-time, and are easily degradable by enzymes [6–8].

Hyperlipidemia is described as an abnormally high amount of lipids and lipoproteins in the blood, and it is thought to be a major risk factor for accelerated atherosclerosis and, as a result, cardiovascular disease. In hyperlipidemia, cholesterol levels, particularly low-density lipoprotein cholesterol (LDL-C), are often increased and serve as the primary therapeutic target [9].

As the result of a lack of LDL receptors in hepatocytes, hyperlipidemia, a disease linked to atherosclerosis, can develop [10,11]. Since injected polymeric NPs are quickly taken up by hepatic Kupffer cells, LDL-absorbing NPs may have improved LDL transport to the liver. At the same time, intrinsic toxicity can also be treated with NPs. These NPs have unique physicochemical properties, which have shown promising drug delivery systems (DDS) to the desired sites in the body. These enhance the drug release, improve the bioavailability and solubility, and minimize toxicity and drug degradation. Due to the expanded contact region for van der Waals attraction, NPs show strong adhesion. It is necessary to comprehend the pharmaceutically relevant properties of NPs to achieve the better development of the novel DDS [12–14]. The major challenge for developing NP as a DDS is controlling the particle size, surface properties, and time-release of the active moiety to get the site-specific action at the desired proportion and dose. Polymeric NPs offer distinct advantages over other nanocarriers; they increase the drugs/protein stability and show beneficial controlled release properties [15–22].

Pravastatin is a cholesterol-lowering agent that belongs to the statin class of drugs. It came from the microbial transformation of mevastatin, which was the first statin ever found. It is a ring-opened dihydroxyacid with a 6'-hydroxyl group that does not need to be activated in the body. When compared to lovastatin and simvastatin, the greater hydrophilicity of pravastatin is expected to give benefits such as reduced penetration across lipophilic membranes of peripheral cells, increased selectivity for hepatic tissues, and a reduction in adverse effects [23]. Pravastatin is structurally identical to 3-hydroxy-3-methylglutaryl (HMG), an endogenous substitute for HMG-coenzyme A reductase (HMG-CoA reductase). Pravastatin, unlike its parent chemical, mevastatin, and statins such as lovastatin and simvastatin, does not require in vivo activation. Its hydrolyzed lactone ring resembles the tetrahedral intermediate of reductase, allowing it to bind with far more affinity than its normal substrate. The bicyclic component of pravastatin binds to the coenzyme A portion of the active site. Pravastatin works in two ways to reduce cholesterol levels. First, it causes modest decreases in intracellular cholesterol pools due to its reversible suppression of HMG-CoA reductase activity. The number of LDL receptors on the cell surfaces increases, resulting in improved receptor-mediated degradation and clearance of circulating LDL. Second, pravastatin inhibits LDL production by inhibiting the hepatic synthesis of VLDL, the LDL precursor [24–27].

2. Results and Discussion

2.1. Experimental Design for Optimization of Nanoparticles

The 3^2 level central composite design (CCD) with results is shown in Tables 1 and 2. All the batches were formulated and evaluated for particle size and entrapment efficiency. The obtained results provided considerable useful information and confirmed the utility of the statistical design for the conduction of the experiments. Independent variables such as polymer amount (mg), the stirring speed, and probe sonication time significantly influenced the observed responses, particle size (nm), and entrapment efficiency (%). The optimized formulation batch was determined by systematic analysis of data using design expert software.

Table 1. Design table for optimized batch formulation.

Std	Run	Factor 1 A: Polymer Amount (mg)	Factor 2 B: Stirring Speed (rpm)	Factor 3 C: Sonication Time (min)	Response 1 Drug Entrapment %	Response 2 Particle Size (nm)
1	10	50	1000	5	53.6	556
2	11	100	1000	5	56	602
3	17	50	1500	5	58.95	604
4	16	100	1500	5	47	614
5	15	50	1000	10	49.5	556
6	5	100	1000	10	47	655
7	7	50	1500	10	55	500
8	6	100	1500	10	54.6	498
9	14	33	1250	7.5	51	543
10	13	117	1250	7.5	50.33	543
11	20	75	830	7.5	50.68	578
12	1	75	1670	7.5	56.95	502
13	4	75	1250	3	58.2	566
14	12	75	1250	12	54	601
15	8	75	1250	7.5	60	486
16	18	75	1250	7.5	60	486
17	3	75	1250	7.5	60	486
18	9	75	1250	7.5	60	486
19	2	75	1250	7.5	60	486
20	19	75	1250	7.5	60	486

Table 2. Analysis of variance (ANOVA) for factorial model at $p < 0.05$ level of significance.

Source	Sum of Squares	df	Mean Square	F-Value	p-Value	
Model	356.02	9	39.56	9.40	0.0008	Significant
A—Polymer amount	13.50	1	13.50	3.21	0.1035	
B—Stirring speed	29.27	1	29.27	6.96	0.0248	
C—Sonication time	19.97	1	19.97	4.75	0.0544	
AB	18.76	1	18.76	4.46	0.0609	
AC	5.53	1	5.53	1.31	0.2784	
BC	35.07	1	35.07	8.34	0.0162	
A^2	165.68	1	165.68	39.38	<0.0001	
B^2	74.72	1	74.72	17.76	0.0018	
C^2	31.10	1	31.10	7.39	0.0216	
Residual	42.07	10	4.21			
Lack of Fit	42.07	5	8.41			
Pure Error	0.0000	5	0.0000			
Cor Total	398.09	19				

Pravastatin-loaded CSNPs (PS-CS-NPs) were prepared by an ionic gelation method using a probe sonicator. Pravastatin is a BCS class-III drug having high solubility and low permeability. The influence of the amount of chitosan, stirring speed, and probe sonication time at different concentrations and time, respectively, was investigated in the ionic gelation method with a fixed quantity of tripolyphosphate (TPP) and amount of drug. In the prescreening study, chitosan and poloxamer were selected as polymer and stabilizer/gelling agents, respectively. This study found that chitosan and poloxamer gave a desired particle size for all the batches, which was found to be 400–3155 nm with a lower polydispersity index (PDI). The zeta potential ensures the stability of the formulation. The stirring speed was taken with the help of the literature search. The dependent variable, i.e., entrapment efficiency obtained at various levels of three independent variables (A, B, C), was observed. The polynomial equation was obtained in terms of actual factors for entrapment efficiency (%). The correlation coefficient (R^2) value of the polynomial equation was found to be 0.9566, indicating a good fit (Table 2).

The entrapment efficiency for all the batches was calculated, and the wide variation was observed, i.e., the values ranged from 47 to 60%. Hence, the obtained results indicate that the entrapment efficiency value was strongly affected by the variables used in this study. It can be found that A, B, (A^2), and (C^2) are the significant model, as the p values of the above independent variable are less than 0.05. Here, the coefficient of independent variable B, (A^2), and (C^2) has the negative value. An increase in stirring speed or average stirring speed and probe sonication time increases the entrapment efficiency as it increases the viscosity, which leads to a decrease in the particle size of the nanoparticles. As the viscosity of the external aqueous phase decreases, effective diffusion of the organic phase is hindered, leading to smaller droplet formed, which affects the mean size of the particle, and hence, the entrapment efficiency is high. This result is also confirmed by the contour plot for the entrapment efficiency in Figure 1a,b. The desirability of optimized batch is shown in Figure 1c overlay plot.

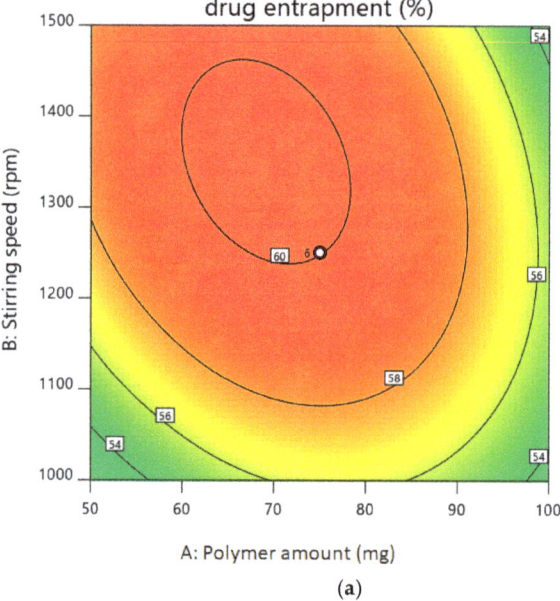

Figure 1. *Cont.*

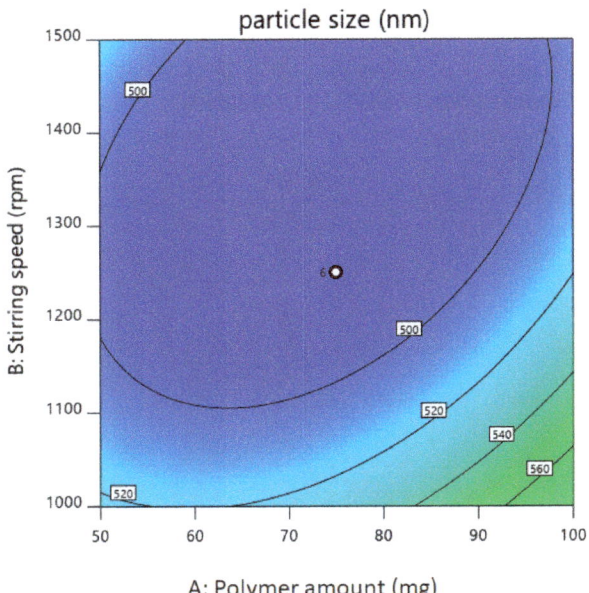

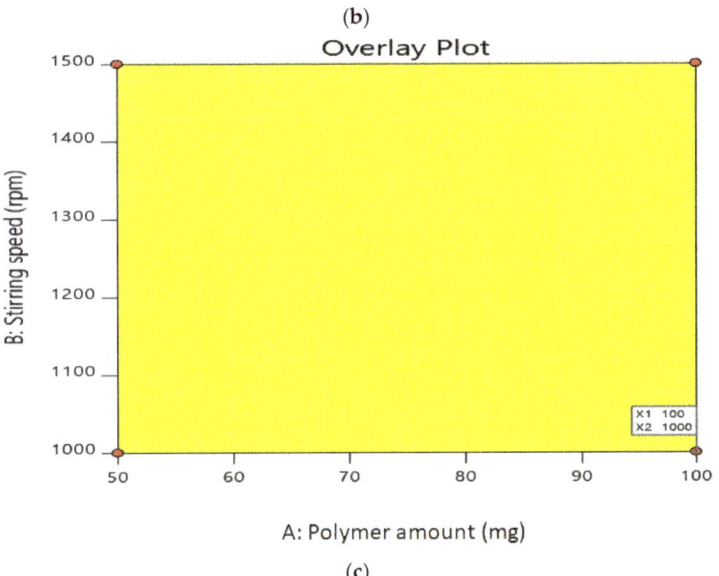

Figure 1. (**a**). Contour plot of drug entrapment. (**b**). Contour plot of particle size. (**c**). Overlay plot for confirmation.

2.2. Characterization of Optimized Nanoparticles

2.2.1. Particle Size

The optimized batch (N20) had a Z-average particle size of 486 nm and PDI of 0.303, indicating that the NPs were distributed uniformly. Figure 2 showed the particle size distribution pattern of the optimized NPs formulation.

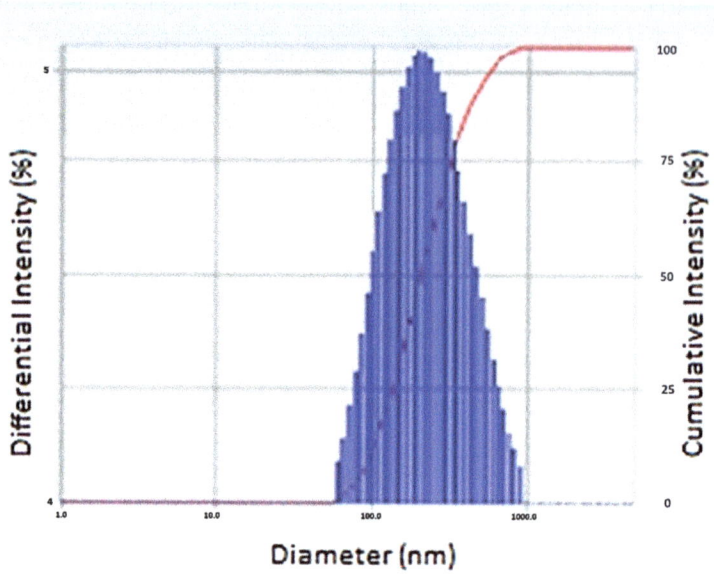

Figure 2. Particle size of optimized nanoparticles.

2.2.2. Entrapment Efficiency

The entrapment efficiency of the optimized batch was found to be 50%. Hence, the results indicate that the measured value obtained for entrapment efficiency was as expected. It is confirmed that at a 95% level of confidence, the values obtained for the optimized batch was in range. The predicted values for entrapment efficiency and particle size are shown in Table 3.

Table 3. Confirmation table of optimized nanoparticles.

Response	Predicted Mean	Predicted Median	Std Deviation	N	SE Prediction	95% Low	Mean	95% High
Entrapment efficiency	49.4784	49.4784	2.05155	1	2.65049	49.368	49.4784	55.384
Particle size	647.994	647.994	23.6573	1	30.5699	579.88	647.994	716.108

2.2.3. Scanning Electron Microscopy (SEM) of Nanoparticles

The SEM image showed that the chitosan nanoparticles have a regular shape (Figure 3).

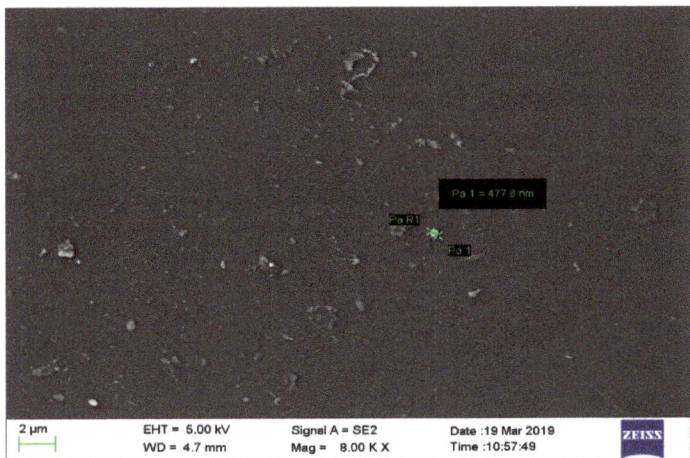

Figure 3. Scanning Electron Microscopy image of chitosan nanoparticles showing regular shape.

2.2.4. In Vitro Drug Release

The optimized PS-CS-NG formulation was subjected for in vitro drug release behavior. It was performed by the dialysis bag method, and the optimized formulation (N20) showed the highest drug release (59.63%) in 24 h (Figure 4). This showed the uniform distribution of drug in the formulation.

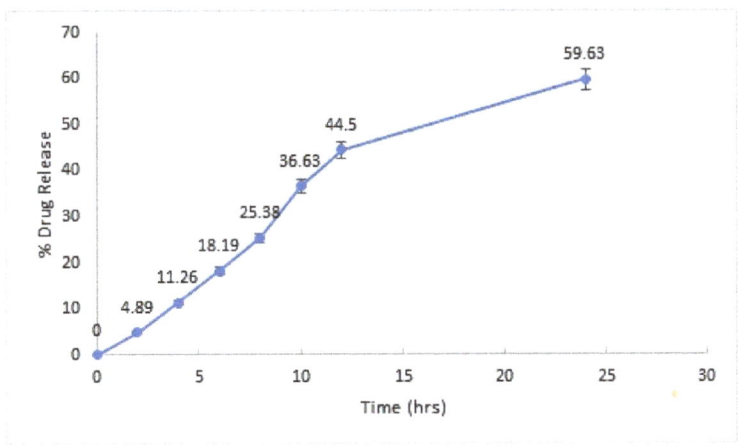

Figure 4. In vitro drug release profile of optimized PS-CS-NG formulation.

2.2.5. Swelling Ratio

At pH 6.8 and 7.4, the swelling ratio for nanogel preparation was determined to be 259% and 382%, respectively. The high swelling ratio showed that the nanogels had a good capacity for accessing and keeping water molecules within the polymer network, and it was clear that this capacity was lower in pH 6.8 buffers than in pH 7.4 buffers. The hydrophilic nature of CS and the nature of bonds inside the matrix structure determine the equilibrium water content of nanogels. Furthermore, the hydrogen bonds between water molecules and functional groups present in nanogel structures, such as hydroxyl groups, ether, amine, and unhydrolyzed acetamide linkages, are the most important controlling factors during swelling. Since the functional groups or matrix structure of CS would alter significantly across the two buffers, a lower swelling ratio in pH 6.8 buffer was most likely attributable to weaker hydrogen bonding than in pH 7.4 buffer. These findings imply that the pH of CS-based nanogels is a significant factor in their swelling and that a high pH is desired.

2.2.6. SEM Image of Structural Network of Nanogel

As a result of the uneven structure of chitosan, SEM image revealed that the NPs have a regular and uniform quasi-shape in the structural network of nanogel (Figure 5).

2.2.7. Effect on Surface Morphology of Erythrocytes

On interaction with aqueous solution of plain drug as compared to normal saline as well as nanogel formulation, the erythrocytes showed the changes in shape and their surface morphology, as depicted in Figure 6. Based on the effects on erythrocytes' surface morphology and hemolytic toxicity investigations, it can be concluded that PS-CS-NGs demonstrated promising performance and can be developed as a carrier system for further biomedical application in other routes of drug delivery.

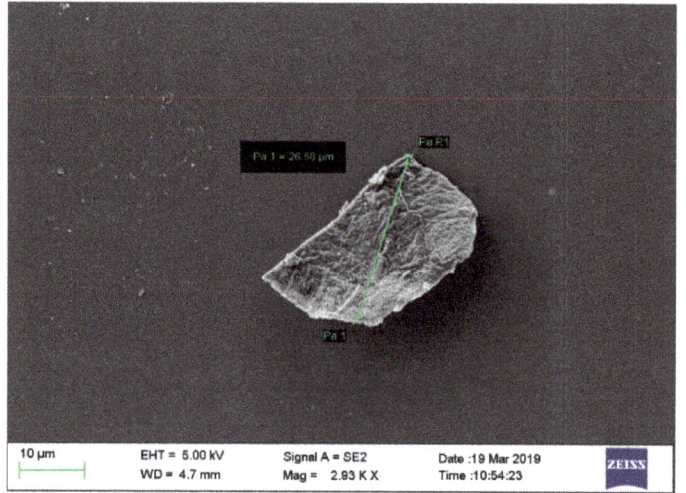

Figure 5. Scanning electron microscopy image of structural network of nanogel revealing a regular and uniform quasi-shape of NPs.

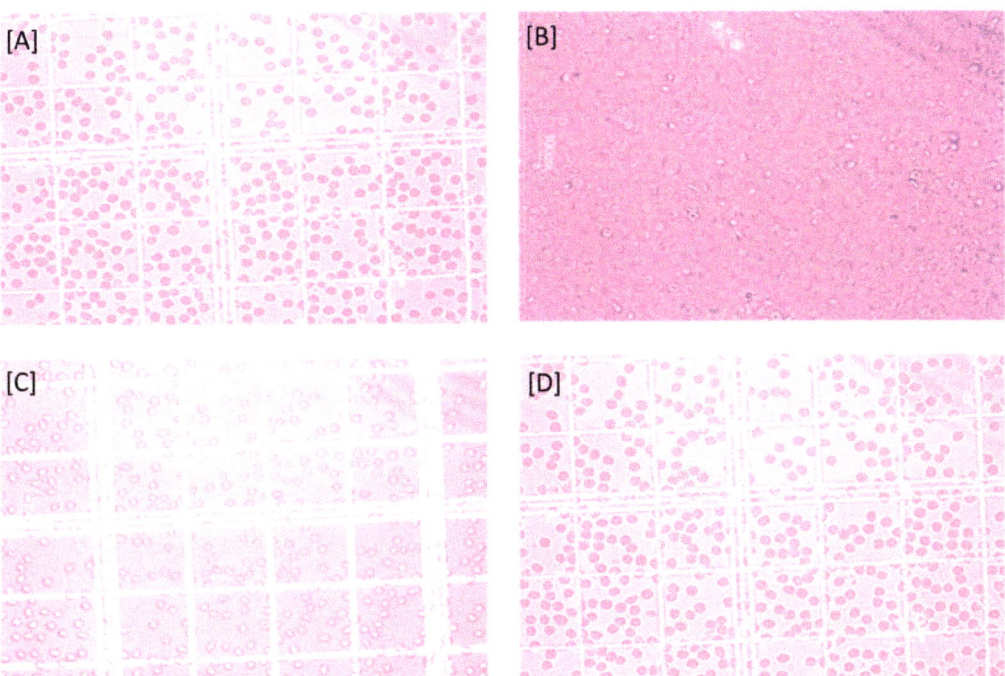

Figure 6. Erythrocytes on interaction with (**A**) Saline solution, (**B**) Water, (**C**) Pure drug solution, and (**D**) Nanogel formulation.

2.2.8. Hemolytic Toxicity

The percentage hemolytic toxicity level of the samples was found to be within the range. The drug-containing nanogel (PS-CS-NG) formulation was found to be less toxic (6.08%) than the pure drug solution (7.2%).

2.2.9. Pharmacokinetic Studies

The plasma concentration–time profile obtained from a single oral administration of pravastatin oral solution and pravastatin oral nanogel to Swiss albino rats was used to calculate the pharmacokinetic parameters (Table 4). According to these findings, the optimized pravastatin oral nanogel formulation has higher bioavailability than the pravastatin oral solution.

Table 4. Summary of values of peak plasma concentration (C_{max}), time of C_{max} attainment (t_{max}), Area under curve (AUC), and Mean residence time (MRT) of pravastatin oral solution and pravastatin oral nanogel.

S. No.	Time (h)	Pravastatin Oral Solution (μg/mL)	Pravastatin Oral Nanogel (μg/mL)
1	0	0	0
2	1	68	14
3	2	47	26
4	3	23	39
5	4	11	79
6	5	0	65
7	6		47
8	7		31
9	8		9
10	10		0
	C_{max}	68 μg/mL	79 μg/mL
	t_{max}	1 h	4 h
	AUC last	140.707 (μg·h/mL)	302.022 (μg·h/mL)
	AUC total	155.85 (μg·h/mL)	312.912 (μg·h/mL)
	T_{half}	0.98 h	0.83 h
	MRT	2.35 h	4.62 h

Some of the major observations and conclusion that could be drawn with the experimental work are represented in Table 5.

Table 5. Concluding features of experimental work for optimized formulation.

Parameters	Result	Inference
Particle size	486.2 nm	Desired and acceptable size
Polydispersity index	0.303	Uniform distribution
Zeta potential	43.4 mV	Positively and evenly distributed
Entrapment efficiency	50%	Nanoparticles leads for higher entrapment of drug
Drug release	59.63% (24 h)	Sustained release of the drug was obtained
SEM studies	-	Regular shape
Compatibility study	Characteristic peak is obtained	Overlay plot confirms the characteristic peaks of drug
Toxicity study	Less toxicity is exhibited	Safer to use
Pharmacokinetic study	Higher bioavailability	

3. Conclusions

The main aim of the present investigation was to develop and optimize the novel pravastatin nanogel using central composite design. The optimized batch showed the desired characteristics and in vitro drug release of the nanogel formulation showing the sustained release. Forced Degradation Studies (FDS) have also indicated about the stability of drug in oxidative and neutral medium (Figures S1 and S2). Pravastatin belongs to cholesterol inhibitor antihyperlipedemia. It inhibits the cholesterol absorption and acts on the brush border of the small intestine. As the drug belongs to BCS class III, inadequate permeability has always been a difficult hurdle to overcome, as it is a critical aspect in a drug's bioavailability after oral administration. Since a rising number of newly produced Statin series medication candidates in pre-clinical development belong to BCS class III, formulation options to circumvent this issue are in high demand. Over the last decade, drug formulation as NPs/nanogel has received a lot of attention as one of the numerous strategies to improve a product's permeability/dissolution rate features with the goal of increasing its oral bioavailability. The hypothesis underlying dissolution rate improvement, when considering drug particle size reduction to the nanometer range, is that the resulting drug-loaded nanogel has a substantially larger effective surface area. Ionic gelation is one of the easiest and most effective strategies for reducing drug particle size to the nanoscale range among the different available technologies.

4. Materials and Methods

4.1. Materials

Gift sample of Pravastatin Sodium drug was received from Biocon Pvt. Ltd., Bengaluru, India. Chitosan was purchased from HiMedia Pvt. Ltd., Mumbai, India. Glacial acetic acid was purchased from Merck, Mumbai, India. Pentasodium tripolyphosphate and Poloxamer were purchased from Sigma Aldrich, Mumbai, India. All other used chemicals and reagents were of analytical grade.

4.2. Methods

4.2.1. Preparation of Pravastatin-Loaded Chitosan Nanogel

Pravastatin-loaded chitosan nanoparticles (PS-CS-NPs) were prepared by the ionic gelation method [28]. Briefly, chitosan was dissolved in 10 mL of 0.5% v/v acetic acid with constant stirring. Tripolyphosphate (TPP) or sodium triphosphate and pravastatin dissolved in 10 mL of deionized water was added drop wise under constant magnetic stirring. The sample was kept for constant magnetic stirring for 2 h. The obtained solution was turbid, and the NPs were obtained. An appropriate amount of poloxamer (5% solution) was added in above prepared NPs and kept for stirring for about half an hour. This led to the formation of nanogel [29].

4.2.2. Experimental Design for Optimization of Nanogel

Pravastatin-loaded chitosan nanogels (PS-CS-NGs) were prepared by an ionic gelation method using the quality by design (QbD) approach [28,29]. For obtaining the desired product quality, critical quality attributes (CQA) were identified from the quality target product profile (QTPP). The CQAs of the PS-CS-NGs are listed in Table 6. During the QbD-based development of PS-CS-NGs, an initial risk assessment (RA) effort was also carried out, which involved evaluation of those formulation variables that could have an impact on CQAs. Table 7 shows the level of initial risk assessment for individual formulation variables.

Table 6. Critical Quality Attributes (CQA) parameters.

CQAs	Polymer Amount	Stirring Speed	Sonication Time
Particle size	Low	Medium	High
% Entrapment efficiency	Low	Medium	High

Table 7. Quality by Design (QbD) details.

Profile	Target	Justification
Dosage form	Nanogel	Novel dosage form for targeted drug delivery
Dosage design	Sustained release oral nanogel	For increasing residence time of pravastatin
Therapeutic indication	Antihyperlipidemia	Pravastatin acts by inhibition of cholesterol producing enzymes
Route of administration	Oral	Most suitable route of administration and can be well absorbed in intestine
Particle size	10–1000 nm	Drug absorption and uniform biodistribution
Zeta potential	−200 to 200 mV	Needed to ensure stability
Entrapment efficiency	>50%	Nanogel entraps higher amount of drug

Particle size and its uniform distribution are the ultimate benchmark for the novel DDS. On the basis of the preliminary trials and risk assessment, a 3^2 level CCD was employed to study the effect of Polymer amount (X1), Stirring speed (X2), and Sonication time (X3) as independent variables (Table 8) while Particle size (Y1) and Entrapment efficiency (Y2) were the dependent variables. Nanogel was prepared as per the experiment design matrix generated by the software (Design-Expert® Software Version 11).

Table 8. Independent variables for Quality by Design (QbD).

Independent Variables Coded Values	Low (−1)	High (+1)
A = Polymer amount (mg)	50	100
B = Stirring speed (rpm)	1000	1500
C = Sonication time (min)	4	8

4.2.3. Characterization and Evaluation of Optimized Nanoparticles/Nanogel

Nanoparticles are characterized based on their size, morphological characteristics, and surface charge. The size distribution, average particle diameter, and surface charge influence the physical stability, redispersibility, and in vivo performance of NPs.

Particle Size and Zeta Potential

The particle size distribution and morphology are the primary determinants of NP characterization. Particle size was determined at 25 °C by the photon correlation spectroscopy technique [30]. This analysis measures the particle size of particles suspended in liquids in the range of 0.6 nm to 10 μm with sample suspension concentrations from 0.00001 to 40%. All the data presented are produced under identical production conditions. The interaction of NPs with the biological environment, as well as their electrostatic interaction with bioactive chemicals, is determined by their surface charge and intensity. To maintain particle stability and avoid aggregation, zeta potential levels (high zeta potential values, either positive or negative) are attained.

Entrapment Efficiency

The obtained formulation was dissolved in methanol and centrifuged in cooling microfuge at 20 °C for 15 min. After centrifugation, the supernatant was collected and filtered. Then, 0.2 mL of this stock solution was diluted with water up to 1 mL. The un-entrapped pravastatin was estimated and analyzed by the UV method (Perkin Elmer series 200) at 238 nm. Drug entrapped was estimated by subtracting un-entrapped drug from the total drug. Entrapment efficiency was further calculated from the entrapped drug.

Morphology Observation

With direct observation of the NPs, the electron microscopy-based approach evaluates their size, shape, and surface morphology. The solution of NPs was first converted into a dry powder, which was then deposited on a sample holder before being sputter-coated with a conducting metal (such as gold). The entire sample was scanned with a focused fine beam of electrons for analysis, and surface characteristics were determined.

Fourier-Transform Infrared Spectroscopy

FT-IR spectroscopy was used to investigate drug–polymer interactions. FT-IR studies were performed on both the pure drug and the excipients (Figures S3 and S4). Physical mixes were also scanned in the wavelength range of 400–4000 cm^{-1} in an FT-IR spectrophotometer, and the spectra were recorded.

In Vitro Drug Release

The most adaptable and widely used approach for determining drug release from nanosized dosage forms is dialysis. The dialysis bag was filled with 10 mL of optimized nanogel formulation and was immersed immediately in the 50 mL of phosphate buffer saline (PBS) pH 7.4. The sampling for drug release was performed at definite time intervals. The sample (5 mL) was withdrawn at each time interval, and the withdrawn volume was replaced by fresh PBS to maintain sink condition. The samples were analyzed UV spectrophotometrically.

Swelling Ratio

The swelling ratio is a measurement of the capability of a nanogel to absorb water. All samples reached the equilibrium water content within 1 h, and the swelling ratio values were determined at pH 6.8 and pH 7.4 [31].

Hemolytic Toxicity

Hemolytic toxicity was determined using the previously described method [32,33]. In a nutshell, whole human blood was collected in a HiAnticlot blood collection vial from a healthy donor. The blood was washed with PBS pH 7.4, which was followed by separation of the erythrocytes by centrifugation at 3000 rpm for 5 min, whereby the supernatant was pipette off repeatedly (n = 3), and the erythrocytes were suspended in normal saline solution to obtain 10% hematocrit. Then, 1 mL of RBC suspension was incubated with distilled water (taken as 100% hemolytic standard) and normal saline (taken as blank spectrophotometric estimation). In this study, Afterwards, 1 mL of erythrocytes suspension and 1 mL of formulation were taken in separate tubes and the volume made up to 10 mL with normal saline. Similarly, 1 mL of drug solution was mixed with 9 mL of normal saline and interacted with erythrocytes suspension. The tubes were allowed to stand for 1 h at 37 °C with intermittent shaking. The tubes were centrifuged for 15 min at 3000 rpm, and the absorbance of supernatants was measured at 308 nm, which was used to estimate the percentage of hemolysis using 100% hemolytic standard obtained with distilled water diluted; similarly, the percent hemolysis was calculated for each sample by using the following Equation (1):

$$\text{Hemolysis} = AB_s / AB_{100} \times 100 \qquad (1)$$

Effect on Surface Morphology of Erythrocytes

Erythrocyte suspension was treated with drug and nanogel formulation at a specified concentration (1 mg/mL), and the morphological status of erythrocytes was assessed under an optical microscope (Leica, DMLB, Heerbrugg, Switzerland) [32,33].

Pharmacokinetics Studies

After a one-week acclimatization period, albino rats weighing 120 ± 10 g were employed in the experiment. All animal experiments were carried out in accordance with the protocol approved by Institutional Animal Ethics Committee (IAEC) of NMIMS School of Pharmacy & Technology Management, Shirpur (Maharashtra), India (SPTM-IAEC/Dec-18/03/03; dated 16 December 2018). A pharmacokinetics investigation was carried out using an optimized nanogel formulation and an in-house made immediate release solution. Each rat received a dose of pravastatin sodium corresponding to 10 mg per kg of body weight. Oral solution was made by dissolving an equivalent amount of drug in water in order to provide the drug dose based on the animal's body weight. The animals were placed into two groups, each with three animals. An i.m. injection of a 1:5 mixture of xylazine (1.9 mg/kg) and ketamine (9.3 mg/kg) was used to lightly anaesthetize the animals. A solution providing the required dose based on the rat's body weight was given orally to one group. Another group received an optimized nanogel with the required dose based on the rat's body weight. A 26-gauge needle was used to take 1 mL of blood from the marginal ear vein every 1 h for up to 10 h. To separate plasma, blood samples were centrifuged at 8000 rpm for 10 min at 15 °C (REMI Pvt. Ltd., Vasai, India). Until further examination, plasma samples were kept at −20 °C. The plasma samples were analyzed for the pravastatin using HPLC method (Figure S5).

Supplementary Materials: The following supporting information can be downloaded at: https://www.mdpi.com/article/10.3390/gels8020081/s1, Figure S1: Hydrolytic degradation chromatogram of pravastatin sodium, Figure S2: Oxidative degradation chromatogram of pravastatin sodium, Figure S3: FTIR Spectrum of pravastatin sodium, Figure S4: Characteristics peaks of compatibility study, Figure S5: HPLC chromatogram of pravastatin sodium.

Author Contributions: Conceptualization, G.K.S., S.T. and V.M.; writing—original draft preparation, G.K.S., S.T., V.M., A.A., P.N. and M.M.T.; writing—review and editing, G.K.S., Y.M., V.M., A.A., P.N. and M.M.T.; supervision, G.K.S. and V.M.; Data curation, S.T., Y.M., P.N. and M.M.T.; Software, S.T.; Validation, V.M., A.A. and M.M.T.; Resources, A.A. All authors have read and agreed to the published version of the manuscript.

Funding: The researchers would like to thank the Deanship of Scientific Research (DSR), Qassim University for funding the publication of this project.

Institutional Review Board Statement: The animal study protocol was approved by Institutional Animal Ethics Committee (IAEC) of NMIMS School of Pharmacy & Technology Management, Shirpur (Maharashtra), India (SPTM-IAEC/Dec-18/03/03; dated 16.12.2018) for studies involving animals.

Informed Consent Statement: Not applicable.

Data Availability Statement: Data are contained within the article or supplementary material.

Acknowledgments: The authors acknowledge DSR, Qassim University for financial support.

Conflicts of Interest: The authors declare no conflict of interest.

References

1. Ghaywat, S.D.; Mate, P.S.; Parsutkar, Y.M.; Chandimeshram, A.D.; Umekar, M.J. Overview of nanogel and its applications. *GSC Biol. Pharm. Sci.* **2021**, *16*, 040–061. [CrossRef]
2. Ansari, S.; Karimi, M. Novel developments and trends of analytical methods for drug analysis in biological and environmental samples by molecularly imprinted polymers. *TrAC Trends Anal. Chem.* **2017**, *89*, 146–162. [CrossRef]
3. Sivaram, A.J.; Rajitha, P.; Maya, S.; Jayakumar, R.; Sabitha, M. Nanogels for delivery, imaging and therapy. *Wiley Interdiscip. Rev. Nanomed. Nanobiotechnol.* **2015**, *7*, 509–533. [CrossRef] [PubMed]
4. Vashist, A.; Kaushik, A.; Vashist, A.; Bala, J.; Nikkhah-Moshaie, R.; Sagar, V.; Nair, M. Nanogels as potential drug nanocarriers for CNS drug delivery. *Drug Discov. Today* **2018**, *23*, 1436–1443. [CrossRef] [PubMed]
5. Anooj, E.S.; Charumathy, M.; Sharma, V.; Vibala, B.V.; Gopukumar, S.T.; Jainab, S.B.; Vallinayagam, S. Nanogels: An overview of properties, biomedical applications, future research trends and developments. *J. Mol. Struct.* **2021**, *1239*, 130446. [CrossRef]
6. Grimaudo, M.A.; Concheiro, A.; Alvarez-Lorenzo, C. Nanogels for regenerative medicine. *J. Control. Release* **2019**, *313*, 148–160. [CrossRef] [PubMed]
7. Cuggino, J.C.; Blanco, E.R.; Gugliotta, L.M.; Igarzabal, C.I.; Calderón, M. Crossing biological barriers with nanogels to improve drug delivery performance. *J. Control. Release* **2019**, *307*, 221–246. [CrossRef]
8. Yin, Y.; Hu, B.; Yuan, X.; Cai, L.; Gao, H.; Yang, Q. Nanogel: A versatile nano-delivery system for biomedical applications. *Pharmaceutics* **2020**, *12*, 290. [CrossRef]
9. Korani, S.; Korani, M.; Bahrami, S.; Johnston, T.P.; Butler, A.E.; Banach, M.; Sahebkar, A. Application of nanotechnology to improve the therapeutic benefits of statins. *Drug Discov. Today* **2019**, *24*, 567–574. [CrossRef]
10. Hill, M.F.; Bordoni, B. Hyperlipidemia. In *StatPearls*; StatPearls Publishing: Treasure Island, FL, USA, 2021. Available online: https://www.ncbi.nlm.nih.gov/books/NBK559182/ (accessed on 30 September 2021).
11. Nelson, R.H. Hyperlipidemia as a risk factor for cardiovascular disease. *Prim. Care Clin. Off. Pract.* **2013**, *40*, 195–211. [CrossRef]
12. Ludwig, A. The use of mucoadhesive polymers in ocular drug delivery. *Adv. Drug Deliv. Rev.* **2005**, *57*, 1595–1639. [CrossRef] [PubMed]
13. Barbu, E.; Verestiuc, L.; Iancu, M.; Jatariu, A.; Lungu, A.; Tsibouklis, J. Hybrid polymeric hydrogels for ocular drug delivery: Nanoparticulate systems from copolymers of acrylic acid-functionalized chitosan and N-isopropylacrylamide or 2-hydroxyethyl methacrylate. *Nanotechnology* **2009**, *20*, 225108. [CrossRef] [PubMed]
14. Kao, H.J.; Lo, Y.L.; Lin, H.R.; Yu, S.P. Characterization of pilocarpine-loaded chitosan/Carbopol nanoparticles. *J. Pharm. Pharmacol.* **2006**, *58*, 179–186. [CrossRef] [PubMed]
15. Järvinen, K.; Järvinen, T.; Thompson, D.O.; Stella, V.J. The effect of a modified β-cyclodextrin, SBE4-β-CD, on the aqueous stability and ocular absorption of pilocarpine. *Curr. Eye Res.* **1994**, *13*, 897–905. [CrossRef] [PubMed]
16. Yu, D.-G. Preface. *Curr. Drug Deliv.* **2021**, *18*, 2–3. [CrossRef]
17. Xu, H.; Xu, X.; Li, S.; Song, W.-L.; Yu, D.-G.; Annie Bligh, S.W. The Effect of Drug Heterogeneous Distributions within Core-Sheath Nanostructures on Its Sustained Release Profiles. *Biomolecules* **2021**, *11*, 1330. [CrossRef]
18. Kang, S.; He, Y.; Yu, D.-G.; Li, W.; Wang, K. Drug–zein@ lipid hybrid nanoparticles: Electrospraying preparation and drug extended release application. *Colloids Surf. B Biointerfaces* **2021**, *201*, 111629. [CrossRef]

19. Borujeni, S.H.; Mirdamadian, S.Z.; Varshosaz, J.; Taheri, A. Three-dimensional (3D) printed tablets using ethyl cellulose and hydroxypropyl cellulose to achieve zero order sustained release profile. *Cellulose* **2020**, *27*, 1573–1589. [CrossRef]
20. Kaur, M.; Sudhakar, K.; Mishra, V. Fabrication and biomedical potential of nanogels: An overview. *Int. J. Polym. Mater. Polym. Biomater.* **2019**, *68*, 287–296. [CrossRef]
21. Kesharwani, P.; Jain, A.; Srivastava, A.K.; Keshari, M.K. Systematic development and characterization of curcumin-loaded nanogel for topical application. *Drug Dev. Ind. Pharm.* **2020**, *46*, 1443–1457. [CrossRef]
22. Sharma, A.; Garg, T.; Aman, A.; Panchal, K.; Sharma, R.; Kumar, S.; Markandeywar, T. Nanogel—An advanced drug delivery tool: Current and future. *Artif. Cells Nanomed. Biotechnol.* **2016**, *44*, 165–177. [CrossRef]
23. Endo, A. A historical perspective on the discovery of statins. *Proc. Jpn. Acad. Ser. B* **2010**, *86*, 484–493. [CrossRef] [PubMed]
24. Murphy, C.; Deplazes, E.; Cranfield, C.G.; Garcia, A. The role of structure and biophysical properties in the pleiotropic effects of statins. *Int. J. Mol. Sci.* **2020**, *21*, 8745. [CrossRef]
25. Manzoni, M.; Rollini, M. Biosynthesis and biotechnological production of statins by filamentous fungi and application of these cholesterol-lowering drugs. *Appl. Microbiol. Biotechnol.* **2002**, *58*, 555–564.
26. Cuggino, J.C.; Molina, M.; Wedepohl, S.; Igarzabal, C.I.; Calderón, M.; Gugliotta, L.M. Responsive nanogels for application as smart carriers in endocytic pH-triggered drug delivery systems. *Eur. Polym. J.* **2016**, *78*, 14–24. [CrossRef]
27. Shidhaye, S.S.; Thakkar, P.V.; Dand, N.M.; Kadam, V.J. Buccal drug delivery of pravastatin sodium. *AAPS PharmSciTech* **2010**, *11*, 416–424. [CrossRef]
28. Desai, K.G. Chitosan nanoparticles prepared by ionotropic gelation: An overview of recent advances. *Crit. Rev. Ther. Drug Carr. Syst.* **2016**, *33*, 107–158. [CrossRef]
29. Wang, H.; Deng, H.; Gao, M.; Zhang, W. Self-assembled nanogels based on ionic gelation of natural polysaccharides for drug delivery. *Front. Bioeng. Biotechnol.* **2021**, *9*, 703559. [CrossRef]
30. Asadian-Birjand, M.; Bergueiro, J.; Rancan, F.; Cuggino, J.C.; Mutihac, R.C.; Achazi, K.; Dernedde, J.; Blume-Peytavi, U.; Vogt, A.; Calderón, M. Engineering thermoresponsive polyether-based nanogels for temperature dependent skin penetration. *Polym. Chem.* **2015**, *6*, 5827–5831. [CrossRef]
31. Durán-Lobato, M.; Carrillo-Conde, B.; Khairandish, Y.; Peppas, N.A. Surface-modified P (HEMA-co-MAA) nanogel carriers for oral vaccine delivery: Design, characterization, and in vitro targeting evaluation. *Biomacromolecules* **2014**, *15*, 2725–2734. [CrossRef]
32. Mishra, V.; Jain, N.K. Acetazolamide encapsulated dendritic nano-architectures for effective glaucoma management in rabbits. *Int. J. Pharm.* **2014**, *461*, 380–390. [CrossRef] [PubMed]
33. Mishra, V.; Gupta, U.; Jain, N.K. Influence of different generations of poly (propylene imine) dendrimers on human erythrocytes. *Die Pharm.* **2010**, *65*, 891–895.

Article

Synthesis and Characterization of Conjugated Hyaluronic Acids. Application to Stability Studies of Chitosan-Hyaluronic Acid Nanogels Based on Fluorescence Resonance Energy Transfer

Volodymyr Malytskyi [1,2,*], Juliette Moreau [1], Maïté Callewaert [1], Céline Henoumont [3], Cyril Cadiou [1], Cécile Feuillie [4], Sophie Laurent [3,5], Michael Molinari [4] and Françoise Chuburu [1,*]

1. Institut de Chimie Moléculaire de Reims, University of Reims Champagne Ardenne, CNRS, ICMR UMR 7312, 51097 Reims, France; juliette.moreau@univ-reims.fr (J.M.); maite.callewaert@univ-reims.fr (M.C.); cyril.cadiou@univ-reims.fr (C.C.)
2. Institut Parisien de Chimie Moléculaire, Sorbonne Université, CNRS, IPCM UMR 8232, 4 Place Jussieu, 75252 Paris, France
3. NMR and Molecular Imaging Laboratory, University of Mons UMons, B-7000 Mons, Belgium; celine.henoumont@umons.ac.be (C.H.); sophie.laurent@umons.ac.be (S.L.)
4. Center for Microscopy and Molecular Imaging, Rue Adrienne Bolland 8, B-6041 Charleroi, Belgium; cecile.feuillie@u-bordeaux.fr (C.F.); michael.molinari@u-bordeaux.fr (M.M.)
5. Institut de Chimie et Biologie des Membranes et des Nano-Objets, CNRS UMR 5248, University of Bordeaux, IPB, 33600 Pessac, France
* Correspondence: volodymyr.malystkyi@sorbonne-universite.fr (V.M.); francoise.chuburu@univ-reims.fr (F.C.)

Abstract: Hyaluronic acid (HA) was functionalized with a series of amino synthons (octylamine, polyethylene glycol amine, trifluoropropyl amine, rhodamine). Sodium hyaluronate (HAs) was first converted into its protonated form (HAp) and the reaction was conducted in DMSO by varying the initial ratio ($-NH_2$ (synthon)/COOH (HAp)). HA derivatives were characterized by a combination of techniques (FTIR, ^{1}H NMR, 1D diffusion-filtered ^{19}F NMR, DOSY experiments), and degrees of substitution (DS_{HA}) varying from 0.3% to 47% were determined, according to the grafted synthon. Nanohydrogels were then obtained by ionic gelation between functionalized hyaluronic acids and chitosan (CS) and tripolyphosphate (TPP) as a cross-linker. Nanohydrogels for which HA and CS were respectively labeled by rhodamine and fluorescein which are a fluorescent donor-acceptor pair were subjected to FRET experiments to evaluate the stability of these nano-assemblies.

Keywords: nanohydrogels–hyaluronic acid–HA-mPEG$_{2000}$; fluorinated and fluorescent HA conjugates–hyaluronic acid degree of substitution–diffusion ordered spectroscopy (DOSY)–1D diffusion-filtered ^{19}F NMR–atomic force microscopy–FRET experiments–hyaluronidase–nanohydrogel stability

1. Introduction

Since its first isolation in 1934 from the vitreous humor of bovine eyes, hyaluronic acid (**HA**) has been used in many applications and research areas [1]. This unbranched glycosaminoglycan which is composed of repeating units of disaccharides N-acetyl glucosamine (GlcNAc) and D-glucuronic acid (GlcA) linked together through alternating β-1,3 and β-1,4 glycosidic bonds, is a negatively charged polymer [2] at physiological pH (3 < pKa (COOH groups on the D-glucuronic acid residues) < 4). **HA** molecules strongly bind to water molecules and become heavily hydrated to form a viscous gel. This property is at the origin of the viscoelastic character and the control of tissue hydration [3,4] and, as a primary component of extracellular matrix (ECM) vitreous humor and synovial fluid of vertebrates, functions as a scaffold for the organization of these biofluids. **HA** has become a

carrier of great interest not only owing to its advantages such as biodegradability, biocompatibility, but also to its intrinsic targeting properties, based on the selective interactions with receptors, such as CD44 or hyaluronan receptors for endocytosis (HARE) [5]. For these reasons, exogenous **HA** has been investigated as a drug delivery system for therapeutics and diagnostics [3–14].

To improve its properties and target its applications, **HA** can be subjected to chemical modifications. To do this, synthetic approaches are mainly based on (a) the functionalization of a carboxylic acid group by peptide coupling [15,16], esterification [17], or Ugi condensation reaction [18] (b) the functionalization of hydroxyl groups by alkylation [19] or acylation [20,21] or (c) a partial oxidative degradation of the polymer [22]. It is worth mentioning that between these methods, peptidic coupling is the most commonly used due to the accessibility of amine functions for the introduction of various side-groups and due to the robust nature of the amide bond formed. The commercially available bio-extracted sodium hyaluronate (**HAs**) is a water-soluble polyanionic polymer. Its chemical modification by peptidic coupling reaction is typically carried out using conventional coupling agents such as EDC/NHS in an aqueous medium [23,24]. However, these latter form in situ activated intermediates that can be hydrolyzed by water molecules prior to their reaction with amines [25–27]. This usually determines low yields of grafting in water and requires the use of a significant excess of coupling agents and amines. Another drawback of this strategy is the difficulty to evidence the formation of the amide bond because the amide protons are often invisible when NMR is performed in protic solvents. An alternative is to carry out the reaction in organic aprotic solvents and under anhydrous conditions. Palumbo et al. [28] recently showed the possibility to manipulate hyaluronic acid in a pure organic solvent, such as DMSO, by the transformation of **HAs** into its tetrabutylammonium (TBA$^+$) salt and its further activation using 4-NPBC which is completely unstable in water. However, the application of functionalized polysaccharides in the nanomedicine field subsequently requires their solubility in water, and therefore an additional step to make cation exchange again (TBA$^+$ to Na$^+$) is necessary. The alternative is to use a protonated form of hyaluronic acid (**HAp**) because this polymer is simultaneously soluble both in DMSO and in water unlike the **HAs** form, and the peptide coupling in DMSO allows the straightforward determination of the degree of substitution by NMR. To our knowledge, only two examples of peptidic coupling using such an approach can be found in the literature to date [29,30].

In this context and in order to extend the scope of this method we have systematically reinvestigated **HAp** functionalization, in which the level of **HAp** substitution was varied (by increasing the initial synthon/COOH HA molar ratio), characterized by FTIR and quantified by a combination of NMR techniques. The method was developed from the model functionalization reaction between **HAp** and n-octylamine, using HATU (1-[bis(dimethylamino)methylene]-1H-1,2,3-triazolo [4,5-b]pyridinium 3-oxid hexafluorophosphate) as a coupling agent, a method which allows to graft different ligands of interest for **HA** such as fluorinated synthons and PEG moieties (stealthiness) and fluorescent tags (rhodamine for fluorescence imaging). Moreover, we have demonstrated that nanohydrogels (NGs) obtained by ionic gelation between **HA** and chitosan (**CS**) in the presence of tripolyphosphate (TPP) as a cross-linking agent are particularly well suited to encapsulate gadolinium chelates (**GdCAs**) and tremendously increase the efficiency of these paramagnetic MRI probes [31–34]. Ionic gelation relies on the development of electrostatic interactions between the negative charges of hyaluronic acid and the positive charges of chitosan. Therefore, special attention must be paid to the HA degree of substitution (DS$_{HA}$) in order to ensure that after functionalization, there are enough negative charges left on the functionalized HA for the establishment of these interactions (which must remain sufficient for the ionic gelation to still lead to the formation of functionalized nanoparticles). Furthermore, when it comes to nanogels obtained by ionic gelation, the question of their stability is raised. In order to answer this question, we will use the functionalization of **HAs** developed herein with rhodamine (**HA-Rhod**) to elaborate nanogels with recently fluorescein-labeled chitosan (**CS-Fluo**) and (i) evaluate by fluorescence spectroscopy the occurrence of a Förster resonance transfer (FRET) signal within nanogels and (ii) test the conditions of degradation of the edifice, in particular in the presence of enzymes [35,36].

2. Results and Discussion

2.1. Chemical Functionalization of Hyaluronic Acid

The functionalization of HA by peptidic coupling was run in DMSO under anhydrous conditions in order to avoid any hydrolysis of the intermediates which could be detrimental to the performance of functionalization [25–27]. This implies beforehand to improve the solubility of HA in this solvent. For that, the commercially available sodium hyaluronate **HAs** was converted into its protonated form **HAp** by column exchange chromatography, according to the procedure of Vasi et al. [20]. After complete dehydration by lyophilization, **HAp** was used in a peptidic coupling reaction in anhydrous DMSO using **HATU** as a coupling agent. A series of amines have been used in the general synthetic method as illustrated in Scheme 1.

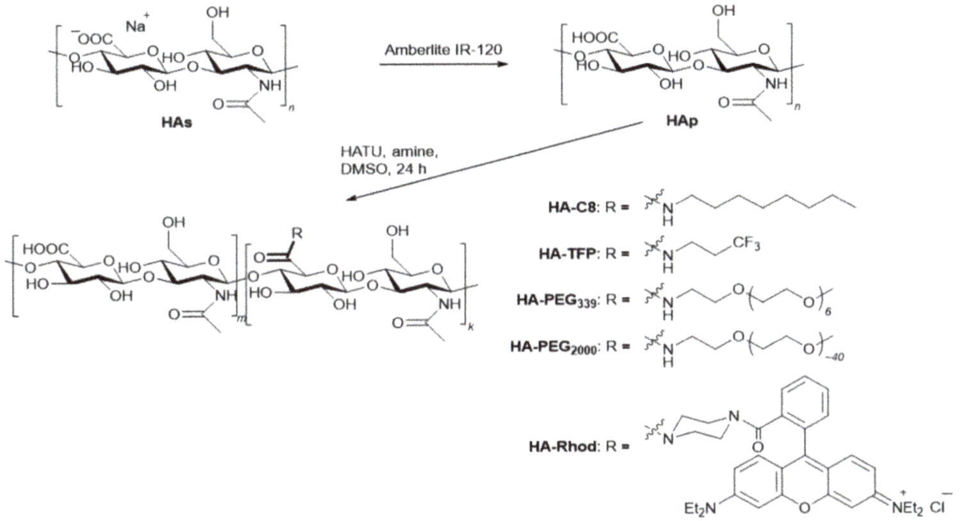

Scheme 1. Syntheses of functionalized HA described in the paper.

Five amino synthons were used to functionalize **HAp**, *n*-octylamine **C8-NH$_2$** (as a model to set up the synthesis and characterization protocols), 1,1,1-trifluoropropylamine **TFP-NH$_2$**, 2,5,8,11,14,17,20-heptaoxadocosan-22-amine **PEG$_{339}$-NH$_2$**, methoxy-poly(ethylene)glycol-amine **PEG$_{2000}$-NH$_2$**, and amine-functionalized rhodamine **Rhod-NH**. The molar ratios ($-$NH$_{2 \text{ synthons}}$ to $-$COOH $_{\text{HA}}$) were initially fixed at 10% (condition a), 20% (condition b), 50% (condition c), and 100% (condition d), respectively, for each amino synthon. The reactions were carried out under ambient conditions for one day. Functionalized polymers were precipitated from the organic solution, purified by ultrafiltration to eliminate all the unreacted low-molecular-weight compounds, and freeze-dried prior to their characterization by FTIR and NMR methods.

2.1.1. Grafting of *n*-Octylamine on Hyaluronic Acid

n-Octylamine (**C8-NH$_2$**) is a simple and accessible product that allows the introduction of an alkyl chain into a polysaccharide backbone. This latter can be easily identified by ^{1}H NMR and FTIR analysis and as such, can be used as a model to develop the conditions for **HA** functionalization and characterization of functionalized **HA**. At the same time and from an applicative point of view, hydrogels obtained with HA derivatized with such alkyl chains are known to exhibit improved viscoelastic properties and increased resistance to enzymatic hydrolysis [28]. After functionalization, Haps functionalized by octylamine (**HA-C8** polymers) revealed a modification of their solubility: **HA-C8a**, **HA-C8b**, and **Ha-C8c** were water soluble, **HA-C8c** was also soluble in chloroform, while **HA-C8d** was insoluble in water. After purification by ultrafiltration (**HA-C8a**, **HA-C8b**, and **HA-C8c**) or centrifugation (**HA-C8d**), all the samples were characterized by FTIR spectroscopy (Figure 1). A clear change in HA carbonyl vibration patterns was observed. As the amount of amine increased, a clear growth in the amide I band (ν(C=O), centered at 1644 cm^{-1}) was observed. It was accompanied by the same enhancement of the amide II band (ν(C–N), centered at 1557 cm^{-1}), confirming the increase in the quantity of amide groups in samples **HA-C8a-d** and therefore, a rise in the degree of functionalization. At the same time, the intensity of the HA carboxylic acid band (ν(C=O), centered at 1732 cm^{-1}) consistently decreased. The increasing quantity of grafted alkyl chains was also confirmed by the C–H vibration band enhancement. Two distinct peaks were rising at 2857 cm^{-1} and 2925 cm^{-1}, corresponding, respectively, to the symmetrical and the asymmetrical stretching vibrations of −CH$_2$ groups. The last band also exhibited a shoulder around 2953 cm^{-1} which could be attributed to the asymmetrical C–H stretching vibrations of the −CH$_3$ group.

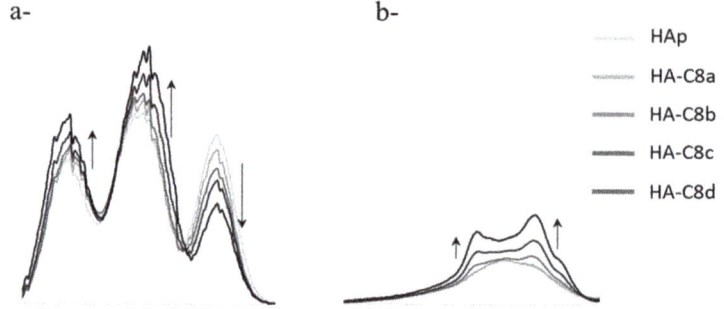

Figure 1. FTIR spectra of octylamine functionalized HA samples: (**a**) carbonyl vibrations region (1500–1800 cm^{-1}) and (**b**) C–H stretching vibrations region (2700–3000 cm^{-1}).

In order to evaluate the grafting degree and then, the degree of substitution of HA by octylamine, **HA-C8** polymers were subjected to a ^{1}H NMR analysis (318 K, DMSO d_6, Figure 2).

In these spectra (**HA-C8a** to **HA-C8d**), the rise of the alkyl chain protons at 0.87 ppm (CH$_3$), 1.26, 1.42, and 3.02 ppm (CH$_2$) was clearly observed, with the increasing quantity of amine. However, contrary to the commonly accepted opinion, they did not necessarily correspond to grafted synthons but can also come from octylamine (or octylammonium) associated via electrostatic interactions to a polyanionic polymer such as **HAp** [37]. The integration of one of these signals in comparison to one of the HA signals, however, led to the determination of the association rate (AR) of the C8 chain to HA (Table 1, entry 1). The most straightforward method to quantify the degree of substitution of HA (DS$_{HA}$), and then the grafting efficiency, consisted in integrating amide proton peaks (Table 1, entry 2). Indeed, while more and more carboxyl groups were modified, the splitting of the acetamide signal at 7.4 ppm was observed as well as the occurrence of a new signal at 7.9 ppm, that corresponded to the newly formed amide. Such an observation was only possible using DMSO-d_6 as a solvent and not in a solvent for which NH proton signals cannot be detected due to their exchangeable nature.

These results were corroborated by DOSY experiments (see Supplementary Materials, Figure S1). As described in the experimental part, the fitting of the diffusion curves of **HA-C8** extracted from DOSY experiments allowed to determine the fraction of C8 covalently grafted to HA. For **HA-C8a** and **HA-C8b**, a biexponential curve was obtained, and its fitting with Equation (1) allowed to extract the percentage of grafted C8 over the total amount of C8 (71% and 92% for **HA-C8a** and **HA-C8b**, respectively). For HA-C8c and HA-C8d, a monoexponential curve was obtained, and its fitting gave a diffusion coefficient equal to that of HA, proving that 100% of C8 was covalently grafted to HA.

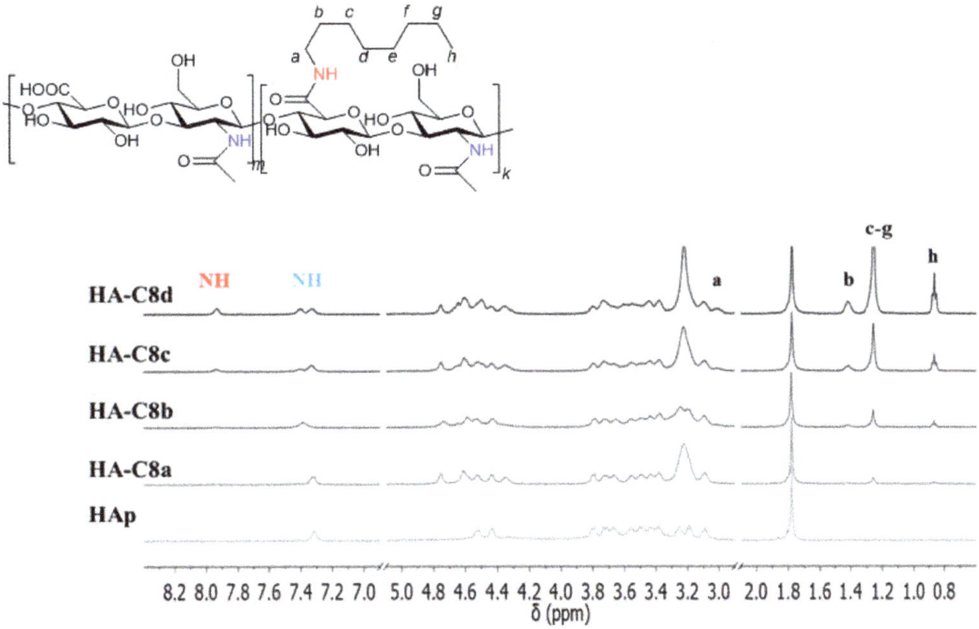

Figure 2. ^{1}H NMR spectra of octylamine functionalized HA: **Hap, HA-C8a, HA-C8b, HA-C8c,** and **HA-C8d**) (500 MHz, 318 K, DMSO d_6, and DMSO peak is omitted for clarity).

Table 1. Degrees of substitution of HA functionalized with *n*-octylamine.

	HA-C8a	HA-C8b	HA-C8c	HA-C8d
% mol (amine/COOH)$_{initial}$	10%	20%	50%	100%
Association rate (AR) [1]	4.0%	8.4%	23.4%	47.0%
Degree of substitution (DS$_{HA}$) [2]	1.9%	7.2%	22.4%	42.9%
Grafting degree (GD) [3]	71%	92%	100%	100%
Degree of substitution (DS$_{HA}$) [4]	2.8%	7.7%	23.4%	47.0%

[1] Found by integration of ^{1}H NMR signal at 0.87 ppm normalized to acetamide CH$_3$ peak at 1.78 ppm; [2] found by integration of ^{1}H NMR signal at 7.9 ppm normalized to acetamide NH peaks at 7.4 ppm; [3] determined from DOSY experiments; [4] DS = AR × GD.

The DS values obtained for **HA-C8** by both methods (by integrating NH peaks or by integrating aliphatic peaks corrected with DOSY analysis) were found to be close (±2–4%) and linearly dependent on the amount of amine initially introduced in the reaction medium (see Supplementary Materials, Figure S2).

2.1.2. Grafting of Polyethyleneglycol Oligomers on Hyaluronic Acid

Polyethyleneglycol (PEG) synthons are often used to improve nanostructure stealthiness in biological media [38–41]. In this context, the grafting of two PEG amines was tested, one bearing exclusively seven ethylene glycol residues of (**PEG$_{339}$-NH$_2$**) and the other being a mixture of larger oligomers with an average molar mass of 1834 Da (**PEG$_{2000}$-NH$_2$**), i.e., approximately 40 ethylene-oxy residues (Scheme 1). HA functionalization was subsequently followed by FTIR and ^{1}H NMR spectroscopies, according to increasing initial quantities of PEG amines (**HA-PEG$_{339}$a-d** and **HA-PEG$_{2000}$a-d** polymers, respectively).

The FTIR spectra of **HA-PEG$_{339}$a-d** and **HA-PEG$_{2000}$a-d** confirmed the successful modification of HA with oligomer PEG chains (see Supplementary Materials, Figures S3 and S4, and related commentaries).

As previously demonstrated for **HA-C8** polymers, ^{1}H NMR spectroscopy in DMSO-d_6 allowed the determination of the DS$_{HA}$ with PEG moieties. The signals corresponding to oligomeric ethylene glycol units and terminal methoxy groups were clearly observed at 3.51 ppm and 3.24 ppm respectively (see Supplementary Materials, Figures S5 and S6) but, due to an overlap with peaks of **HA** backbone, these signals cannot be used to quantify the extent of functionalization. In these conditions, DS$_{HA}$ were determined by the integration of amide proton signals at 8.10 ppm for each copolymer (Table 2).

Table 2. Degrees of substitution of HA functionalized with oligo-ethyleneglycol-amines.

	HA-PEG$_{339}$				HA-PEG$_{2000}$			
Entries	a	b	c	d	a	b	c	d
% mol (amine/COOH)$_{initial}$	10%	20%	50%	100%	10%	20%	50%	100%
Degree of substitution (DS$_{HA}$) [1]	2.4%	7.6%	15.7%	38.9%	1.3%	7.0%	14.1%	32.0%

[1] Found by the integration of ^{1}H NMR signal at 8.1 ppm normalized to acetamide NH peaks at 7.4 ppm.

The DS$_{HA}$ obtained for **HA-PEG** polymers were in the same order of magnitude as the ones determined for **HA-C8** polymers It is interesting to notice that (i) the variation of DS$_{HA}$ according to initial amounts of pegylated amine introduced in the preparation was again linear (see Figures S7 and S8) and (ii) the DS$_{HA}$ measured with PEG synthons were in the same order of magnitude as the ones determined with the lipophilic C8 chain.

2.1.3. Grafting of Trifluoropropylamine on Hyaluronic Acid

The introduction of fluorinated groups is considered as a solution in pharmaceutical chemistry for improving the lipophilicity of active substances and their subsequent accumulation in lymph nodes [37,42]. For this reason, the synthetic method developed herein was extended to the introduction of fluorinated groups on HA by means of the commercially available 1,1,1-trifluoropropylamine (**TFP-NH$_2$**, Scheme 1). After synthesis and purification, the successful grafting was evidenced by FTIR (see, Figure S9) and ^{1}H NMR (see Supplementary Materials, Figure S10).

Thanks to the presence of terminal CF$_3$ groups, evidence of grafting was also obtained in ^{19}F NMR spectroscopy (Figure 3).

As soon as the **TFP-NH$_2$** compound is grafted onto HA, a strong-field shift of CF$_3$ signal occurs, from 76.4 ppm (ungrafted **TFP-NH$_2$**) to 75.9 ppm (**HA-TFP**) (Figure 3a). One-dimensional (1D) diffusion-filtered ^{19}F spectra concomitantly recorded (Figure 3b and see also Supplementary Materials Figure S11) showed that for all the **HA-TFP** samples, the application of a 95% diffusion filter was accompanied by the permanence of the ^{19}F signal, while for **TFP-NH$_2$** the application of the same filter induced the disappearance of the signal. Under these applied filtering conditions, species that quickly diffuse are removed, while the 2% gradient condition is not able to discriminate between low and rapid diffusion species. Consequently, for all **HA-TFPa-d** samples, the peak at 75.9 ppm was unambiguously assigned to the signal of TFP grafted to the HA backbone.

The quantification of DS$_{HA}$ in **HA-TFPa-d** samples was then performed by ^{1}H NMR spectroscopy. As previously noticed for HA-PEG copolymers, the ^{1}H signals of the newly grafted −(CH$_2$)$_2$ chain were masked by the peaks of the polymer. Therefore, the quantification of DS$_{HA}$ was again performed by integrating the ^{1}H signal of the amidic proton associated with the newly formed peptide bond at 8.25 ppm (Table 3). Compared to C8- and PEG-functionalized HA, a slight low-field shift of the amidic proton signal was observed due to the electron-withdrawing effect of the trifluoromethyl group. Obtained DS values were gathered in Table 3 (see also Supplementary Materials, Figure S12).

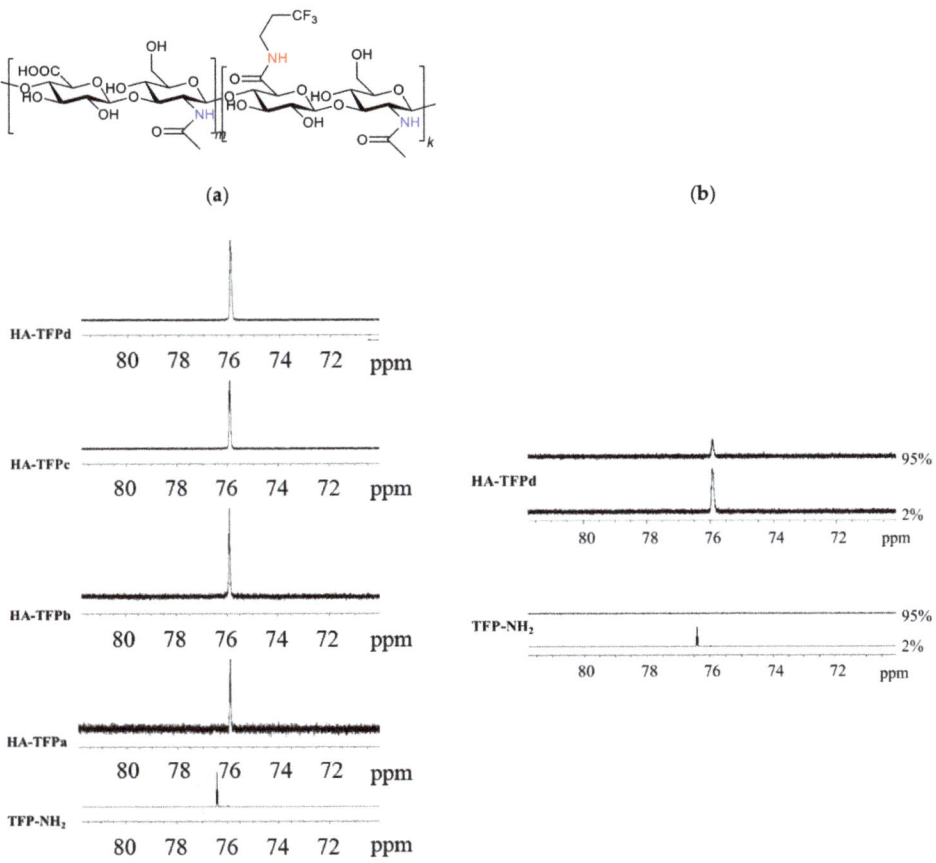

Figure 3. (a) ^{19}F NMR spectra of trifluoropropyl functionalized HA: **HA-TFPa**, **HA-TFPb**, **HA-TFPc** and **HA-TFPd** for initial molar ratios (NH$_2$/COOH) = 10 (bottom), 20, 50, and 100 (top) %, respectively; **TFP-NH$_2$** ^{19}F spectrum is given as a reference (470.64 MHz, 318 K, DMSO-d_6). (b) One-dimensional (1D) diffusion-filtered ^{19}F NMR spectra of **HA-TFPd** and **TFP-NH$_2$** with a gradient g of 2% and 95%.

Table 3. Degrees of substitution of HA functionalized with trifluoropropyl-amine.

	HA-TFPa	HA-TFPb	HA-TFPc	HA-TFPd
% mol (amine/COOH)$_{initial}$	10%	20%	50%	100%
Degree of substitution (DS$_{HA}$) [a]	5.2%	6.5%	15.6%	29.2%

[a] Found by integration of ^{1}H NMR signal at 8.25 ppm normalized to acetamide NH peak at 7.4 ppm.

2.1.4. Grafting of Rhodamine B Amine on Hyaluronic Acid

An important property for nanomaterials designed for the biomedical field is their ability to be tracked in vivo, particularly by fluorescence imaging. In the current study, this requires the control of the HA grafting reaction by fluorescent synthons. We have chosen as a model fluorophore the rhodamine **Rhod-NH** [43] which was introduced on the HA skeleton by peptidic coupling. (Scheme 1) Therefore, HA backbone functionalization was performed with rhodamine and the corresponding conjugates characterized as above to obtain a precise evaluation of the grafting rate and then of DS$_{HA}$.

The efficiency of the grafting was firstly followed by an FTIR analysis (see Supplementary Materials, Figure S13).

Unlike the four synthons described above, **Rhod-NH** is a secondary amine. Therefore, after its grafting to HA, the amide formed is tertiary and bears no proton, which makes the characterization more challenging. As a result, the ^{1}H NMR spectra of **HA-Rhod** polymers (see Supplementary Materials, Figure S14) showed that there was no newly arising amide proton signal at low fields, and the only clearly distinctive synthon-related peaks were related to the aromatic protons (between 6.4 and 8.1 ppm) and to the methyl protons (at 1.2 ppm).

Although the latter signal at the high field was quite intense (corresponding to 12 H from two diethylamino groups of rhodamine) and well-suitable to quantification (well-separated from the other signals), it did not necessarily correspond to the grafted rhodamine only, as already discussed for **HA-Rhod** polymers. Indeed, at this level, it was not possible to distinguish between associated (by electrostatic interactions) and grafted Rhod synthons. That is why **HA-Rhod** polymers were subjected to DOSY experiments and were carried out to determine the grafting degree of rhodamine synthons on HA chains (Figure 4).

The obtained diffusion curves were clearly nonlinear (Figure 4). A biexponential fitting of these curves with Equation (1) was performed. The first coefficient of 2.40×10^{-10} m^2 s^{-1} corresponded to ungrafted Rhod synthons that quickly diffused, and the second one (1.8×10^{-11} m^2 s^{-1}), corresponding to Rhod synthons that diffused much more slowly, was attributed to grafted functionalized rhodamine moieties. For the latter, the diffusion coefficient was the same as that of HA (see experimental section). This was expected because rhodamine and HA chains have very different molecular weights, and rhodamine grafting should not restrict HA chain mobility. In a second step, this fitting allowed the extraction of the percentage of grafted rhodamine over the total amount of rhodamine (grafting degree (GD) Rhod$_G$/Rhod$_T$, Equation (2), experimental section, and Table 4). Finally, the combination of ^{1}H NMR integration (of the peak at 1.2 ppm) and DOSY analysis allowed to obtain the final DS$_{HA}$ (Table 4).

The variation of DS$_{HA}$ according to initial amounts of rhodamine introduced in the preparation was also linear here (see Supplementary Materials, Figure S15).

2.2. Nanogel Syntheses with Functionalized HA and Characterization—Evaluation of Nanogel Stability by Förster Energy Transfer Experiments (FRET)

2.2.1. Nanogel Synthesis with Functionalized HA and Characterization

Functionalized HA polymers in association with chitosan (CS) were used to produce nanoparticles by physical gelation, in a one-step procedure. This method relied upon the establishment of multivalent electrostatic interactions between HA derivatives (polyanionic) and CS (polycationic). The resulting supramolecular network could be reinforced by cross-linking mediated by small anionic cross-linkers such as sodium tripolyphosphate (TPP) [44]. Functionalized HA with various DS$_{HA}$ were then evaluated for their ability to produce functionalized CS-TPP/HA NPs by ionic gelation. Functionalized CS-TPP/HA nanogels formation was evidenced by DLS. The average hydrodynamic diameters of NPs were determined by dynamic light scattering (DLS, Table 5) recording hydrodynamic diameters and polydispersity index (PDI) of the nanosuspensions. Nanoparticle zeta potential (ζ) which was indicative of their outermost surface charge was determined by ELS.

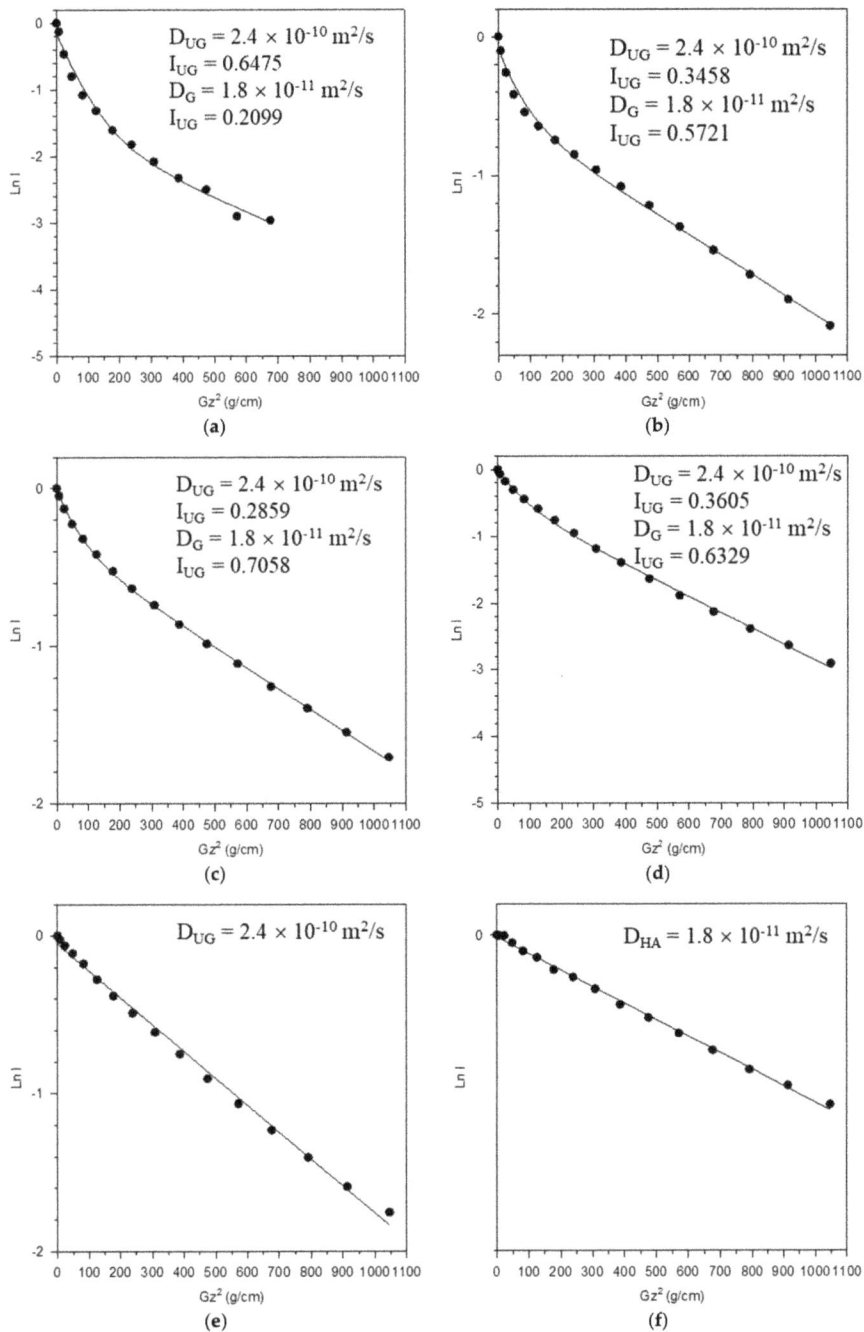

Figure 4. Diffusion curves and diffusion coefficients extracted from DOSY spectra (ref δ (^{1}HRhod) = 1.1 ppm) for (**a**)—**HA-Rhod a**, (**b**)—**HA-Rhod b**, (**c**)—**HA-Rhod c**, and (**d**)—**HA-Rhod d** and (**e**) Rhod-NH alone and (**f**) HA alone as controls. See Equation (1) for I_{UG} and I_G definitions (UG corresponds to ungrafted Rhod synthon, while G corresponds to grafted Rhod synthon).

Table 4. Degrees of substitution of HA functionalized with rhodamine.

	HA-Rhod a	HA-Rhod b	HA-Rhod c	HA-Rhod d
% mol (amine/COOH)$_{initial}$	10%	20%	50%	100%
Association rate (AR) [1]	1.1%	1.9%	4.0%	7.6%
Grafting degree (GD) [2]	25%	62%	71%	64%
Degree of substitution (DS$_{HA}$) [3]	0.3%	1.2%	2.8%	4.9%

[1] Found by integration of ^{1}H NMR signal at 1.2 ppm normalized to acetamide CH$_3$ peak at 1.78 ppm; [2] determined from DOSY experiments; [3] DS = AR × GD.

Table 5. Intensity weighted (Z-average) diameters, polydispersity indexes (PdI), and zeta potential (ζ) of CS-TPP/functionalized HA nanoparticles according to HA degree of substitution (DS$_{HA}$).

Synthon	DS$_{HA}$ [%]	Z-Average ± sd (nm)	PdI ± sd	ζ ± sd (mV)
PEG$_{339}$	2.4	132 ± 2	0.19 ± 0.01	+26 ± 3
	7.6	128 ± 1	0.18 ± 0.02	+23 ± 4
	15.7	138 ± 2	0.19 ± 0.01	+24 ± 3
	38.9	128 ± 1	0.17 ± 0.01	+29 ± 3
PEG$_{2000}$	1.3	141 ± 2	0.18 ± 0.02	+26 ± 4
	7	149 ± 1	0.18 ± 0.01	+21 ± 3
	14.1	137 ± 2	0.19 ± 0.02	+24 ± 4
	32	146 ± 1	0.18 ± 0.01	+24 ± 3
TFB	5.2	153 ± 3	0.18 ± 0.02	+28 ± 3
	6.5	140 ± 2	0.20 ± 0.01	+25 ± 4
	15.6	148 ± 3	0.18 ± 0.02	+23 ± 3
	29.2	146 ± 3	0.20 ± 0.01	+23 ± 3
Rhod	0.3	137 ± 3	0.18 ± 0.01	+22 ± 3
	1.2	147 ± 2	0.18 ± 0.02	+23 ± 4
	2.8	148 ± 3	0.20 ± 0.01	+23 ± 3
	4.9	141 ± 3	0.18 ± 0.02	+21 ± 4
No synthon	0	139 ± 2	0.18 ± 0.01	+22 ± 3

DLS experiments showed the presence of relatively monodisperse nanoassemblies (PDI ≤ 0.35) whose size varied from 130 to 155 nm. For some samples, AFM images in liquid mode corroborated the formation of nanoparticles by evidencing nanoassemblies of lower size (30–70 nm) and the presence of some aggregates (see Supplementary Materials, Figure S16). Such differences between DLS and AFM measurements have already been observed for nanogels [45] and attributed to the fact that in DLS, because of the presence of aggregates, the response could be biased by the use of mathematical models of signal processing. For **CS-TPP/HA-Rhod** nanogels, the confocal image and the associated fluorescence spectrum exhibited the expected features for the **CS-TPP/HA-Rhod** NGs, confirming the fact that the NGs are fluorescent (see Supplementary Materials, Figure S17).

2.2.2. Evaluation of Nanogels Stability by FRET Experiments

As shown, nanogels can be readily obtained by an ionotropic gelation process between functionalized hyaluronic acid solutions and chitosan ones, in the presence of tripolyphosphate (TPP) as a crosslinker [44]. We have previously demonstrated that these nanogels are very helpful to boost the performance of gadolinium chelates (GdCAs) used as contrast agents in MRI [31,32,34]. There remains a need for knowledge of the stability of these nanoassemblies and the synthesis of **HA-Rhod** polymers can be used to evaluate it by FRET. More precisely, FRET experiments have allowed to evaluate the molecular proximity of both polymers thanks to a fluorescent donor–acceptor pair. For this purpose, nanogels were synthesized by mixing **HA-Rhod** and **CS-Fluo** partners according to the conditions used for the synthesis of nanogels that encapsulate GdCAs. In these conditions, the ratio [A]/[D] was equal to 0.5. Since the degree of substitution of each polymer was low (DS_{CS} = 1%, DS_{HA} = 4.9%), the properties of each partner were not perturbed (i.e., the number of positive and negative charges carried by CS and HA, respectively) and the **CS-Fluo-TPP/HA-Rhod** nanogel formation was evidenced by DLS and ELS measurements (Z-ave = 115 nm, PDI = 0.21, ζ = 26 mV). The emission spectrum of **CS-Fluo-TPP/HA-Rhod** nanogel was then recorded after excitation at 470 nm, i.e., at the excitation wavelength of the donor dye, and compared to the ones of **CS-Fluo-TPP/HA** and **CS-TPP/HA-Rhod** nanogels (Figure 5a). The **CS-Fluo-TPP/HA-Rhod** nanogel fluorescence spectrum exhibited two signals at 525 and 591 nm attributed to fluorescein and rhodamine emissions respectively. By comparison to the **CS-TPP/HA-Rhod** nanogel luminescence spectrum recorded under similar conditions (after excitation at fluorescein wavelength at 470 nm), it was noticeable that the emission intensity of rhodamine signal at 591 nm in the **CS-Fluo-TPP/HA-Rhod** nanogel was greatly exalted. This was the fingerprint of an energy transfer between fluorescein and rhodamine and this FRET signal confirmed the close proximity of **CS-Fluo** and **HA-Rhod** within the **CS-Fluo-TPP/HA-Rhod** nanogel structure. This signal was persistent over a period of one month in PBS (a longer analysis period has not been tested), illustrating the stability of the nanogels under these conditions. Furthermore, FRET properties of **CS-Fluo-TPP/HA-Rhod** nanogels were tested in the presence of hyaluronidase enzyme (HA-ase). **CS-Fluo-TPP/HA-Rhod** nanogels were incubated at 37 °C in the presence of hyaluronidase HYAL-1 at a concentration of 60 ng·mL^{-1}, which is the HYAL-1 concentration in human serum [46]. No changes in the FRET spectrum were detected (Figure 5b), which highlighted the stability of **CS-Fluo-TPP/HA-Rhod** nanogels under these physiological conditions.

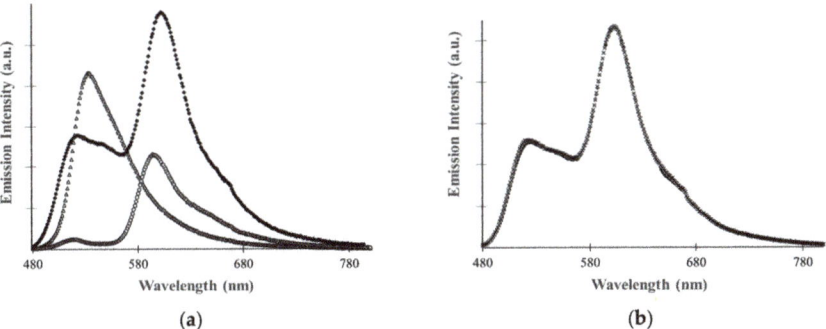

Figure 5. (a) Emission spectrum of **CS-Fluo-TPP/HA-Rhod** nanogels in PBS (◆), by comparison of **CS-Fluo-TPP/HA** (△) and **CS-TPP/HA-Rhod** (○) in **CS-Fluo-TPP/HA-Rhod** nanogels in PBS ($\lambda_{ex} = \lambda_{ex}^{D}$ = 470 nm, CS-Fluo being the donor dye and HA-Rhod being the acceptor dye). (b) Fluorescence measurements of **CS-Fluo-TPP/HA-Rhod** nanogels ([A]/[D] ratio = 0.5) without (◆) and in the presence of hyaluronidase (×).

3. Conclusions

To conclude, our objective in this work was to obtain functionalized HA with stealth, lipophilic, or fluorescent properties and to test their ability to form nanogels by ionic gelation with chitosan. Successful HA grafting was obtained in DMSO through a peptidic coupling between the amino-terminal group of the grafted synthons and the carboxylic moieties of protonated HA. In DMSO, the identification of the amidic function was most often straightforward and allowed the determination of the HA degree of substitution (DS_{HA}). When this identification was not possible, a combination of ^{1}H NMR and DOSY experiments was used. A series of functionalized HAs were then described and DS_{HA} seemed to cap to about 30–50% according to the grafted function except for rhodamine synthon for which DS_{HA} did not exceed 5%. This was probably due to the fact that the reactive rhodamine nitrogen atom was secondary, more sterically hindered, and then less reactive towards the peptidic coupling strategy.

Ionic gelation from all HA conjugates, whatever DS_{HA}, proved to be efficient to provide CS-TPP/functionalized HA nanohydrogels having morphological characteristics compatible with biomedical applications. Ionic gelation was then used to synthesize nanohydrogels combining fluorescein-labeled chitosan and HA-Rhod. FRET experiments performed with the corresponding nanoassemblies that carried this fluorescent donor–acceptor pair allowed to demonstrate the close proximity of CS and HA polymers within the nanogel matrix. In the presence of physiological amounts of hyaluronidase, no modification of the FRET signal was observed which allowed to conclude a good stability of these nanohydrogels in a biological medium, which was a prerequisite to their use in biomedical applications.

4. Materials and Methods

4.1. Materials

Hyaluronic acid sodium salt (**HAs**, from *Streptococcus equi* M_W ~1.5–1.8 × 10^6 Da), 1-[bis(dimethylamino)methylene]-1H-1,2,3-triazolo[4,5-b]pyridinium 3-oxid hexafluoro-phosphate (**HATU**) and hyaluronidase (HYAL-1 from bovine testes, 407 UI·mg^{-1}) were purchased from Sigma-Aldrich (Saint Louis, MO, USA). Amino synthons involved in this work were *n*-octylamine **C8-NH₂** (Alfa Aesar, Kandel, Germany), 1,1,1-trifluoropropylamine **TFP-NH₂** (Sigma-Aldrich, Saint Louis, MO, USA), 2,5,8,11,14,17,20-heptaoxadocosan-22-amine **PEG₃₃₉-NH₂**, and methoxy-poly(ethylene)glycol-amine **PEG₂₀₀₀-NH₂** (both synthons purchased from Iris Biotech GmbH, Marktredwitz, Germany). Amine-functionalized rhodamine **Rhod-NH** was prepared from Rhodamine B (purchased from Sigma Aldrich, Saint Louis, MO, USA) following a literature-inspired method [43]. For calculations, **HAs** repetitive unit molecular mass was considered to be M_W in average (**HAs**) = 401 g·mol^{-1}. Amberlite™ IR 120 ion exchange resin was purchased from Fluka (Buchs, Switzerland). Vivaspin® 20 ultrafiltration tubes (MWCO 10,000 Da) were purchased from Sartorius (Göttingen, Germany). Ultrafiltration experiments were realized with an Allegra X-30 Centrifuge (Beckman-Coulter) (7500 rpm, between 45 and 60 min, at room temperature). DMSO-d_6 was purchased from Eurisotop. The syntheses of organic nanoparticles by the ionic gelation method were performed using chitosan (**CS**, Sigma-Aldrich, low viscosity, deacetylation degree 86% determined by ^{1}H NMR spectroscopy [47–49]) and sodium tripolyphosphate (TPP, Acros Organics). Sterile water for injections (Laboratoire Aguettant, Lyon, France) was systematically used for nanoparticle preparations and analyses. All products were used as received, without further purification.

Polymers and copolymers (**HAp, HA-C8, HA-TFP, HA-PEG₃₃₉, HA-PEG₂₀₀₀, HA-Rhod**) were characterized by means of FTIR (Thermo Scientific™ Nicolet™ iS5 spectrometer equipped with ATR iD5 accessory), ^{1}H, and ^{19}F NMR spectroscopies (Bruker Avance III (^{1}H—500 MHz, ^{19}F—470.6 MHz) spectrometer) at 318K with DMSO-d_6 as the solvent. The diffusion coefficients of different materials (**HAp, HA-C8, HA-Rhod, C8-NH₂, Rhod-NH**) were determined by DOSY experiments (diffusion ordered spectroscopy) on an Avance II 500 spectrometer (Bruker).

4.2. Syntheses and Purifications of Functionalized HA

HA functionalization occurs in two steps: first, the protonation of carboxylate groups to make it soluble in DMSO [50,51] and second, peptidic coupling with the amine, using HATU as the coupling agent.

4.2.1. Conversion of Sodium Hyaluronate HAs into Its Protonated Form HAp

At neutral pH, HA is in the form of a sodium salt and is referred to as sodium hyaluronate (**HAs**). In order to transform it in its protonated form, ion exchange was undertaken similarly as described by Vasi et al. [20]. Sodium hyaluronate (1.00 g), dissolved in demineralized water (400 mL) was slowly eluted through an Amberlite™ IR 120 ion exchange resin (25 mL in dry volume, dispersed in 50 mL of demineralized water) conditioned under HCl form (by addition of 50 mL HCl 1M and then rinsing with 50 mL of demineralized water). The resulting solution was pre-concentrated under reduced pressure and freeze-dried to afford 0.92 g of protonated HA (yield 97%).

4.2.2. General Method of HA Functionalization by Peptidic Coupling Reaction with Amine Synthons in DMSO

The starting compounds were separately dissolved in anhydrous DMSO: protonated HA (**HAp**, 80 mg, 0.21 mmol, 8 mL DMSO), **HATU** (40 mg, 0.105 mmol, 1 mL DMSO), and amine synthon (0.1 mmol, 1 mL DMSO). **HAp** solution was added in four glass vials (2 mL in each vial, 0.05 mmol), equipped with magnetic stirring bars, and previously purged with argon. Increasing quantities of **HATU** were then added in each vial (50, 100, 250, and 500 µL, corresponding to 0.1, 0.2, 0.5, and 1.0 equivalent, respectively) and followed by dilution with anhydrous DMSO (900, 800, 500, and 0 µL, respectively). HA was activated over 15 min and amine solution was added in each vial (50, 100, 250, and 500 µL, corresponding to 0.1, 0.2, 0.5, and 1.0 equivalent, respectively). After 24 h of reaction at room temperature, the solutions were transferred into 50 mL Falcon® tubes and the addition of diethyl ether (27 mL into each tube) provoked the polymer precipitation. Functionalized HAs were then isolated by centrifugation and the corresponding solids were washed once more with diethyl ether (20 mL). After a second centrifugation, the products were dried under reduced pressure. Each product was then dissolved in 5 mL of 0.1 M HCl, transferred into a Vivaspin® 20 tube (with MWCO 10,000 Da), diluted with 7 mL of demineralized water, and centrifuged at 6000 g. After two cycles of ultrafiltration, the final solutions were transferred into 15 mL Falcon® tubes and freeze-dried. The sample was obtained by grafting 1 equiv. of octylamine on HA was not soluble in water. Three cycles of washing by centrifugation were applied in place of ultrafiltration.

4.3. Determination of Functionalized HA Degree of Substitution DS_{HA}

The evaluation of the degree of substitution of HA (DS_{HA}) on four categories of compounds (aliphatic, fluorinated, pegylated, and fluorescent amines (rhodamine)) was performed by ^{1}H NMR methods. The functionalization with octylamine was chosen as a model reaction to develop the method for determining the DS_{HA} in the corresponding **HA-C8** polymers. Then, DS_{HA} was first determined by the integration of the newly formed amide proton ^{1}H NMR signal, normalized to an acetamide NH signal of the HA backbone. This approach allowed to directly give a percentage of HA functionalized COOH groups. At this point, DOSY experiments were used to corroborate this percentage (*vide infra*). Then the integration approach was used to determine DS_{HA} in the case of **HA-TFP**, **HA-PEG$_{339}$**, and **HA-PEG$_{2000}$** polymers. Indeed, for these compounds, the peptidic coupling also generated a secondary amide bond, with a ^{1}H signal that acts as a probe for the functionalization. In the case of **HA-Rhod** polymers, the newly formed amide bond is tertiary. For **HA-Rhod** polymers, DS_{HA} was calculated by comparison of the integrals of distinctive aliphatic protons peaks related to the introduced side groups and a HA acetamide methyl signal; these ratios were further corrected by DOSY analysis as described in one of our previous works [37].

For DOSY experiments, bipolar gradient pulses with two spoil gradients were used to measure the diffusion coefficients (BPP-LED pulse sequence). The value of the gradient pulse length δ was 2 or 4 ms depending on the samples, while the value of the diffusion time Δ was set to 150, 250, or 500 ms depending on the samples. The pulse gradients were incremented in 16 steps from 2% to 95% of the maximum gradient strength (53.5 G/cm) in a linear ramp and the temperature was set at 30 °C. Under these conditions, preliminary DOSY experiments were performed to determine **HA** and octylamine diffusion coefficients

(D_{HA} and D_{C8}, respectively). Values of 1.8×10^{-11} m$^2 \cdot$s^{-1} and 6.0×10^{-10} m$^2 \cdot$s^{-1}, were obtained for HA and octylamine respectively.

Similar DOSY experiments were then performed with **HA-C8** polymers to characterize the diffusion coefficients of ungrafted and grafted C8 chains. The diffusion curves were extracted from **HA-C8** DOSY spectra for two peaks of C8 at 0.8 and 1.2 ppm and were characterized by two contributions: one coming from the ungrafted C8 (C8$_{UG}$) which diffuses fast, and the other coming from the grafted C8 (C8$_G$). Diffusion curves can thus be fitted with a bi-exponential equation taking into account the two contributions (Equation (1)) [52,53].

$$I = I_G \exp[-\gamma^2 g^2 D_G \delta^2 (\Delta - (\delta/3) - (\tau/2))] + I_{UG} \exp[-\gamma^2 g^2 D_{UG} \delta^2 (\Delta - (\delta/3) - (\tau/2))] \quad (1)$$

where I_G and I_{UG} are the intensities at 0% gradient of grafted and not grafted C8, respectively, γ is the gyromagnetic ratio, g is the gradient strength, D_G and D_{UG} are the diffusion coefficients of grafted and ungrafted C8, respectively, δ is the gradient pulse length, Δ is the diffusion time, and τ is the interpulse spacing in the BPP-LED pulse sequence.

Assuming that the **HA-C8** molecular weight must be close to the one of HA (due first to the large difference between C8 and HA molecular weights), one can consider that C8$_G$ (and then HA-C8) has the same diffusion coefficient as HA. During the fitting, D_G and D_{UG} were then fixed to values measured independently on HA and C8, respectively: $D_{HA} = 1.8 \times 10^{-11}$ m$^2 \cdot$s^{-1}, $D_{C8} = 6.0 \times 10^{-10}$ m$^2 \cdot$s^{-1}.

The values of I_G and I_{UG} extracted from the fitting allowed to calculate the percentage of the grafted C8 over the total amount of C8 (C8$_G$/C8$_T$):

$$\frac{C8_G}{C8_T} = \frac{I_G}{I_G + I_{UG}} \times 100 \quad (2)$$

The percentage of C8 grafted to HA chains (DS$_{C8/HA}$) was then calculated from ^{1}H NMR and DOSY experiments:

$$DS_{C8/HA} = \%\frac{C8_G}{HA} = \frac{I(3H, C8)_{0.87\text{ ppm}}}{I(3H, HA)_{1.78\text{ ppm}}} \times \frac{I_G}{I_G + I_{UG}} \times 100 \quad (3)$$

where I represents the integration of the peaks indicated in brackets and I_G and I_{UG} stand for the intensities extracted from the DOSY experiments, for grafted and ungrafted C8, respectively.

The same procedure was applied to characterize **HA-Rhod** polymers and determine their substitution degrees DS. For that, the diffusion coefficient of rhodamine was measured and a value of 2.4×10^{-10} m$^2 \cdot$s^{-1} was obtained, which corresponds to D_{UG} in Equation (1).

4.4. Preparation of Nanoparticles with Functionalized HA Polymers by Ionic Gelation and Characterizations

4.4.1. CS-TPP/Functionalized HA Nanogel Synthesis

Functionalized HAs obtained in this study (**HA-PEG$_{339}$**, **HA-PEG$_{2000}$**, **HA-TFB**, **HA-Rhod**, 3.6 mg) were dissolved in water (3.15 mL) and allowed to stir overnight in the presence of NaOH (1.35 mL of NaOH 0.1 M) for deprotonation of the remaining carboxylic acid groups. Stock solutions of CS were prepared by dissolution of the CS powder (2.5 mg·mL^{-1}) in a 10% (*m/v*) citric acid aqueous solution, or in a 10% (*v/v*) acetic acid solution, and stirred overnight. Insoluble residues were removed by centrifugation at 3800 rpm for 4 min at room temperature. CS-TPP/functionalized HA nanogels were obtained by an ionotropic gelation process. The polyanionic phase (4.5 mL), i.e., functionalized HA (0.8 mg·mL^{-1}) and TPP (2.4 mg·mL^{-1}) were added dropwise to the CS solution (9 mL) under sonication (750 W, amplitude 32%) to obtain nanosuspensions. At the end of the addition, magnetic stirring was maintained for 10 min. The removal of unreacted compounds was achieved by dialysis (Spectrapore®, MWCO 25 kDa, Spectrumlab) against water (3 × 12 h).

4.4.2. Nanogels Characterization by Dynamic Light Scattering (DLS)

Averaged hydrodynamic diameters (Z-ave) of nanoparticles were determined by Dynamic Light Scattering (DLS) with a Zetasizer Nano ZS (Malvern Zetasizer Nano-ZS, Malvern Instruments, Worcestershire, UK). Polydispersity indexes (PdI) were determined by cumulant analysis. Each nanosuspension was analyzed in triplicate at 20 °C at a scattering angle of 173°, after 1/20 dilution in water. Water for injection was used as a reference dispersing medium. ζ-(zeta) potential data were collected through electrophoretic light scattering at 20 °C, 150 V, in triplicate for each sample, after 1/20 dilution in water. The instrument was calibrated with a Malvern—68 mV standard before each analysis cycle.

4.4.3. Atomic Force Microscopy

CS-TPP/functionalized HA nanosuspensions were analyzed by Atomic Force Microscopy (AFM) in solution in order to afford minimum perturbation of the samples [54,55]. 35 µL of each nanosuspension was directly deposited on freshly cleaved mica disks. After 20 min of deposition at ambient temperature, the sample was rinsed several times in distilled water. Nanosuspensions were then imaged in distilled water, under manually operated PeakForce Tapping mode (PFT) on a Brüker Resolve setup (Billerica, MA, USA). The average PeakForce setpoint was set around 100 pN, which was found to be a good compromise to remain in good tracking conditions and to avoid particle damage. MSNL probes (Bruker, Billerica, MA, USA) with an average nominal spring constant of 0.07 N/m were used. For each type of nanogel, three different samples were prepared and at least three different areas were imaged per sample to ensure the reproducibility of the measurements. For image processing, all images were analyzed and particle diameters were estimated using Nanoscope Analysis 1.8 (Bruker, Billerica, MA, USA). For the particle analysis, only individual and well-distinguished nanoparticles were taken into consideration and to obtain reliable statistical results.

The AFM setup is directly coupled to a confocal Zeiss LSM 800 microscope (Oberkochen, Germany) allowing to correlate fluorescent and AFM images.

4.4.4. Evaluation of CS-Fluo-TPP/HA-Rhod Nanogels Stability by Förster Resonance Energy Transfer (FRET) Experiments

CS-Fluo-TPP/HA-Rhod nanogels for which the [A]/[D] ratio was equal to 0.5 were synthesized according to the protocol previously described. **HA-Rhod** (as the acceptor dye 3.6 mg, DS_{HA} = 4.9%) was dissolved in water (3.15 mL) and allowed to stir overnight in the presence of NaOH (1.35 mL of NaOH 0.1 M) as previously described. Chitosan grafted with the fluorescein probe (**CS-Fluo**, as the donor dye, DS_{CS} = 1.0% [34]) was prepared by dissolution of the **CS-Fluo** powder (2.5 mg·mL^{-1}) in a 10% (m/v) citric acid aqueous solution and stirred overnight. Insoluble residues were removed by centrifugation at 3800 rpm for 4 min at room temperature.

For FRET measurements, nanosuspensions were diluted 10-fold in ultrapure water, to be in the concentration range suitable for analysis. At this dilution, nanoparticles exhibited the same morphological characteristics as the raw suspensions, as confirmed by DLS measurements. Fluorescence measurements were conducted on an Edinburg FLS100 spectrophotometer. The fluorescence emission in response to an excitation at 470 nm was recorded between 480 and 800 nm (with $\Delta\lambda_{exc}$ = 1.6 nm and $\Delta\lambda_{em}$ = 1.8 nm), using in a 10 mm thick quartz cuvette (Hëllma) and ultrapure water as a reference. The FRET signal was detected at 600 nm. For experiments in the presence of the enzyme, HA-ase solution at 60 ng·mL^{-1} and nanogels suspensions were pre-heated at 37 °C for 10 min. **CS-Fluo-TPP/HA-Rhod** nanogels and HA-ase were then mixed and incubated at 37 °C for 1 h 30 min. Then, the nanogels in the presence of HA-ase were cooled at 4 °C and characterized by DLS and fluorescence measurements, as previously described.

Supplementary Materials: The following files are available free of charge. The following supporting information can be downloaded at: https://www.mdpi.com/article/10.3390/gels8030182/s1, Figure S1. Diffusion curves extracted from the DOSY experiments recorded on **HA-C8** samples for the peaks of octylamine at 0.8 and 1.2 ppm. Diffusion curves extracted from DOSY experiments run on HA and C8 separately were added for comparison. Figure S2. Evolution of DS_{HA} according to increasing C8-NH$_2$/COOH$_{HA}$ initial ratios. Figure S3. FTIR spectra of PEG$_{339}$ functionalized HA samples: (a) carbonyl stretching vibration region (1480–1820 cm^{-1}) and b) C-H stretching vibration region (2700–3000 cm^{-1}). Figure S4. FTIR spectra of PEG$_{2000}$ functionalized HA samples: (a) carbonyl stretching vibration region (1480–1820 cm^{-1}) and (b) C-H stretching vibration region (2700–3000 cm^{-1}). Figure S5. Structure and ^{1}H NMR spectra of PEG$_{339}$ functionalized HA: **HAp**, **HA-PEG$_{339}$a**, **HA-PEG$_{339}$b**, **HA-PEG$_{339}$c**, and **HA-PEG$_{339}$d**, for initial molar ratios (amine/COOH) = 0 (bottom), 10, 20, 50, and 100 (top) % respectively (500 MHz, 318 K, DMSO-d_6). Figure S6. Structure and ^{1}H NMR spectra of PEG$_{2000}$ functionalized HA: **HAp**, **HA-PEG$_{2000}$a**, **HA-PEG$_{2000}$b**, **HA-PEG$_{2000}$c**, and **HA-PEG$_{2000}$d**, for initial molar ratios (amine/COOH) = 0 (bottom), 10, 20, 50, and 100 (top) % respectively (500 MHz, 318 K, DMSO-d_6). Figure S7. Evolution of DS_{HA} according to increasing PEG$_{339}$-NH$_2$/COOH$_{HA}$ initial ratios. Figure S8. Evolution of DS_{HA} according to increasing PEG$_{2000}$-NH$_2$/COOH$_{HA}$ initial ratios. Figure S9. FTIR spectra of TFP functionalized HA samples: (a) carbonyl stretching vibration region (1480–1820 cm^{-1}) and (b) C-H stretching vibration region (2700–3000 cm^{-1}). Figure S10. Structure and ^{1}H NMR spectra of trifluoropropyl functionalized HA: **HAp**, **HA-TFPa**, **HA-TFPb**, **HA-TFPc**, and **HA-TFPd**, for initial molar ratios (amine/COOH) = 0 (bottom), 10, 20, 50, and 100 (top) % respectively (500 MHz, 318 K, DMSO-d_6). Figure S11. 1D diffusion-filtered ^{19}F NMR spectra of **HA-TFPa**, **HA-TFPb**, **HA-TFPc**, **HA-TFPd**, and **TFP-NH$_2$** with a gradient g of 2% and 95%. Figure S12. Evolution of DS_{HA} according to increasing C8-TFP/COOH$_{HA}$ initial ratios. Figure S13. FTIR spectra of rhodamine functionalized HA samples: (a) carbonyl stretching vibration region (1480–1820 cm^{-1}) and (b) C-H stretching vibration region (2700–3000 cm^{-1}). Figure S14. Structure and ^{1}H NMR spectra of rhodamine B functionalized HA (500 MHz, 318 K, NS = 32); from bottom to up: **HAp**, **HA-Rhoda**, **HA-Rhodb**, **HA-Rhodc**, and **HA-Rhodd**. Figure S15. Evolution of DS_{HA} according to increasing Rhod-NH/COOH$_{HA}$ initial ratios. Figure S16: topographical AFM images of (a) CS-TPP/HA-PEG$_{2000}$, (b) CS-TPP/HA-Rhod, and (c) CS-TPP/HA (control) NGs. Figure S17: Coupled AFM and Confocal images of CS-TPP/HA-Rhod nanogels.

Author Contributions: The manuscript was written through contributions of all authors. Conceptualization, V.M., J.M., C.C., M.C. and F.C.; methodology, V.M., J.M., M.C., C.F. and C.H.; validation, V.M., J.M., C.C., M.M. and C.H.; investigation, V.M., J.M. and M.C.; writing—original draft preparation, V.M. and C.C.; writing—review and editing, F.C.; supervision, S.L. and F.C.; project administration, S.L. and F.C.; funding acquisition, S.L. and F.C. All authors have read and agreed to the published version of the manuscript.

Funding: The work was funded by the "Programme de coopération transfrontalière Interreg France-Wallonie-Vlaanderen" (Nanocardio project (http://nanocardio.eu, accessed on 3 December 2021).

Institutional Review Board Statement: Not applicable.

Informed Consent Statement: Not applicable.

Data Availability Statement: Not applicable.

Acknowledgments: V. Malytskyi is grateful to the "Programme de coopération transfrontalière Interreg France-Wallonie-Vlaanderen" for funding his post-doctoral fellowship. The Center for Microscopy and Molecular Imaging (CMMI, supported by the European Regional Development Fund and the Region Wallone), the Bioprofiling platform (supported by the European Regional Development Fund and the Walloon Region, Belgium) and the PlAneT and the NanoMat' platforms (supported by the European Regional Development Fund, the Region Grand Est, and the DRRT Grand Est) are thanked for their support. Antony Robert, Amandine Destrebecq, and Christelle Kowandy are thanked for their help in ^{1}H NMR spectra recording, ICP OES and SEC measurements respectively.

Conflicts of Interest: The authors declare no conflict of interest.

References

1. Fallacara, A.; Baldini, E.; Manfredini, S.; Vertuani, S. Hyaluronic Acid in the Third Millennium. *Polymers* **2018**, *10*, 701. [CrossRef] [PubMed]
2. Dosio, F.; Arpicco, S.; Stella, B.; Fattal, E. Hyaluronic acid for anticancer drug and nucleic acid delivery. *Adv. Drug Deliver. Rev.* **2016**, *97*, 204–236. [CrossRef] [PubMed]
3. Gallo, N.; Nasser, H.; Salvatore, L.; Natali, M.L.; Campa, L.; Mahmoud, M.; Capobianco, L.; Sannino, A.; Madaghiel, M. Hyaluronic acid for advanced therapies: Promises and challenges. *Eur. Polym. J.* **2019**, *117*, 134–147. [CrossRef]
4. Graça, M.F.P.; Miguel, S.P.; Cabrala, C.S.D.; Correia, I.J. Hyaluronic acid—Based wound dressings: A review. *Carbohydr. Polym.* **2020**, *241*, 116364. [CrossRef]
5. Khan, W.; Abtew, E.; Modani, S.; Domb, A.J. Polysaccharide based nanoparticles. *Isr. J. Chem.* **2018**, *58*, 1315–1329. [CrossRef]
6. Li, M.; Sun, J.; Zhang, W.; Zhao, Y.; Shufen, Z.; Zhang, Z. Drug delivery systems based on CD44-targeted glycosaminoglycans for cancer therapy. *Carbohydr. Polym.* **2021**, *251*, 117103. [CrossRef]
7. Wolf, K.J.; Kumar, S. Hyaluronic Acid: Incorporating the Bio into the Material. *ACS Biomater. Sci. Eng.* **2019**, *5*, 3753–3765. [CrossRef]
8. Vasvani, S.; Kulkarni, P.; Rawtani, D. Hyaluronic acid: A review on its biology, aspects of drug delivery, route of administrations and a special emphasis on its approved marketed products and recent clinical studies. *Int. J. Biol. Macromol.* **2020**, *151*, 1012–1029. [CrossRef]
9. Kirschning, A.; Dibbert, N.; Drager, G. Chemical functionalization of polysaccharides—Towards biocompatible hydrogels for biomedical applications. *Chem. Eur. J.* **2018**, *24*, 1231–1240. [CrossRef]
10. Rho, J.G.; Han, H.S.; Han, J.H.; Lee, H.; Nguyen, V.Q.; Lee, W.H.; Kim, W. Self-assembled hyaluronic acid nanoparticles: Implications as a nanomedicine for treatment of type 2 diabetes. *J. Control. Release* **2018**, *279*, 89–98. [CrossRef]
11. Kim, K.; Choi, H.; Choi, E.S.; Park, M.-H.; Ryu, J.-H. Hyaluronic Acid-Coated Nanomedicine for Targeted Cancer Therapy. *Pharmaceutics* **2019**, *11*, 301. [CrossRef]
12. Prajapati, V.D.; Maheriya, P.M. Hyaluronic acid as potential carrier in biomedical and drug delivery applications. In *Functional Polysaccharides for Biomedical Applications*; Maiti, S., Jana, S., Eds.; Elsevier: Amsterdam, The Netherlands, 2019; pp. 213–265.
13. Kaewruethai, T.; Laomeephol, C.; Pan, Y.; Luckanagul, J.A. Multifunctional Polymeric Nanogels for Biomedical Applications. *Gels* **2021**, *7*, 228. [CrossRef]
14. Yuan, J.; Maturavongsadit, P.; Zhou, Z.; Lv, B.; Lin, Y.; Yang, J.; Luckanagul, J. Hyaluronic acid-based hydrogels with tobacco mosaic virus containing cell adhesive peptide induce bone repair in normal and osteoporotic rats. *Biomater Transl.* **2020**, *1*, 89–98.
15. Jia, X.; Han, Y.; Pei, M.; Zhao, X.; Tian, K.; Zhou, T.; Liu, P. Multi-functionalized hyaluronic acid nanogels crosslinked with carbon dots as dual receptor-mediated targeting tumor theranostics. *Carbohydr. Polym.* **2016**, *152*, 391–397. [CrossRef]
16. Silva Garcia, J.M.; Panitch, A.; Calve, S. Functionalization of hyaluronic acid hydrogels with ECM-derived peptides to control myoblast behavior. *Acta Biomater.* **2019**, *84*, 169–179. [CrossRef]
17. Du, X.; Yin, S.; Wang, Y.; Gu, X.; Wang, G.; Li, J. Hyaluronic acid-functionalized half-generation of sectorial dendrimers for anticancer drug delivery and enhanced biocompatibility. *Carbohydr. Polym.* **2018**, *202*, 513–522. [CrossRef]
18. Crescenzi, V.; Francescangeli, A.; Capitani, D.; Mannina, L.; Renier, D.; Bellini, D. Hyaluronan networking via Ugi's condensation using lysine as cross-linker diamine. *Carbohydr. Polym.* **2003**, *53*, 311–316. [CrossRef]
19. Ramachandran, B.; Chakraborty, S.; Kannan, R.; Dixit, M.; Muthuvijayan, V. Immobilization of hyaluronic acid from Lactococcus lactis on polyethylene terephthalate for improved biocompatibility and drug release. *Carbohydr. Polym.* **2019**, *206*, 132–140. [CrossRef]
20. Vasi, A.-M.; Popa, M.I.; Butnaru, M.; Dodi, G.; Verestiuc, L. Chemical functionalization of hyaluronic acid for drug delivery applications. *Mater. Sci. Eng. C* **2014**, *38*, 177–185. [CrossRef]
21. Wei, K.; Zhu, M.; Sun, Y.; Xu, J.; Feng, Q.; Lin, S.; Wu, T.; Xu, J.; Tian, F.; Xia, J.; et al. Robust Biopolymeric Supramolecular "Host−Guest Macromer" Hydrogels Reinforced by in Situ Formed Multivalent Nanoclusters for Cartilage Regeneration. *Macromolecules* **2016**, *49*, 866–875. [CrossRef]
22. Liu, C.; Liu, D.; Wang, Y.; Li, Y.; Li, T.; Zhou, Z.; Yang, Z.; Wang, J.; Zhang, Q. Glycol chitosan/oxidized hyaluronic acid hydrogels functionalized with cartilage extracellular matrix particles and incorporating BMSCs for cartilage repair. *Artif. Cells Nanomed. Biotechnol.* **2018**, *46*, 721–732. [CrossRef] [PubMed]
23. Kaczmarek, B.; Sionkowska, A.; Kozlowska, J.; Osyczka, A.M. New composite materials prepared by calcium phosphate precipitation in chitosan/collagen/hyaluronic acid sponge cross-linked by EDC/NHS. *Int. J. Biol. Macromol.* **2018**, *107*, 247–253. [CrossRef] [PubMed]
24. Song, H.-Q.; Fan, Y.; Hu, Y.; Cheng, G.; Xu, F.-J. Polysaccharide–Peptide Conjugates: A Versatile Material Platform for Biomedical Applications. *Adv. Funct. Mater.* **2021**, *31*, 2005978. [CrossRef]
25. Nakajima, N.; Ikada, Y. Mechanism of Amide Formation by Carbodiimide for Bioconjugation in Aqueous Media. *Bioconjugate Chem.* **1995**, *6*, 123–130. [CrossRef]
26. D'Este, M.; Eglin, D.; Alini, M. A systematic analysis of DMTMM vs. EDC/NHS for ligation of amines to hyaluronan in water. *Carbohydr. Polym.* **2014**, *108*, 239–246. [CrossRef]
27. Yan, Q.; Zheng, H.-N.; Jiang, C.; Li, K.; Xiao, S.-J. EDC/NHS activation mechanism of polymethacrylic acid: Anhydride versus NHS-ester. *RSC Adv.* **2015**, *5*, 69939–69947. [CrossRef]

28. Palumbo, F.S.; Fiorica, C.; Di Stefano, M.; Pitarresi, G.; Gulino, A.; Agnello, S.; Giammona, G. In situ forming hydrogels of hyaluronic acid and inulin derivatives for cartilage regeneration. *Carbohydr. Polym.* **2015**, *122*, 408–416. [CrossRef]
29. Almeida, P.V.; Shahbazi, M.-A.; Mäkilä, E.; Kaasalainen, M.; Salonen, J.; Hirvonen, J.; Santos, H.A. Amine-modified hyaluronic acid-functionalized porous silicon nanoparticles for targeting breast cancer tumors. *Nanoscale* **2014**, *6*, 10377–10387. [CrossRef]
30. Schneider, A.; Picart, C.; Senger, B.; Schaaf, P.; Voegel, J.; Frisch, B. Layer-by-Layer Films from Hyaluronan and Amine-Modified Hyaluronan. *Langmuir* **2007**, *23*, 2655–2662. [CrossRef]
31. Courant, T.; Roullin, V.G.; Cadiou, C.; Callewaert, M.; Andry, M.C.; Portefaix, C.; Hoeffel, C.; de Goltstein, M.C.; Port, M.; Laurent, S.; et al. Hydrogels Incorporating GdDOTA: Towards Highly Efficient Dual T1/T2 MRI Contrast Agents. *Angew. Chem. Int. Ed.* **2012**, *51*, 9119–9122. [CrossRef]
32. Callewaert, M.; Roullin, V.G.; Cadiou, C.; Millart, E.; Van Gulik, L.; Andry, M.C.; Portefaix, C.; Hoeffel, C.; Laurent, L.; Vander Elst, L.; et al. Tuning the composition of biocompatible Gd nanohydrogels to achieve hypersensitive dual T1/T2 MRI contrast agents. *J. Mater. Chem. B* **2014**, *2*, 6397–6405. [CrossRef]
33. Malytskyi, V.; Moreau, J.; Callewaert, M.; Rigaux, G.; Cadiou, C.; Laurent, S.; Chuburu, F. Organic nanoparticles and gadolinium chelates. In *Materials for Biomedical Engineering: Organic Micro and Nanostructures*, 1st ed.; Grumezescu, A., Holban, A.-M., Eds.; Elsevier: Amsterdam, The Netherlands, 2019; pp. 425–476.
34. Moreau, J.; Callewaert, M.; Malytskyi, V.; Henoumont, C.; Voicu, S.N.; Stan, M.S.; Molinari, M.; Cadiou, C.; Laurent, S.; Chuburu, F. Fluorescent chitosan-based nanohydrogels and encapsulation of gadolinium MRI contrast agent for magneto-optical imaging. *Carbohydr. Polym. Technol. Appl.* **2021**, *2*, 100104.
35. Chib, R.; Raut, S.; Fudala, R.; Chang, A.; Mummert, M.; Rich, R.; Gryczynski, Z.; Gryczynski, I. FRET Based-Metric Sensing of Hyaluronidase in Synthetic Urine as a Biomarker for Bladder and Prostate Cancer. *Curr. Pharm. Biotechnol.* **2013**, *14*, 470–474. [CrossRef]
36. Fudala, R.; Mummert, M.E.; Gryczynski, Z.; Gryczynski, I. Fluorescence detection of hyaluronidase. *J. Photochem. Photobiol. B* **2011**, *104*, 473–477. [CrossRef]
37. Belabassi, Y.; Moreau, J.; Gheran, V.; Henoumont, C.; Robert, A.; Callewaert, M.; Rigaux, G.; Cadiou, C.; Vander Elst, L.; Laurent, S.; et al. Synthesis and characterization of PEGylated and fluorinated chitosans: Application to the synthesis of targeted nanoparticles for drug delivery. *Biomacromolecules* **2017**, *18*, 2756–2766. [CrossRef]
38. Eslami, P.; Rossi, F.; Fedeli, S. Hybrid Nanogels: Stealth and Biocompatible Structures for Drug Delivery Applications. *Pharmaceutics* **2019**, *11*, 71. [CrossRef]
39. Guerrini, L.; Alvarez-Puebla, R.; Pazos-Perez, N. Surface Modifications of Nanoparticles for Stability in Biological Fluids. *Materials* **2018**, *11*, 1154. [CrossRef]
40. Hussain, Z.; Khan, S.; Imran, M.; Sohail, M.; Shah, S.W.A.; de Matas, M. PEGylation: A promising strategy to overcome challenges to cancer-targeted nanomedicines: A review of challenges to clinical transition and promising resolution. *Drug Deliv. Transl. Res.* **2019**, *9*, 721–734. [CrossRef]
41. Jokerst, J.V.; Lobovkina, T.; Zare, R.N.; Gambhir, S.S. Nanoparticle PEGylation for imaging and therapy. *Nanomedicine* **2011**, *6*, 715–728. [CrossRef]
42. Misselwitz, B. MR contrast agents in lymph node imaging. *Eur. J. Radiol.* **2006**, *35*, 375–382. [CrossRef]
43. Kačenka, M.; Kaman, O.; Kikerlová, S.; Pavlů, B.; Jirák, Z.; Jirák, D.; Herynek, V.; Černý, J.; Chaput, F.; Laurent, S.; et al. Fluorescent magnetic nanoparticles for cell labeling: Flux synthesis of manganite particles and novel functionalization of silica shell. *J. Colloid Interface Sci.* **2015**, *447*, 97–106. [CrossRef]
44. Oyarzun-Ampuero, F.A.; Brea, J.; Loza, M.I.; Torres, D.; Alonso, M.J. Chitosan–hyaluronic acid nanoparticles loaded with heparin for the treatment of asthma. *Int. J. Pharm.* **2009**, *381*, 122–129. [CrossRef]
45. Rigaux, G.; Gheran, C.V.; Callewaert, M.; Cadiou, C.; Voicu, S.N.; Dinischiotu, A.; Andry, M.C.; Vander Elst, L.; Laurent, S.; Muller, R.N.; et al. Characterization of Gd loaded chitosan-TPP nanohydrogels by a multi-technique approach combining dynamic light scattering (DLS), asymetrical flow-field-flow fractionation (AF4) and atomic force microscopy (AFM) and design of positive contrast agents for molecular resonance imaging (MRI). *Nanotechnology* **2017**, *28*, 055705.
46. Stern, R.; Jedrzejas, M.J. Hyaluronidases: Their genomic, structures and mechanisms of action. *Chem. Rev.* **2006**, *106*, 818–839. [CrossRef]
47. Buschmann, M.D.; Merzouki, A.; Lavertu, M.; Thibault, M.; Jean, M.; Darras, V. Chitosans for delivery of nucleic acids. *Adv. Drug Deliv. Rev.* **2013**, *65*, 1234–1270. [CrossRef]
48. Hirai, A.; Odani, H.; Nakajima, A. Determination of degree of deacetylation of chitosan by ^{1}H NMR spectroscopy. *Polym. Bull.* **1991**, *26*, 87–94. [CrossRef]
49. Vårum, K.M.; Antohonsen, M.W.; Grasdalen, H.; Smidsrød, O. Determination of the degree of N-acetylation and the distribution of N-acetyl groups in partially N-deacetylated chitins (chitosans) by high-field N.M.R. spectroscopy. *Carbohydr. Res.* **1991**, *211*, 17–23. [CrossRef]
50. Palumbo, F.S.; Pitarresi, G.; Mandracchia, D.; Tripodo, G.; Giammona, G. New graft copolymers of hyaluronic acid and polylactic acid: Synthesis and characterization. *Carbohydr. Polym.* **2006**, *66*, 379–385. [CrossRef]
51. Cho, H.-J.; Yoon, H.Y.; Koo, H.; Ko, S.-H.; Shim, J.-S.; Cho, J.-H.; Park, J.H.; Kim, K.; Kwon, I.C.; Kim, D.-D. Hyaluronic acid-ceramide-based optical/MR dual imaging nanoprobe for cancer diagnosis. *J. Control. Release* **2012**, *162*, 111–118. [CrossRef]

52. Johnson, C.S., Jr. Diffusion Ordered Nuclear Magnetic Resonance Spectroscopy: Principles and Applications. *Prog. Nucl. Magn. Reson. Spectrosc.* **1999**, *34*, 203–256. [CrossRef]
53. Augé, S.; Amblard-Blondel, B.; Delsuc, M.A. Investigation of the diffusion measurement using PFG and tTest r against experimental conditions and parameters. *J. Chim. Phys. Phys.-Chim. Biol.* **1999**, *96*, 1559–1565. [CrossRef]
54. Best, J.P.; Neubauer, M.P.; Javed, S.; Dam, H.H.; Fery, A.; Caruso, F. Mechanics of pH-Responsive Hydrogel Capsules. *Langmuir* **2013**, *29*, 9814–9823. [CrossRef] [PubMed]
55. Cui, J.; Björnmalm, M.; Liang, K.; Xu, C.; Best, J.P.; Zhang, X.; Caruso, F. Super-Soft Hydrogel Particles with Tunable Elasticity in a Microfluidic Blood Capillary Model. *Adv. Mater.* **2014**, *26*, 7295–7299. [CrossRef] [PubMed]

MDPI
St. Alban-Anlage 66
4052 Basel
Switzerland
Tel. +41 61 683 77 34
Fax +41 61 302 89 18
www.mdpi.com

Gels Editorial Office
E-mail: gels@mdpi.com
www.mdpi.com/journal/gels

www.ingramcontent.com/pod-product-compliance
Lightning Source LLC
LaVergne TN
LVHW070636100526
838202LV00012B/820